D1749308

**Hydrogen Storage Technologies**

*Agata Godula-Jopek, Walter Jehle,
and Jörg Wellnitz*

## Related Titles

Zhang, J., Zhang, L., Liu, H., Sun, A., Liu, R.-S. (Eds.)

**Electrochemical Technologies for Energy Storage and Conversion**

2012
ISBN: 978-3-527-32869-7

Hirscher, M. (Ed.)

**Handbook of Hydrogen Storage**

New Materials for Future Energy Storage

2010
ISBN: 978-3-527-32273-2

Stolten, D. (Ed.)

**Hydrogen and Fuel Cells**

Fundamentals, Technologies and Applications

2010
ISBN: 978-3-527-32711-9

Garcia-Martinez, J. (Ed.)

**Nanotechnology for the Energy Challenge**

2010
ISBN: 978-3-527-32401-9

Mitsos, A., Barton, P. I. (Eds.)

**Microfabricated Power Generation Devices**

Design and Technology

2009
ISBN: 978-3-527-32081-3

Barbaro, P., Bianchini, C. (Eds.)

**Catalysis for Sustainable Energy Production**

2009
ISBN: 978-3-527-32095-0

Olah, G. A., Goeppert, A., Prakash, G. K. S.

**Beyond Oil and Gas: The Methanol Economy**

Second, Updated and Enlarged edition

2009
ISBN: 978-3-527-32422-4

Armaroli, N., Balzani, V.

**Energy for a Sustainable World**

From the Oil Age to a Sun-Powered Future

2011
ISBN: 978-3-527-32540-5

*Agata Godula-Jopek, Walter Jehle, and Jörg Wellnitz*

# Hydrogen Storage Technologies

New Materials, Transport, and Infrastructure

WILEY-VCH

WILEY-VCH Verlag GmbH & Co. KGaA

**The Editors**

*Dr.-Ing. Agata Godula-Jopek*
EADS Innovation Works
Dept. IW-EP
81663 München
Germany

*Walter Jehle*
ASTRIUM GmbH
Dept.: TO 51
88090 Immenstaad
Germany

*Prof. Jörg Wellnitz*
Privat-Institut für Technik
und Design e.V.
Marie-Curie-Str. 6
85055 Ingolstadt
Germany

All books published by **Wiley-VCH** are carefully produced. Nevertheless, authors, editors, and publisher do not warrant the information contained in these books, including this book, to be free of errors. Readers are advised to keep in mind that statements, data, illustrations, procedural details or other items may inadvertently be inaccurate.

**Library of Congress Card No.:** applied for

**British Library Cataloguing-in-Publication Data**
A catalogue record for this book is available from the British Library.

**Bibliographic information published by the Deutsche Nationalbibliothek**
The Deutsche Nationalbibliothek lists this publication in the Deutsche Nationalbibliografie; detailed bibliographic data are available on the Internet at <http://dnb.d-nb.de>.

© 2012 Wiley-VCH Verlag & Co. KGaA, Boschstr. 12, 69469 Weinheim, Germany

All rights reserved (including those of translation into other languages). No part of this book may be reproduced in any form – by photoprinting, microfilm, or any other means – nor transmitted or translated into a machine language without written permission from the publishers. Registered names, trademarks, etc. used in this book, even when not specifically marked as such, are not to be considered unprotected by law.

**Composition** MPS Limited, Chennai
**Printing and Binding** Markono Print Media Pte Ltd, Singapore
**Cover Design** Adam Design, Weinheim

**Print ISBN:** 978-3-527-32683-9
**ePDF ISBN:** 978-3-527-64995-2
**ePub ISBN:** 978-3-527-64994-5
**mobi ISBN:** 978-3-527-64993-8
**oBook ISBN:** 978-3-527-64992-1

Printed in Singapore
Printed on acid-free paper

# Contents

| | | |
|---|---|---|
| **1** | **Introduction** *1* | |
| 1.1 | History/Background *1* | |
| 1.2 | Tanks and Storage *4* | |
| | | |
| **2** | **Hydrogen – Fundamentals** *11* | |
| 2.1 | Hydrogen Phase Diagram *13* | |
| 2.2 | Hydrogen in Comparison with Other Fuels *14* | |
| 2.3 | Hydrogen Production *16* | |
| 2.3.1 | Reforming Processes in Combination with Fossil Fuels (Coal, Natural Gas, and Mineral Oil) *18* | |
| 2.3.1.1 | Steam Reforming of Natural Gas *19* | |
| 2.3.1.2 | Partial Oxidation and Autothermal Reforming of Hydrocarbons *20* | |
| 2.3.1.3 | HyPr-RING Method to Produce Hydrogen from Hydrocarbons *21* | |
| 2.3.1.4 | Plasma-Assisted Production of Hydrogen from Hydrocarbons *23* | |
| 2.3.1.5 | Coal Gasification *25* | |
| 2.3.2 | Water-Splitting Processes (Hydrogen from Water) *27* | |
| 2.3.2.1 | Electrolysis of Water with Electricity from Renewable and Nonrenewable Energy Sources (Low-Temperature Water Splitting) *27* | |
| 2.3.2.2 | Different Types of Electrolyzers *33* | |
| 2.3.2.3 | High-Temperature Water Splitting in Combination with High-Temperature Nuclear Energy and Solar Energy *42* | |
| 2.3.3 | Hydrogen from Biomass *45* | |
| 2.3.3.1 | Thermochemical Processes *47* | |
| 2.3.3.2 | Biological Processes *47* | |
| 2.3.4 | Hydrogen from Aluminum *50* | |
| 2.3.5 | Outlook *51* | |
| 2.4 | Hydrogen Storage Safety Aspects *53* | |
| 2.4.1 | Hydrogen Properties Related to Safety *55* | |
| 2.4.2 | Selected Incidents with Hydrogen *61* | |
| 2.4.3 | Human Health Impact *62* | |
| 2.4.4 | Sensors *63* | |
| 2.4.5 | Regulations, Codes, and Standards (RCS) *63* |

| | | |
|---|---|---|
| 2.4.6 | Safety Aspects in the Hydrogen Chain from Production to the User | *65* |
| 2.4.6.1 | Hydrogen Production | *66* |
| 2.4.6.2 | Hydrogen Refuelling Stations | *67* |
| 2.4.6.3 | Storage/Transportation (Compressed/Liquid/Metal Hydride) | *68* |
| 2.4.6.4 | Garage for Repairing Cars | *70* |
| 2.4.7 | Safety Aspects of Hydrogen Vehicles | *70* |
| 2.4.8 | Safe Removal of Hydrogen | *73* |
| | References | *73* |
| | | |
| **3** | **Hydrogen Application: Infrastructural Requirements** | *81* |
| 3.1 | Transportation | *81* |
| 3.2 | Filling Stations | *86* |
| 3.3 | Distribution | *87* |
| 3.4 | Military | *89* |
| 3.5 | Portables | *92* |
| 3.6 | Infrastructure Requirements | *93* |
| | References | *96* |
| | Further Reading | *96* |
| | | |
| **4** | **Storage of Pure Hydrogen in Different States** | *97* |
| 4.1 | Purification of Hydrogen | *97* |
| 4.2 | Compressed Hydrogen | *98* |
| 4.2.1 | Properties | *98* |
| 4.2.2 | Compression | *98* |
| 4.2.2.1 | Mechanical Compressors | *100* |
| 4.2.2.2 | Nonmechanical Compressor | *101* |
| 4.2.3 | Materials | *106* |
| 4.2.3.1 | Hydrogen Embrittlement | *106* |
| 4.2.3.2 | Hydrogen Attack | *107* |
| 4.2.3.3 | Hydrogen Permeation | *107* |
| 4.2.3.4 | Used Structural Materials | *108* |
| 4.2.3.5 | Used Materials for Sealing and Liners | *109* |
| 4.2.3.6 | High Pressure Metal Hydride Storage Tank | *109* |
| 4.2.4 | Sensors, Instrumentation | *110* |
| 4.2.5 | Tank Filling | *110* |
| 4.2.6 | Applications | *111* |
| 4.2.6.1 | Storage in Underground | *111* |
| 4.2.6.2 | Road and Rail Transportation | *112* |
| 4.2.6.3 | Vehicles | *112* |
| 4.3 | Liquid/Slush Hydrogen | *114* |
| 4.3.1 | Properties | *114* |
| 4.3.2 | Ortho Para Conversion | *114* |
| 4.3.3 | Liquefaction | *116* |

| | | |
|---|---|---|
| 4.3.3.1 | Linde Process | *116* |
| 4.3.3.2 | Claude Process | *117* |
| 4.3.3.3 | Collins Process | *117* |
| 4.3.3.4 | Joule–Brayton Cycle | *118* |
| 4.3.3.5 | Magnetic Liquefaction | *118* |
| 4.3.3.6 | Thermoacoustic Liquefaction | *120* |
| 4.3.4 | Hydrogen Slush | *120* |
| 4.3.5 | Boil-Off | *121* |
| 4.3.5.1 | Zero Boil-Off Solutions | *122* |
| 4.3.6 | Materials | *123* |
| 4.3.6.1 | Tank Material | *123* |
| 4.3.6.2 | Insulation | *123* |
| 4.3.6.3 | Braze Materials | *124* |
| 4.3.7 | Sensors, Instrumentation | *124* |
| 4.3.8 | Applications | *125* |
| 4.3.8.1 | Storage | *125* |
| 4.3.8.2 | Sea Transportation | *126* |
| 4.3.8.3 | Road and Rail Transportation | *126* |
| 4.3.8.4 | Vehicles | *127* |
| 4.3.8.5 | Aircraft | *130* |
| 4.3.8.6 | Rockets | *131* |
| 4.3.8.7 | Solar Power Plants | *131* |
| 4.4 | Metal Hydrides | *131* |
| 4.4.1 | Classical Metal Hydrides | *135* |
| 4.4.1.1 | Intermetallic Hydrides (Heavy Metal Hydrides) | *135* |
| 4.4.1.2 | Magnesium-Based Hydrides | *137* |
| 4.4.2 | Light Metal Complex Hydrides | *139* |
| 4.4.2.1 | Alanates | *139* |
| 4.4.2.2 | Amides-Imides ($Li_3N$–$Li_2NH$–$LiNH_2$) | *143* |
| 4.4.2.3 | Borohydrides | *146* |
| 4.4.3 | Application | *149* |
| 4.4.4 | Outlook | *163* |
| | References | *166* |
| | | |
| **5** | **Chemical Storage** | *171* |
| 5.1 | Introduction | *171* |
| 5.2 | Materials and Properties | *172* |
| 5.3 | Hydrogen Storage in Hydrocarbons | *173* |
| 5.4 | Hydrocarbons as Hydrogen Carrier | *177* |
| 5.5 | Application: Automotive | *178* |
| 5.6 | Ammonia | *181* |
| 5.6.1 | Properties | *181* |
| 5.6.2 | Application Areas of Ammonia | *182* |
| 5.6.3 | Production | *184* |
| 5.6.3.1 | Production from Nitrogen and Hydrogen | *184* |

| | | |
|---|---|---|
| 5.6.3.2 | Production from Silicon Nitride | 184 |
| 5.6.4 | Methods for Storing Ammonia | 185 |
| 5.6.4.1 | Liquid Dry Ammonia | 185 |
| 5.6.4.2 | Solid-State Ammonia Storage | 185 |
| 5.6.5 | Use of Ammonia as Fuel in High-Temperature Fuel Cells | 186 |
| 5.6.6 | Hydrogen from Ammonia | 187 |
| 5.6.6.1 | Ammonia Electrolysis | 187 |
| 5.6.6.2 | Catalytic Decomposition | 187 |
| 5.6.7 | Hydrogen from Ammonia and Metal Hydride | 189 |
| 5.6.8 | Energetic Consideration | 190 |
| 5.7 | Borohydrides | 191 |
| 5.7.1 | Sodium Borohydride | 191 |
| 5.7.1.1 | Direct Use of Sodium Borohydride as Fuel in a PEM-Based Fuel Cell | 191 |
| 5.7.1.2 | Hydrogen Generation by Hydrolytic Release | 192 |
| 5.7.2 | Ammonia Borane | 193 |
| | References | 194 |
| | | |
| **6** | **Hydrogen Storage Options: Comparison** | **197** |
| 6.1 | Economic Considerations/Costs | 197 |
| 6.2 | Safety Aspects | 200 |
| 6.2.1 | Safety Rules and Regulations | 200 |
| 6.2.2 | Safety Equipment | 205 |
| 6.3 | Environmental Considerations: Waste, Hazardous Materials | 209 |
| 6.4 | Dimension Considerations | 212 |
| 6.5 | Sociological Considerations | 216 |
| 6.6 | Comparison with Other Energy Storage System | 218 |
| | References | 222 |
| | | |
| **7** | **Novel Materials** | **225** |
| 7.1 | Silicon and Hydropolysilane (HPS) | 225 |
| 7.2 | Carbon-Based Materials – General | 228 |
| 7.2.1 | Carbon Nanotubes (CNT), Activated Carbon (AC), Graphite Nanofibers | 229 |
| 7.2.2 | Other High-Surface Area Materials | 233 |
| 7.2.3 | Zeolites | 234 |
| 7.2.4 | Metal-Organic Frameworks (MOFs) | 235 |
| 7.2.5 | Covalent Organic Frameworks (COF) | 236 |
| 7.3 | Microspheres | 239 |
| 7.3.1 | Methods for Discharging | 244 |
| 7.3.2 | Resume | 245 |
| | References | 246 |

**Index** 249

# Preface

Energy is an essential component of life. After coal and oil, hydrogen will play a very important role as energy carrier. Hydrogen has the advantage that there are several options for producing it, including use of renewable energy sources. Hydrogen, as energy carrier, can also be used for energy storage and can be easily converted to electric energy and heat. Hydrogen could also be used for mobility as its use promises low impact to the environment. This technology is environment friendly.

During the past years, the necessity to review alternative energy sources has become increasingly important. Environmental issues related to global warming, climate change, and the need to reduce $CO_2$ emissions stimulated the interest in looking into new technologies that may overcome the effect. There are numerous applications in which hydrogen can be used directly, as in road transport or space vehicles or in stationary applications.

The key to a new pathway in mobility will be definitely laid in the mid- and long-term use of hydrogen for general transportation. Hydrogen is the lightest and most energy-efficient element in the world, which is available as an almost unlimited source on our planet. From the perspective of the authors and the hydrogen community, storage of hydrogen plays the key role for the future transportation solutions for passenger cars, trucks, and general transportation. The solution and the cost-effective feasibility of hydrogen storage in reliable tank systems will be the gateway to $CO_2$-free emission driving. This book should provide experts' view to the scientific community to understand the issue of hydrogen storage and to show case solutions for storage. In addition, this book describes some of the significant developments that have emerged in this field. The reader should be able to understand the complexity of storage systems and to design own solutions for transportation vehicles of his choice. We are very happy to have this extraordinary book in our hands today, because we think this is the main contribution to the next-generation transportation.

June 2012

*Agata Godula-Jopek*
*Walter Jehle*
*Jörg Wellnitz*

# 1
# Introduction

## 1.1
## History/Background

Storage of hydrogen is still one of the key issues of the usage of hydrogen itself for vehicle transportation. Main activities on these fields were recorded in the early times of space flight, whereas launching-systems/liquid propelled systems were driven by hydrogen and oxygen fuel.

The development of lightweight tank systems played – from the beginning – a very important role [1], as well as the adjusted issues of pipes, hoses, and dressings.

In Figure 1.1 a typical setup of a pressurized hydrogen or oxygen tank is given, designed with a filament winding method from the early 1950s and 1960s of the last century.

From these issues the storage of hydrogen can be mainly divided into two major technology approaches:

- Pressure storage, $CH_2$, using high pressurized $H_2$ in special tank systems in order to store an amount of n-kg of mass for the use in vehicles.
- Liquid storage, $LH_2$, where the gaseous agent is liquefied below 50 K with moderate pressure (less than 10 bar) and held in a thermo-insulated tank setup.

In this chapter the emphasis will lie on the background of pressurized storage of hydrogen with a special focus on ground transportation, automotive, and tracking.

In Figure 1.2 the basic storage capacities of $LH_2$ and $CH_2$ are given with respect to cost and manufacturability margins.

The early developments of hydrogen tanks are closely linked with space flight programs such as Mercury, Gemini, Delta, and the Apollo program, only to mention NASA projects.

The demands of space flights as a part of early application in transportation are very high, on the other hand cost and manufacturing issues played a minor role in this field of application. Due to technical boundary conditions, mainly cylinder- or elliptical-cylinder-tanks were designed which were the best fit for the fuselage of launchers. This layout was mainly driven by static determinations of the current

*Hydrogen Storage Technologies: New Materials, Transport, and Infrastructure*, First Edition.
Agata Godula-Jopek, Walter Jehle, and Jörg Wellnitz.
© 2012 Wiley-VCH Verlag GmbH & Co. KGaA.
Published 2012 by Wiley-VCH Verlag GmbH & Co. KGaA

**Figure 1.1** $H_2$ and $O_2$ tanks manufactured with filament winding technique in Lockheed–Martin laboratories.

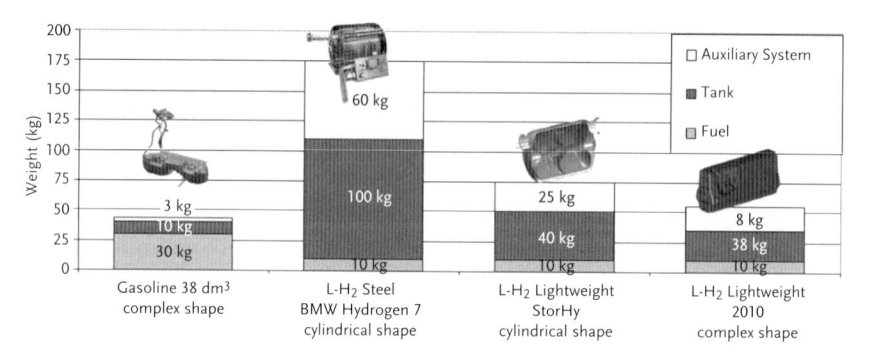

**Figure 1.2** $CH_2$ and $LH_2$ storage outlines with respect to design, cost, and weight margins.

limited ultimate load-cases where the burst and rupture strength of the structure were taken into account for layout. In addition to that, dressing systems were mainly driving the weight-penalty of the structure, by demanding special in- and out-design-features as an interface to composite materials.

Figure 1.3 explains the problem field of interfaces and connectors between dressing-systems and the homogeneous and monolithic tank structure.

The usage of tank systems for ground transportation or vehicles had already been introduced in the early 19th century within the use of hydrogen carriages. Figure 1.4 shows the earliest hydrogen application ever recorded on a carriage system by De Rivaz from 1808.

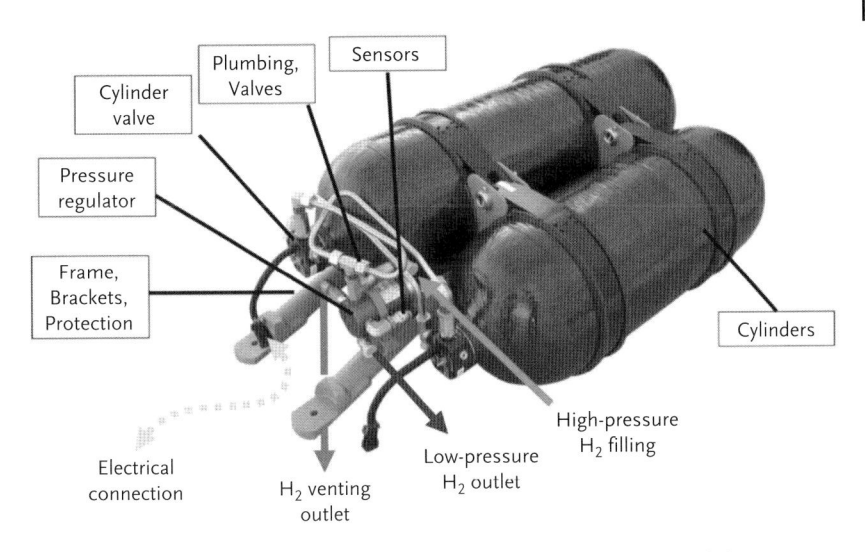

**Figure 1.3** Problem field of interconnection between dressings and the monolithic composite.

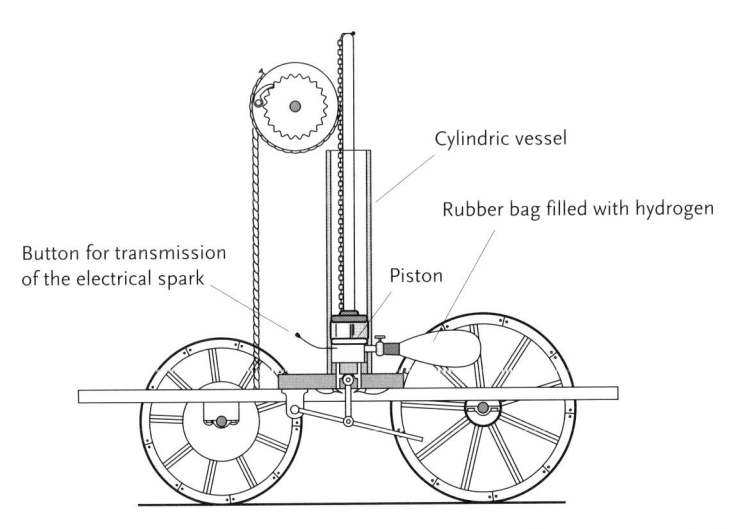

**Figure 1.4** The earliest application of hydrogen in a ground transportation vehicle — carriage — the earliest predecessor of cars.

The development of tank systems for automotive application in the second half of the last century was mainly driven by BMW automotive tank systems, which were using internal combustion engines (ICEs) as the main propulsion system (Figure 1.5). This project, among other parallel developments such as Ford, Man, Mazda, see [2], was one of the main drivers of hydrogen storage, as well as this, BMW has decided to use the storage system $LH_2$ from the very beginning in order to carry more kilograms of hydrogen and therefore to extend the range.

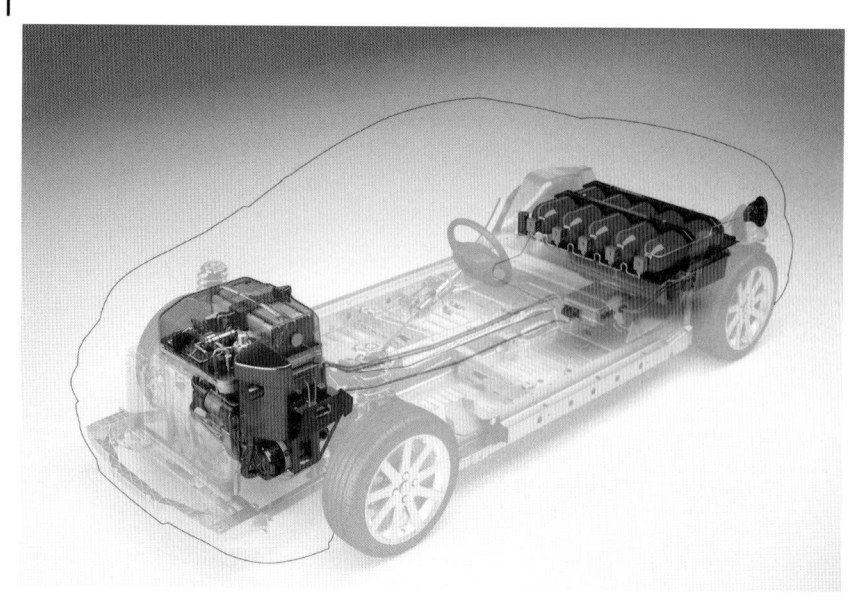

**Figure 1.5** Tank system with the use of $CH_2$ for PSA automotive application using a cylindrical tank rack system in a conventional passenger car.

The recent projects founded by the European government – mainly StorHy (2004–2008) – were focused on the cost critical and production critical issues of the storage of hydrogen and have also lined out major criteria for a practical use of hydrogen in passenger car vehicles.

In [3], a brief overlook at the results of the StorHy project is given, the main aspects of the developments of this project are also used by the authors as an input for future strategies of hydrogen storage.

## 1.2
### Tanks and Storage

The storage of a gaseous agent in tank systems under pressure or high pressure can be linked to the conventional task of layout of pressurized vessels in engineering mechanics.

Mainly the gas, as the fuel for the propulsion system, will be put under high pressure and stored in mostly rotationally symmetric tank systems. The pressure especially for hydrogen will vary between 100 and 750 bar, in particular driven by the demand of the vehicle. This pressure range allows between 0.5 and 3 kg of $H_2$ to be carried. Figure 1.6 shows a conventional tank system by the manufacturer Dynetek with the daily operational pressure of 350 bar and the burst pressure of about 1000 bar.

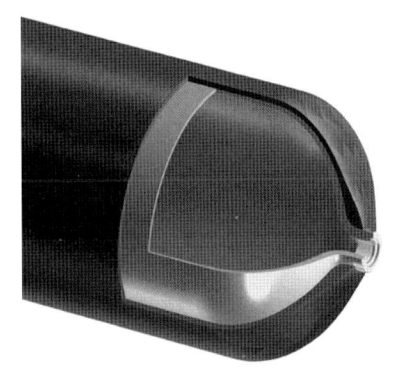

**Figure 1.6** Conventional CH$_2$ pressurized tank system for automotive usage at 350 bar.

This tank is designed with an aluminum liner on the inside to avoid critical H$_2$ permeation and a filament winded carbon fiber hull is used to cope with the high circumferential stresses.

With a storage mass of about 1 kg of hydrogen a typical range of about 100 km can be reached using an internal combustion engine (ICE) and about 150 km on conventional fuel cell applications (F/C). This would automatically demand a typical required mass of at least 3 kg hydrogen for a passenger car, in order to establish practical useful ranges for the customer.

Pressurized gas in tank systems will lead to high circumferential normal stresses, which can be calculated using the pressure-vessel theory. Pressure vessels based on a rotationally symmetric topology can be calculated with the so-called half membrane theory, which will include the membrane stress state and a set of transfer forces for the static equilibrium balance. The normal forces and the shear stresses can be calculated as a function of the metric of the tank system with the conventional orthotropic shell theory [4].

In Figure 1.7 the stresses are shown as a function of the cutting reactional forces of the shell under internal pressure load.

With the help of the statical determination by balancing all forces and moments, the main stresses of the vessel under internal pressure load can be calculated. The investigation of this phenomenon shows that all designed tank systems following this strategy would be inwardly statically determined, whereas the determined stresses are only a function of metric and wall thickness.

If the loads of the shell mid place are described by the load vector

$$\mathbf{q} = q^\alpha \mathbf{a}_\alpha + q^3 \mathbf{a}_3$$

that is relating to their unit of area, the membrane equilibrium state of the shell element follows – under the condition of neglected shear forces $Q^\alpha = 0$ – to

$$N^{\alpha\beta}\big|_\alpha + q^\beta = 0$$

$$N^{\alpha\beta} b_{\alpha\beta} + q^3 = 0$$

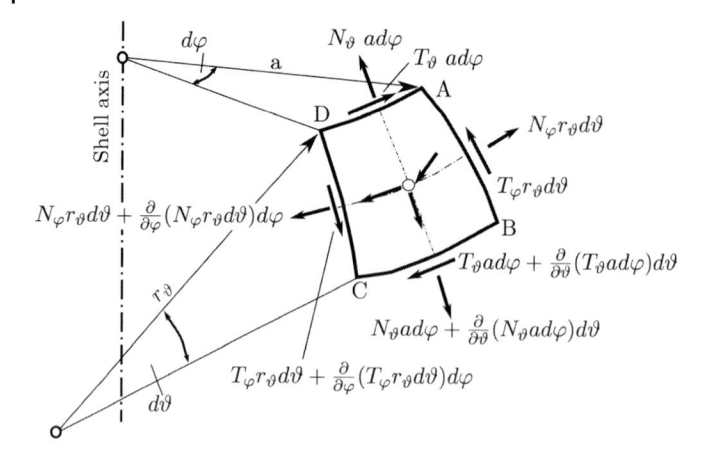

**Figure 1.7** Cutting reactional forces of a shell element under internal pressure.

The partial derivative of the covariant basis vectors

$$\mathbf{a}_{\alpha,\beta} = \frac{\partial \mathbf{a}_\alpha}{\partial \Theta \beta} = \Gamma^\rho{}_{\alpha\beta}\mathbf{a}_\rho + \Gamma^3{}_{\alpha\beta}\mathbf{a}_3$$

is here reflected in the covariant derivative $n^{\alpha\beta}||_\alpha$. The term

$$\Gamma^3{}_{\alpha\beta}\,\mathbf{a}_3 = \mathbf{a}_{\alpha,\beta}\cdot\mathbf{a}_3$$

results in the consideration of the curvature in the equilibrium of forces in normal direction. Further deduction leads to the balance of forces in component notation

$$\overline{N}^{\alpha\beta}\big|_\alpha + \bar{q}^\beta = 0$$

$$\overline{N}^{\alpha\beta}x^3\big|_{\alpha\beta} + \overline{N}^{\alpha\beta}\big|_\alpha x^3\big|_\beta + \bar{q}^3 = 0$$

Showing the main circumferential stresses as a function of $pR/t$ the layout of the vessel system can mainly be focused on this simple formula. Following the metric of cylinder/paraboloid shells versus spherical shells the critical stresses in spheres are only half of the main axial stresses in the other shells.

Because of the character of spherical shells storage, the advantage of this geometry is not only in stresses but also in having a high mass of hydrogen in a very limited amount of space which means the smallest room. This was the reason why spherical shells were mainly used for space-flight applications as well as for submarine and sea-vessel usages.

The complexity of the dressing system and the necessity of putting in and out hoses and wirings lead to so-called edge-design, which can cause the introduction of sharp bending cutting reactional bending moment gradients. Figure 1.8 shows a typical peak-stress environment on a conventional tank system filled with a liquid source.

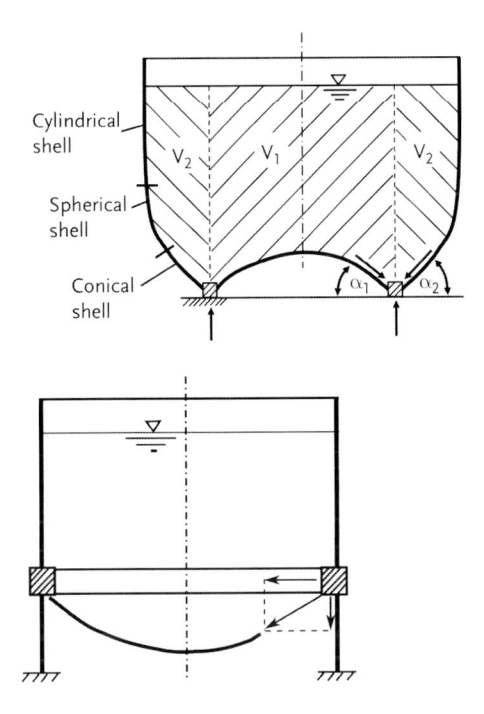

**Figure 1.8** Typical bending moment peak-gradient on the transition phase between the cylinder and bottom wall.

With the calculation of the main-stress-state of the tank system, the strength of the structure can be derived mainly by using conventional failure criteria, as shown by Von Mises and Tresca. Using these criteria would lead to the insight that — for the fatigue loading of tanks with storage pressure of more than 200 bar — a conventional steel design is no longer feasible. The yield strength of modern steel alloys, even for multiphase steels lies in a region of 700–1100 MPa.

Comparing the strength allowable with the calculated main stresses, low or negative margins of safety for homogeneous monolithic tanks using isotropic materials is shown. In order to gain the required strength of the to-be-designed tank structure monolithic materials such as steel or aluminum alloys on a sheet metal base are no longer sufficient. High stresses due to internal pressure loading up to 750 bar require the usage of new fiber material such as carbon or polyamide fibers. This requirement leads to the introduction of high tensile carbon fibers, which can provide strength values up to 2000–3000 MPa. In addition to that, low rupture strain characteristics of the fibers will lead to a reliable layout of the new tank generation. The combination of carbon fiber with resin material such as epoxy or PUR will lead to a new generation of carbon fiber reinforced plastics (CFRP).

In Table 1.1 the basic material properties of new fiber generations are given. It is clearly visible that only carbon or PA fibers can cope with the demand of high circumferential daily fatigue stresses of the hydrogen tank.

Table 1.1 Tensile strength values and Young's moduli of modern fiber generations.

| Fiber | Density (g cm$^{-3}$) | Young's modulus (longitudinal) (GPa) | Tensile strength (MPa) | Elongation at fracture (%) | Coefficient of thermal expansion ($10^6\ °C^{-1}$) |
|---|---|---|---|---|---|
| Carbon | 1.74–1.81 | 230–800 | 2150–4500 | 0.4–2.1 | −1.1– −0.5 |
| Glass | 2.14–2.54 | 55–90 | 1650–4500 | 3.0–4.0 | 3.5–7.2 |
| Aramid | 1.44 | 135–185 | 2800–3500 | 2.1–4.3 | −2 |
| Basalt | 2.65–2.75 | 89–100 | 3000–4800 | 3.2 | – |
| Zylon® (PBO) | 1.55 | 180–270 | 5800 | 2.5–3.5 | −6 |
| Dyneema® (PE) | 0.97 | 89–172 | 2700–3600 | 3.6 | −12.1 |
| Innegra® (PA) | 0.84 | 18 | 590 | 7.2–8.6 | – |
| Vectran® | 1.40 | 103 | 1100–3200 | 3.3 | – |
| Natural | 1.00–1.50 | 22–55 | 390–700 | – | – |
| Silicon carbide | 2.55 | 176–400 | 2450–2950 | 0.6–1.9 | 3.1 |
| Quartz | 2.20 | 78 | 3300–3700 | – | 0.5 |
| Aluminum oxide | 2.70–3.90 | 150–380 | 1700–3100 | 0.6–1.1 | 3–8 |

CFRP with thermoset resin systems would have high permeation rates for hydrogen molecules. These values lay more than $10^3$ times higher than conventional metal alloys such as aluminum or steel. For this reason an inner- or outer-lining system has to be introduced in order to minimize the leakage rate. Such lining systems are mainly used for ground transportation or car usages and will be produced by 3D-rollforming. The aluminum vessel can be used as a mold for filament winding processes or tape-layup CFRP production techniques. Following that the carbon fiber reinforcement will strengthen the aluminum liner to provide high rupture strength in the matter of a circumferential wrapped filament. The permeation rate of aluminum sheet metal is very low, so aluminum is a favorite material for any containment issues for hydrogen.

The key issue for the implementation of pressure tanks for compressed hydrogen is strongly linked with the application of carbon fibers. This is mainly driven by the requirements of the strength of the structure and the subsequent prevention of burst cases. Following this philosophy a 750 bar tank has to have a design using high tenacity (HT) carbon fibers, which will be an issue of cost and availability. The carbon fiber market worldwide is saturated, carbon fiber raw material costs will exceed 20–25 Euro/kg, worldwide availability of carbon fibers is around 50,000 tons per annum. These boundary conditions will restrict the must-production usage of CFRP for hydrogen tanks for conventional cars.

The solution of this problem is the key to a new generation of hydrogen tank systems, whereas new promising materials for high strength fibers or filaments are currently being investigated. The most promising approach is the usage of basalt-fibers and/or super-light-weight PA-fibers such as Innegra. Table 1.1 indicates new material families for the usage of filament winding processes for

**Figure 1.9** Tank assembly of the Formula H racing car, developed by the University of Applied Sciences Ingolstadt and RMIT Melbourne, see [6].

**Figure 1.10** Formula H racing car at a glance.

pressure tanks. In the European project StorHy the pathway to a new generation of pressure tanks was given, also considering its current limitations to costs and availability as well as for all aspects of safety, see [5].

The usage of conventional 200 bar tanks, which may carry 1 kg of hydrogen, can be very cost efficient when used for small cars or racing projects, such as HyRacer, Formula H, and so on (Figures 1.9 and 1.10). These applications offer a wide range of easy-to-use hydrogen tank systems with ICE and conventional mechanical pressure-regulators and -reducers.

# 2
# Hydrogen — Fundamentals

A lot has been written and said about hydrogen because of its independent discovery by Antoine Lavoisier and Henry Cavendish in the eighteenth century. In 1766, Henry Cavendish discovered a gas he called "inflammable air." Nearly a century later, in 1839, Swiss professor Christian Friedrich Schönbein [1] and London lawyer William Robert Grove described the first idea of fuel cell effect, generating electricity and heat by electrochemical conversion of hydrogen and oxygen from air [2].

Introducing hydrogen into world still dependent on fossil fuels is not easy and certainly will face many challenges. However, it has to be underlined that more recently there is a clear trend to become less dependent on oil and fossils toward reduced carbon consumption and increased use of hydrogen, to develop the use of alternative fuels, and to use renewable energies. There is also a big issue related to climate changes and pollution. It has been found that carbon-rich fuel is responsible for the global warming due to greenhouse gas emission to the atmosphere. As stated by Jeremy Rifkin [3],

> There are rare moments in history when a generation of human beings are given a new gift with which to rearrange their relationship to one another and the world around them. This is such a moment.... Hydrogen is a promissory note for humanity's future on Earth. Whether that promise is squandered in failed ventures and lost opportunities or used wisely on behalf of our species and our fellow creatures is up to us.

Hydrogen being the simplest and the most abundantly available element in the universe (0.9 wt%) is present everywhere. Being an energy carrier, hydrogen (a means of storing and transporting energy) is not an energy source itself, but it can only be produced from other sources of energy, such as fossil fuels (natural gas, coal, and petroleum), renewable sources (biomass, wind, solar, and geothermal), or nuclear power (using the energy stored in fissile uranium) by means of several different energy-conversion processes.

Hydrogen is nontoxic, colorless, odorless, and tasteless gas, causing no problems when inhaled with ambient air. It is environment friendly and nonpollutant;

releasing hydrogen has no effect on atmosphere (no greenhouse gas effect) or water (under normal atmospheric conditions, hydrogen is a gas with a very low solubility in water of 0.01911 dm$^3$ dm$^{-3}$ at 25 °C, 0.1 MPa). Hydrogen is highly combustible, therefore a proper ventilation and sensing must be assured when hydrogen diffuses into nonflammable concentrations. During combustion, only water vapor is produced.

Thermodynamic data of hydrogen at 298.15 K and 0.1 MPa are presented in Table 2.1 [4].

Selected physical properties of hydrogen based on Van Nostrand [5] are as follow (Table 2.2).

The energy content of hydrogen is 33.3 kWh kg$^{-1}$, which corresponds to 120 MJ kg$^{-1}$ (lower heating value, LHV) and 39.4 kWh kg$^{-1}$ corresponding to 142 MJ kg$^{-1}$ (upper heating value, HHV). The difference between the HHV and LHV is the molar enthalpy of vaporization of water, which is 44.01 kJ mol$^{-1}$. HHV is achieved when the water steam is produced as a result of hydrogen combustion, whereas LHV is achieved when the product water (i.e., steam) is condensed back to liquid.

**Table 2.1** Basic thermodynamic data of hydrogen.

| State | $\Delta H°$ (kJ mol$^{-1}$) | $\Delta G°$ (kJ mol$^{-1}$) | $S°$ (J K$^{-1}$ mol$^{-1}$) | $C_p$ (J K$^{-1}$ mol$^{-1}$) |
|---|---|---|---|---|
| Gas H | 218.0 | 203.3 | 114.7 | 20.8 |
| H$_2$ | 0 | 0 | 130.7 | 28.8 |

**Table 2.2** Selected physical properties of hydrogen, reprinted by permission.

| Parameter | Value | Unit |
|---|---|---|
| Molecular weight | 2.016 | mol |
| Melting point | 13.96 | K |
| Boiling point (at 1 atm) | 14.0 | K |
| Density solid at 4.2 K | 0.089 | g$^{-3}$ |
| Density liquid at 20.4 K | 0.071 | g cm$^{-3}$ |
| Gas density (at 0 °C and 1 atm) | 0.0899 | g L$^{-1}$ |
| Gas thermal conductivity (at 25 °C) | 0.00044 | cal cm s$^{-1}$ cm$^{-2}$ °C$^{-1}$ |
| Gas viscosity (at 25 °C and 1 atm) | 0.0089 | Centipoise |
| Gross heat of combustion (at 25 °C and 1 atm) | 265.0339 | kJ g$^{-1}$ mol$^{-1}$ |
| Net heat of combustion (at 25 °C and 1 atm) | 241.9292 | kJ g$^{-1}$ mol$^{-1}$ |
| Autoignition temperature | 858 | K |
| Flammability limit in oxygen | 4–94 | % |
| Flammability limit in air | 4–74 | % |

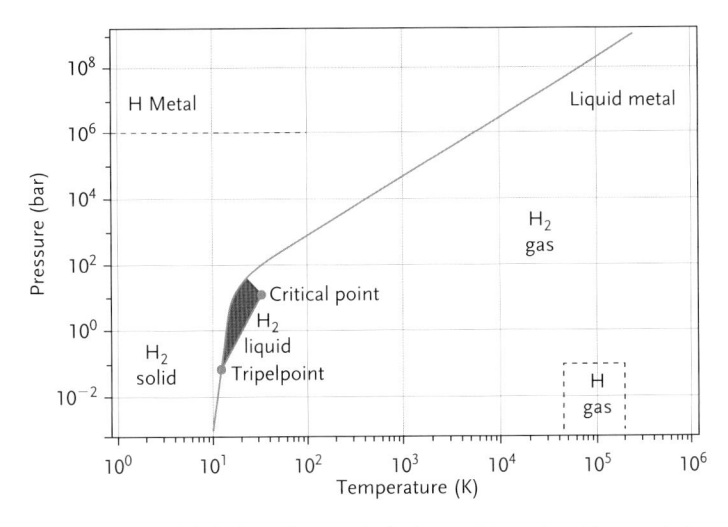

**Figure 2.1** Simplified phase diagram for hydrogen [7], reprinted by permission.

## 2.1
### Hydrogen Phase Diagram

Hydrogen phase diagram is presented in Figure 2.1. The phase diagram of any substance shows the areas of the pressure and temperature where the different phases of this substance are thermodynamically stable. The borders between these areas (phase equilibrium lines) determine the pressure and temperature conditions, $p$ and $T$, where two phases are in equilibrium. The critical point is defining the critical conditions for the physical state of the system separating the states with different properties (liquid–gas), which cannot distinguish between the two states (liquid and gas). For example, for pure gaseous substances, critical point means the critical temperature (and corresponding critical pressure – maximal pressure over the liquid) above which gas cannot condense, independent of its size.

From the diagram, it can be seen that the molecule of hydrogen can be found in different states/forms depending on the temperature and pressure conditions. Critical parameters of hydrogen are: $T_c=33.25$ K ($-239.9\,^\circ$C), $p_c=1.28$ MPa, and $V_c=64.99$ cm$^3$ mol$^{-1}$ [6].

At low temperatures, hydrogen exists as a solid with density of 0.089 g cm$^{-3}$ at 4.2 K.

Liquid hydrogen with density of 0.071 g cm$^{-3}$ at 20.4 K exists between the solid line and the line from the triple point and the critical point at 33.25 K. Hydrogen is a gas at higher temperatures with a density of 0.0899 g L$^{-1}$ at 0 $^\circ$C and 1 atm. At ambient temperature of 25 $^\circ$C (297.15 K), hydrogen gas can be described by the following van der Waals equation:

$$p = \frac{nRT}{V-nb} - a\left(\frac{n}{V}\right)^2 \tag{2.1}$$

## 2 Hydrogen – Fundamentals

**Table 2.3** Hydrogen storage methods, parameters, [7] reprinted by permission.

| Storage method | Gravimetric density (mass %) | Volumetric density (kg $H_2$ $m^{-3}$) | T (°C) | P (bar) |
|---|---|---|---|---|
| High pressure gas cylinders | 13 | <40 | 25 | 800 |
| Liquid hydrogen in cryogenic tanks | Size dependent | 70.8 | −252 | 1 |
| Adsorbed hydrogen | ≈2 | 20 | −80 | 100 |
| Adsorbed hydrogen on interstitial sites | ≈2 | 150 | 25 | 1 |
| Complex compounds | <18 | 150 | >100 | 1 |
| Metal and complexes together with water | <40 | >150 | 25 | 1 |

where:

$a$ (dipole interaction or repulsion constant) and $b$ (volume occupied by hydrogen molecules) are called van der Waals coefficients. For hydrogen, the coefficients $a$ and $b$ are 0.2476 atm $L^2\,mol^{-2}$ ($a$) and $2.661 \times 10^{-2}$ $L\,mol^{-1}$ ($b$), respectively [6],
$R$ is the gas constant (8.134 $JK^{-1}\,mol^{-1}$),
$n$ is the number of moles,
$p$ is the gas pressure,
$V$ is the volume of the gas.

In order to store hydrogen gas, the big volume of gas has to be reduced significantly (under ambient conditions, 1 kg of hydrogen has a volume of 11 $m^3$). Either a certain work has to be performed to compress hydrogen or the temperature has to be decreased below the critical temperature of 33.25 K, or the repulsion must be reduced by the interaction of hydrogen with another material [7]. Hydrogen can be stored as a gas or a liquid. In addition, other methods of storing hydrogen in compounded form such as in metal hydrides, chemical hydrides, carbon materials, glass microspheres are also possible.

Following parameters for storing methods have been given [7] (Table 2.3).

## 2.2
### Hydrogen in Comparison with Other Fuels

Table 2.4 shows the properties of hydrogen compared with other fuels.

From Table 2.4, it is clear that hydrogen yields much higher energy per unit weight than any other fuel. Hydrogen has a high energy-to-weight ratio (around three times more than gasoline, diesel, or kerosene) but is less flammable than these fuels. The required concentration for hydrogen to combust in air is four times higher than gasoline; hydrogen concentrations in air below 4%, and above

**Table 2.4** Mass and volume energy density of hydrogen in comparison with other fuels.

| Fuel | Gravimetric energy density | | Volumetric energy density | | Flammability limits (vol%) | Explosive limits (vol%) | Fraction of heat in radiative form |
|---|---|---|---|---|---|---|---|
| | (MJ kg$^{-1}$) | (kWh kg$^{-1}$) | (MJ L$^{-1}$) | (kWh L$^{-1}$) | | | |
| Hydrogen compressed 200 bar | 120 | 33.3 | 2.1 | 0.58 | 4–75 | 18.3–59.0 | 17–25 |
| Hydrogen liquid | 120 | 33.3 | 8.4 | 2.33 | | | |
| Methanol | 19.7 | 5.36 | 15.7 | 4.36 | 6–36.5 | 6–36 | 17 |
| Petrol | 42 | 11.36 | 31.5 | 8.75 | 1–7.6 | 1.1–3.3 | 30–42 |
| Diesel | 45.3 | 12.58 | 35.5 | 9.86 | | 0.6–7.5 | |
| Kerosene | 43.5 | 12.08 | 31.0 | 8.6 | | 0.7–5 | |

75%, will not burn. In comparison, gasoline concentrations of only 1% are flammable in air (flammability limits of petrol are 1–7.6 vol.%). Flammability range is highest for hydrogen, but as long as it stays in the area of proper ventilation, it is difficult to reach the limit. In addition, hydrogen has a relatively high ignition temperature of 858 K compared to an ignition temperature as low as 501 K for gasoline. Once ignited, even the flame temperature for hydrogen is lower than that for gasoline $-2318$ K (2470 K for gasoline) [8]. Due to low density, hydrogen does not cumulate near ground but dissipates in the air, unlike gasoline and diesel fuel. Hydrogen and methanol with regard to safety, economics, and emissions aspects have been evaluated by Adamson and Pearsons [9]. Comparative risks analysis in case of accident, self-assessed by the authors in enclosed and ventilated areas showed that both hydrogen and methanol are safer than petrol; there is no winner between the first two fuels, but in certain situations hydrogen may be at higher risk than methanol. Hake *et al.* [10] compared different fuels and fuel storage systems of exemplary passenger car including safety characteristics of gasoline, diesel, methanol, methane, and hydrogen and concluded that "there are no technical safety or health aspects that generally exclude the introduction of hydrogen as a fuel." The risk of hydrogen with the infrastructure exists and is different when compared to diesel or gasoline but not higher. In addition, it has to be noted that hydrogen is environmental friendly, producing water vapor as the only waste, compared to other fuels like gasoline. High energy-to-weight ratio and clean by-products are main factors for the automobile industry that make hydrogen a future fuel source. Assuming proper storage and infrastructure, hydrogen can be used in many applications as an alternative fuel or even the fuel of the future.

## 2.3
**Hydrogen Production**

There are several ways of producing hydrogen. Most of the technologies are well technically and commercially developed, but some of them are competing to a certain extent with existing energy technologies. Presently, most of the industrial hydrogen is produced from fossil fuels (natural gas, oil, and coal), mainly by steam reforming of natural gas as the leading process on a large scale, partial oxidation of hydrocarbons, and coal gasification, contributing significantly to $CO_2$ gas emission to the atmosphere as a result. It is reported that each kilogram of hydrogen produced by the steam reforming emitted 13.7 kg of equivalent $CO_2$ [11]. A special attention is paid to the renewable production options that include water electrolysis using renewable power (e.g., wind, solar, hydroelectric, and geothermal), biomass gasification, photoelectrochemical and biological processes, and high-temperature thermochemical cycles. The use of biomass for producing hydrogen instead of fossil fuels contribute to the reduction of the atmospheric emissions, but biomass-related processes are used on a minor scale hydrogen production and still in the demonstration phase. Solar hydrogen generated by water electrolysis with solar cells, direct photocatalytic water, photobiological water splitting, or solar thermal processes represents a highly desirable, clean, and abundant source of hydrogen. Water electrolysis is said to be rather efficient, above 70%, but is expensive and estimated at more than $20/GJ assuming a cost of about $0.05\,kWh^{-1}$ [12]. Water electrolysis is suitable in combination with photovoltaics (PVs) and wind energy and is in direct relation with the availability of electricity. In general, hydrogen production methods (without photovoltaics) can be connected with the nuclear reactor, providing heat and electricity for the process. It is estimated that one of the most promising thermochemical cycles for large-scale hydrogen production is the iodine–sulfur (I–S) cycle (General Atomic) with thermal to hydrogen efficiency of 52% and UT-3 cycle developed at the University of Tokyo with efficiency of around 50%. General pathways of hydrogen production by different processes and from different primary energy sources are presented in Figure 2.2.

The end users of produced hydrogen can be classified in various sectors as presented in Figure 2.3.

In fact a big part of hydrogen produced is used in industry for various syntheses and polygeneration processes and turbines or in households.

Hydrogen can be used in transportation sector to power vehicles by using conventional engines (Diesel, Otto), gas turbines in vehicles. One of the most important advantages is that hydrogen engines emit less pollutants to atmosphere compared to gasoline engines. Combustion product is mainly water vapor and some nitrogen oxides. Certain interest has been risen by aeronautic sector by

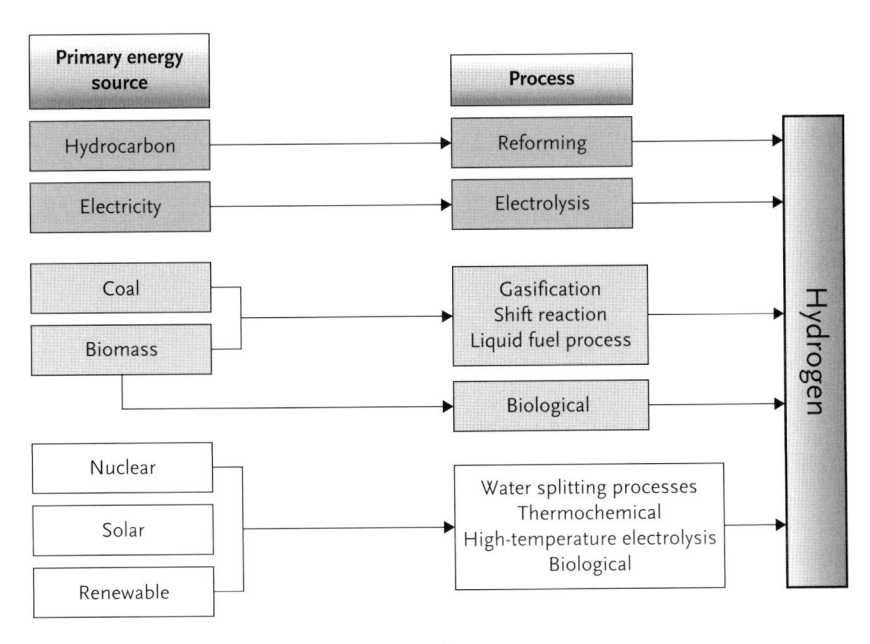

**Figure 2.2** General pathways of hydrogen production.

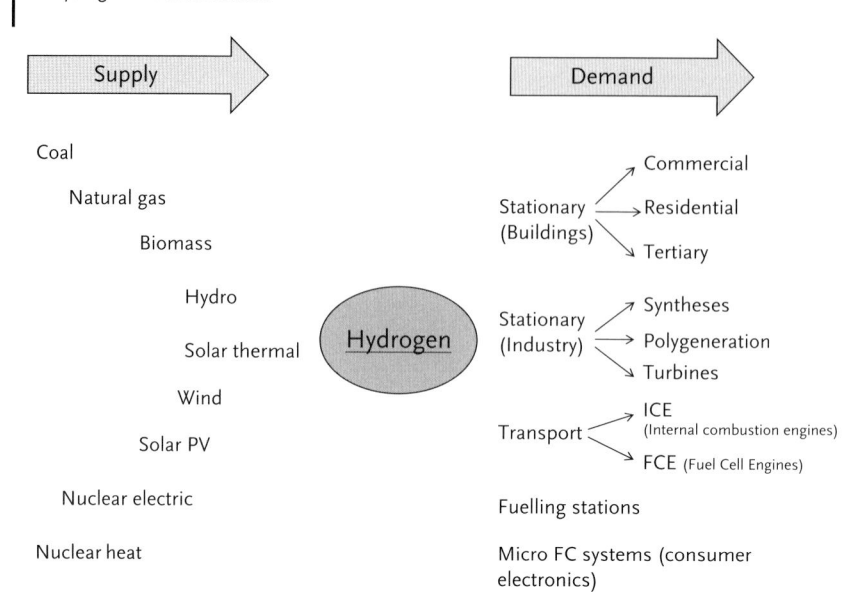

**Figure 2.3** Hydrogen by means of different sources and its potential application in stationary, transportation, and portable sectors.

using liquid hydrogen for direct combustion in gas turbines. Considerable attention has been paid for using hydrogen as a fuel for fuel cells, which are of high and still increasing interest of stationary, portable, and transport groups. Fuel cells are recently one of the most attractive and promising hydrogen utilization technologies. It must be noted that a special attention must be given toward issues related to hydrogen gas transmission, storage and distribution systems, as well as end-user infrastructure. Integrity, durability, safety aspects, standards and norms, and the public acceptance come along with hydrogen application in wide understanding.

The application of hydrogen for different sectors is described above in Section 1.2.

## 2.3.1
### Reforming Processes in Combination with Fossil Fuels (Coal, Natural Gas, and Mineral Oil)

Fossil fuels, coal, oil, and natural gas, are a nonrenewable source of energy. Although combustion of fossil fuel produces significant amount of greenhouse and toxic gases, such as $CO_2$, $SO_2$, $NO_x$, and other pollutants, contributing to the global warming and acid rain, they are still a significant source of hydrogen and electricity production by means of different processes. However, it has to be mentioned that an effort is being made to look for clean and renewable alternatives.

### 2.3.1.1 Steam Reforming of Natural Gas

Steam reforming of natural gas is currently the cheapest, well technically, and commercially established way to produce hydrogen, mainly used in petrochemical and chemical industries. The cost of hydrogen production from steam reforming strongly depends on the costs and availability of the natural gas feedstock. Steam reforming is highly endothermic (heat is absorbed) gas phase conversion process and requires high reaction temperatures, typically above 600 K (823 °C) in the presence of Fe- or Ni-based catalysts conventionally supported on $AL_2O_3$ and $MgAl_2O_4$ and pressure of about 3 MPa. As a result, a syngas containing CO and hydrogen is generated. The efficiency of steam reforming is around 65–70%.

General steam-reforming reaction can be described by the following equation:

$$C_nH_m + nH_2O \rightarrow nH_2 + \left(\frac{(n+m)}{2}\right)CO \quad \Delta H^0_{298} > 0 \quad (2.2)$$

In addition to carbon monoxide and hydrogen, a certain amount of unreacted steam can be found in the reformate gas as well as some unreacted fuel and carbon dioxide. $CO_2$ is formed by the following water gas shift reaction (WGSR):

$$CO + H_2O \rightarrow CO_2 + H_2 \quad \Delta H^0_{298} = -40.4 \text{ kJ mol}^{-1} \quad (2.3)$$

WGSR increases the hydrogen concentration of reformate; methane is usually formed in large amount. It occurs in two stages, as high-temperature shift (HTS) at around 350 °C and low-temperature shift (LTS) at around 200 °C. Ideally, WGS reactions should reduce the CO level down to less than 5000 ppm. The reaction is moderately exothermic with low level of CO at low temperatures but with favorable kinetics at higher temperatures. In industrial applications, catalysts based on Fe–Cr oxide are used for HTS and Cu–ZnO–$Al_2O_3$ for LTS to achieve a good performance under steady-state conditions [13]. Hydrogen produced by steam-reforming process may require additional purification like desulfurization and $CO_2$ removal. Currently, most of the hydrogen plants use pressure swing adsorption, producing very pure hydrogen (99.9%). In modern hydrogen plants, reforming temperatures can operate above 900 °C and steam-to-carbon (S/C) ratio below 2.5, even below 2.0. Under advanced steam-reforming conditions (high temperature of reforming condition combined with low steam ratio), it is possible to operate hydrogen plants with efficiency less than 3 Gcal/1000 $Nm^3$ of hydrogen. The theoretical efficiency of the hydrogen production from methane by steam reforming is 10.8 GJ/1000 $Nm^3$ $H_2$ when starting from water vapor and 11.8 GJ/1000 $Nm^3$ $H_2$ starting from liquid water [14]. Haldor Topsøe developed two types of convection reformers to improve the thermal efficiency of steam reforming, process gas heated reformer (Haldor Topsøe exchange reformer, HTER) and the convective flue gas heated reformer (Haldor Topsøe convective reformer, HTCR). Using HTCR technology, the reforming efficiencies increased to 80% with capacities up to around 25 000 $Nm^3$ $h^{-1}$ [15].

Steam reforming of light hydrocarbons is also a well-established industrial process. One of the benefits of using methanol is that the reforming reaction can be carried out at lower temperature – the endothermic steam reforming can be carried out at around 300 °C over a Co–Zn catalyst [16], whereas about 800 °C is required for hydrocarbons. Theoretically steam reforming of methanol can produce 75% hydrogen concentration at 100% $CO_2$ selectivity; in practice, it is greater than 70% with various catalysts [13]. Higher range of temperature (850–1500 °C) is required for the steam reforming of ethanol because of its C—C bond [17].

#### 2.3.1.2 Partial Oxidation and Autothermal Reforming of Hydrocarbons

Partial oxidation reaction is much faster than steam reforming. It is exothermic reaction with oxygen at moderately high pressure with or without catalyst depending on the feedstock and selected process. Noncatalytic Texaco process (TGP) operates at temperatures in the range of 1200–1500 °C and pressure above 3 MPa. Catalytic partial oxidation uses lower temperatures around 1000 °C, but production of pure hydrogen is less efficient and more costly than steam reforming. Catalysts include supported nickel (NiO–MgO), nickel-modified hexa-aluminates, platinum group metals Pt, Rh, Pd/alumina, on ceria-containing supports or on titania [13].

Partial oxidation is the conversion of fuels under oxygen-deficient conditions according to the following formula [18]:

$$C_xH_yO_z + \frac{(x-z)}{2}(O_2 + 3.76N_2) \rightarrow xCO + \frac{y}{2}H_2 + 3.76\frac{(x-z)}{2}N_2 \quad (2.4)$$

A definite advantage of partial oxidation is that only fuel and feed air are needed for the reaction; there are no evaporation processes. Typical by-product of the reaction is methane and coke formation. The disadvantage of partial oxidation process is catalyst deactivation due to coke deposition and carbon monoxide. Coke formation may be formed due to the following reaction:

$$H_2 + CO \rightarrow H_2O + C \quad \Delta H^0_{298} = -131 \text{ kJ mol}^{-1} \quad (2.5)$$

It is to be noted that the amount of CO is higher when compared with steam-reforming reaction (clean-up step may be needed in case of connecting with fuel cells). Kolb [18] reports two reaction mechanisms for partial oxidation: (i) reaction begins with catalytic combustion followed by reaction of lower rate (steam reforming, $CO_2$ reforming, and water–gas shift) and (ii) direct POX at very short residence time.

Autothermal reforming (ATR) or oxidative steam reforming combines the endothermic steam-reforming process with the exothermic partial oxidation reaction. It is a combination of both processes in which the energy generated by the partial oxidation reaction provides the energy for the steam-reforming reaction. These systems can be very productive, fast starting, and have been demonstrated with methanol, gasoline, and natural gas. Autothermal reforming has several advantages including improved heat integration, faster start-up, and lower operating temperatures. Catalyst choice for autothermal reforming depends on the fuel and operating temperature. For example, for methanol, Cu-based catalysts similar to commercial methanol synthesis catalysts and LTS catalysts are used. For higher hydrocarbons, catalysts containing Pt, Rh, Ru, and Ni supported or deposited on oxides have been reported; highest selectivity in reforming of iso-octane was obtained on Pt formulation ($Pt/Ce_{0.8}Sm_{0.15}Gd_{0.05}O_2$) [13].

The autothermal reaction can be described as follows [18]:

$$C_xH_yO_z + n(O_2 + 3.76N_2) + (x - 2n - z)H_2O$$
$$\rightarrow xCO + \left(x - 2n - z + \frac{y}{2}\right)H_2 + 3.76N_2 \quad (2.6)$$

Air addition should be in limited amounts to prevent coke formation on the catalyst. Usually, an optimum atomic O/C (oxygen to carbon) ratio exists for each fuel under thermally neutral conditions to achieve optimum efficiency. The maximum efficiency available from autothermal process for various hydrocarbons was given by Hagh [19] at fixed S/C ratio and pressure. Simulation on autothermal reforming on low and high molecular weight hydrocarbons (LHCM and HHCM, respectively) showed 80% efficiency for LHCM at 700 °C at S/C 3.5 and O/C ratio 0.28% and 80% efficiency for HHCM at the same temperature, S/C 3.5, O/C 0.5 [20]. Önsan [17] reported that higher S/C ratios and reactor inlet temperature favor hydrogen production. Less fuel has to be burned for the reaction and as a consequence lower O/C ratio can be used. Optimal S/C ratios have been reported for different fuels: 4 for methane, 1.5 for methanol, 2.0 for ethanol, and 1.3 for surrogate gasoline. Increase of S/C ratio favors hydrogen production [17]. CO formation is depressed at higher S/C ratios especially at higher temperatures. Higher operational temperatures enhance CO formation and reduce hydrogen production.

### 2.3.1.3 HyPr-RING Method to Produce Hydrogen from Hydrocarbons

Innovative hydrogen production method, HyPr-RING (hydrogen production by reaction-integrated novel gasification), using hydrocarbons and water was proposed by Lin et al. [21]. General idea of the concept is the integration of water carbon reaction, water–gas shift reaction, and $CO_2$ absorption reaction in a single low-temperature reactor to produce clean and highly concentrated hydrogen, as shown in Figure 2.4.

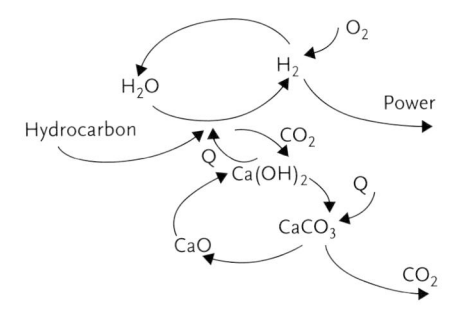

**Figure 2.4** General concept of the HyPr-RING process [21], reprinted by permission.

Hydrocarbons are mixed with water in the reactor operating at temperature 650–700 °C and high pressure (10–100 MPa). Following reactions occur in the reactor:

a. Reaction of carbon with water

$$C + H_2O \rightarrow CO + H_2 \quad \Delta H^0_{298} = 132 \text{ kJ mol}^{-1} \quad (2.7)$$

b. Water–gas shift reaction

$$H_2O + CO \rightarrow CO_2 + H_2 \quad \Delta H^0_{298} = -41.5 \text{ kJ mol}^{-1} \quad (2.8)$$

c. Addition of $CO_2$ absorption reaction (this reaction was performed in order to integrate the above two reactions in one reactor)

$$CaO + CO_2 \rightarrow CaCO_3 \quad \Delta H^0_{298} = -178 \text{ kJ mol}^{-1} \quad (2.9)$$

d. Overall reaction for new reaction system (a + b + c)

$$C + 2H_2O + CaO \rightarrow CaCO_3 + 2H_2 \quad \Delta H^0_{298} = -88 \text{ kJ mol}^{-1} \quad (2.10)$$

The experimental results show that all hydrocarbons can produce high concentrations of hydrogen, up to 80% of the product gas. Besides hydrogen, methane was also present in the gas as well as trace amounts of CO and $CO_2$ [21].

The advantages of the HyPr-RING process is that it can be applied to hydrocarbon sources such as coal, heavy oil, biomass, plastic, and organic waste (Figure 2.5). In a single reactor, these materials can produce sulfur-free gas containing up

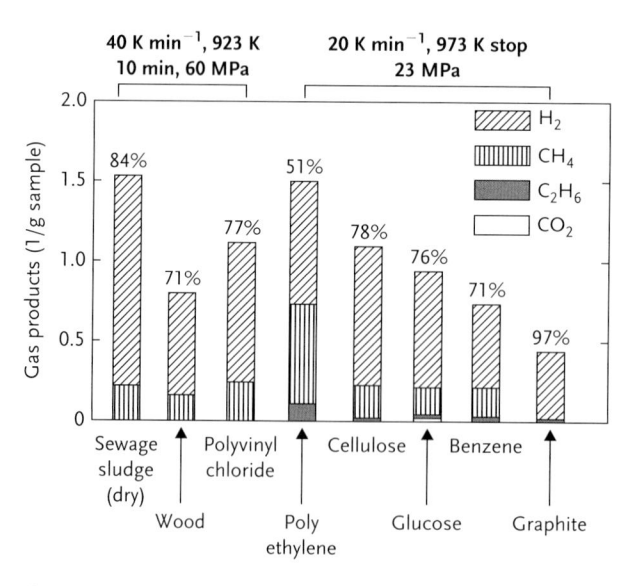

**Figure 2.5** Gas production by HyPr-RING process from various hydrocarbons [22].

to 80% of hydrogen and 20% of methane. Hydrogen produced is of high quality and can be fed directly to the fuel cell [22].

#### 2.3.1.4 Plasma-Assisted Production of Hydrogen from Hydrocarbons

Thermal and nonequilibrium plasma are under consideration as a way for hydrogen production from hydrocarbons. The advantages of thermal plasma chemical methods are very high specific productivity of the apparatus, low investment, and operational costs. The disadvantage is high-energy consumption. The gas and electron temperatures and the energy content are very high; the gas temperature is between 10 000 and 100 000 K. Thermal plasma is an equilibrium plasma because temperatures of the gas and electrons are nearly equal.

Nonthermal plasmas are known as a very high energy density media, able to accelerate the reactions at low temperatures, nonequilibrium properties, and low-power requirements. In addition, if active species generated by the nonthermal plasma can promote many cycles of chemical transformation, high productivity of plasma can be combined with low-energy consumption of conventional catalysts (plasma catalysis). Nonthermal plasma reforming shows interesting results of efficiency, conversion rate, and hydrogen production (Figure 2.6). As an alternative for hydrocarbons reforming for the development of fuel cells in vehicles, nonequilibrium plasma technique has been implemented over the past two decades. Various plasma types have been used: plasmatron, gliding arc, dielectric barrier discharge, corona, microwave, and pulsed discharge [23]. Hydrocarbon conversion in gliding arc showed following results: 3 kWh Nm$^{-3}$ (cost 2.5 eV mol$^{-1}$) of syngas for steam reforming, 0.11 kWh Nm$^{-3}$ (cost 0.09 eV mol$^{-1}$) of syngas for partial oxidation in oxygen, 0.3 kWh Nm$^{-3}$ (cost 0.25 eV mol$^{-1}$) of syngas for steam–oxygen conversion, and 0.98 kWh Nm$^{-3}$ (cost 0.82 eV mol$^{-1}$) of syngas for air–steam conversion in the presence of nickel catalyst [24]. In combination with the pulse microwave discharge by small addition of microwave energy (up to 10% of thermal energy input), a significant increase of conversion degree was observed, resulting in reduction of hydrogen energy costs. In this process, hydrocarbons were first preheated up to 427–727 °C in a conventional heat exchanger followed by nonequilibrium pulse microwave discharge. In case of methane decomposition, conversion degree increased from 7% to 18% ($H_2$ plasma energy cost 0.1 eV mol$^{-1}$), ethane decomposition from 19% to 26% ($H_2$ plasma energy cost 0.25 eV mol$^{-1}$), ethanol decomposition from 23% to 62% ($H_2$ plasma energy cost 0.1 eV mol$^{-1}$), methane–steam decomposition from 10% to 16% ($H_2$ plasma energy cost 0.35 eV mol$^{-1}$), and ethanol–steam decomposition conversion rate increased from 41% to 58% ($H_2$ plasma energy cost $\sim$0.1 eV mol$^{-1}$) [24]. Reforming of hydrocarbons (methane) and alcohols (methanol and ethanol) in mixtures with $CO_2$ or $H_2O$ has been performed in dielectric barrier discharge [25]. No higher hydrocarbons were formed during the reforming of the fuels; the majority of the products were hydrogen and carbon monoxide. For methanol and ethanol, 100% conversion was obtained for relatively high flow of the reactants. Calculated energy consumed in the discharge showed that 1 mol of hydrogen can be produced by energy approximately 4 kWh for methane, 0.6 kWh for methanol, and 0.3 kWh for ethanol [25].

**24** | *2 Hydrogen – Fundamentals*

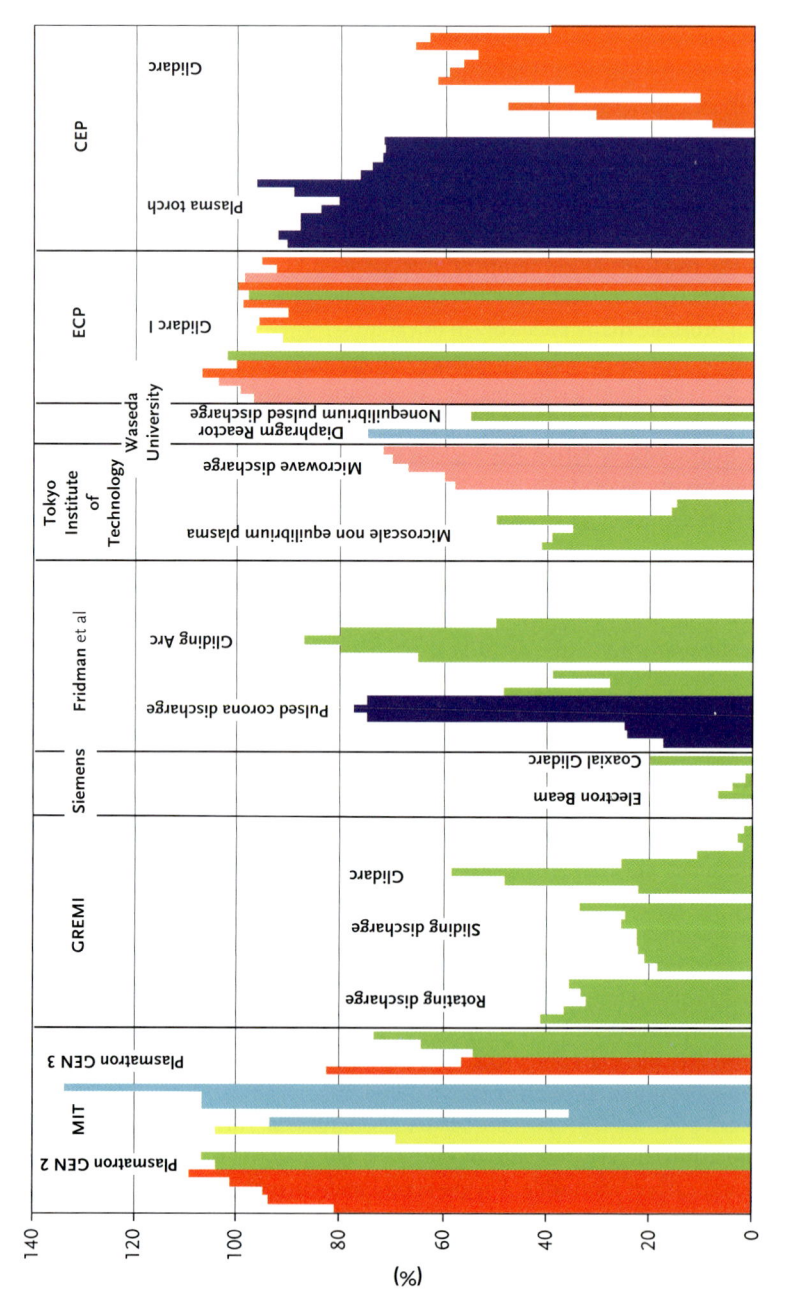

**Figure 2.6** Conversion rates of different nonequilibrium plasma-assisted technologies [23], by kind permission of MINES ParisTech – CEP, Plasma Group.

Novel sliding discharge reactor for hydrogen generation from hydrocarbons has been introduced by GREMI, University of Orleans, France [26]. This reactor was used for steam reforming of methane and propane at atmospheric pressure. The main products of the plasma process were hydrogen (50%), carbon monoxide (up to 30%), and nonconsumed methane or propane. For the same experimental conditions, methane conversion rate was higher than those of propane.

Major works concerning nonequilibrium plasma-assisted reforming have been presented and compared, among others at the Plasma Science and Fusion Center at MIT, USA; Drexel Plasma Institute, USA; Siemens AG, Germany; Kurchatov Institute, Russia; ECP, France; Waseda University, Japan, and so on. A comparison between nonequilibrium plasma reforming reactors (efficiency, conversion rate) for hydrogen production from various hydrocarbons showed that arc discharge-based technologies meet in the best way performances due to their relative simplicity of the set up, high-energy densities, and their ability to create a large reactive volume [23]. The GAT reactor reached the top value with 79% (Figure 2.7). The conversion rates and energetic efficiencies of different nonequilibrium plasma-assisted technologies are presented below.

### 2.3.1.5 Coal Gasification

Systems based on gasification can utilize coal, petroleum coke, biomass, municipal, and hazardous wastes. In principle, the process is similar to partial oxidation of heavy oils and has three main steps, (i) conversion of coal feedstock in the presence of oxidant (typically oxygen or air and steam) to syngas at high temperatures of 1000–1500 °C in gasification reactor, (ii) catalytic shift reaction, (iii) and purification of the produced hydrogen, mainly residual carbon and ash. Depending on the gasification technology, certain amount of water, carbon dioxide, and methane can be present in the syngas including traces components, for example hydrogen cyanide (HCN), hydrogen chloride gas (HCl), hydrogen sulfide ($H_2S$), and carbonyl sulfide (COS). Generated syngas can be used directly to produce electricity or be further processed to pure hydrogen for hydrocracking of petroleum or ammonia production. A lot of research has been performed on integrating the gasifier with a combined cycle gas and steam turbine (IGCC – integrated gasification combined cycle) and a fuel cell (IGFC – integrated gasification fuel cell). IGFC systems tested in the United States, Japan, and Europe with gasifiers by Texaco, Eagle, and Lurgi showed energy conversion at efficiency of 47.5% (HHV), higher than the efficiency of conventional coal gasifiers [27]. Most of the carbon components are removed before combustion when the gas turbine is used before conversion of the feedstock in air or oxygen/steam step. The largest worldwide IGCC Power Plant with 318 $MW_{el}$ is build in Puertollano, Spain by Elcogas. Worldwide, there are 117 operating plants, 385 gasifiers with a total production capacity of around 45 000 $MW_{th}$ [28]. Major gasification technologies are developed by Texaco, E-Gas, Shell, Kellogg, British Gas/Lurgi, KRW, and PRENFLO [27]. In gasification process, other fuels like biomass may also be used (see Section 2.3).

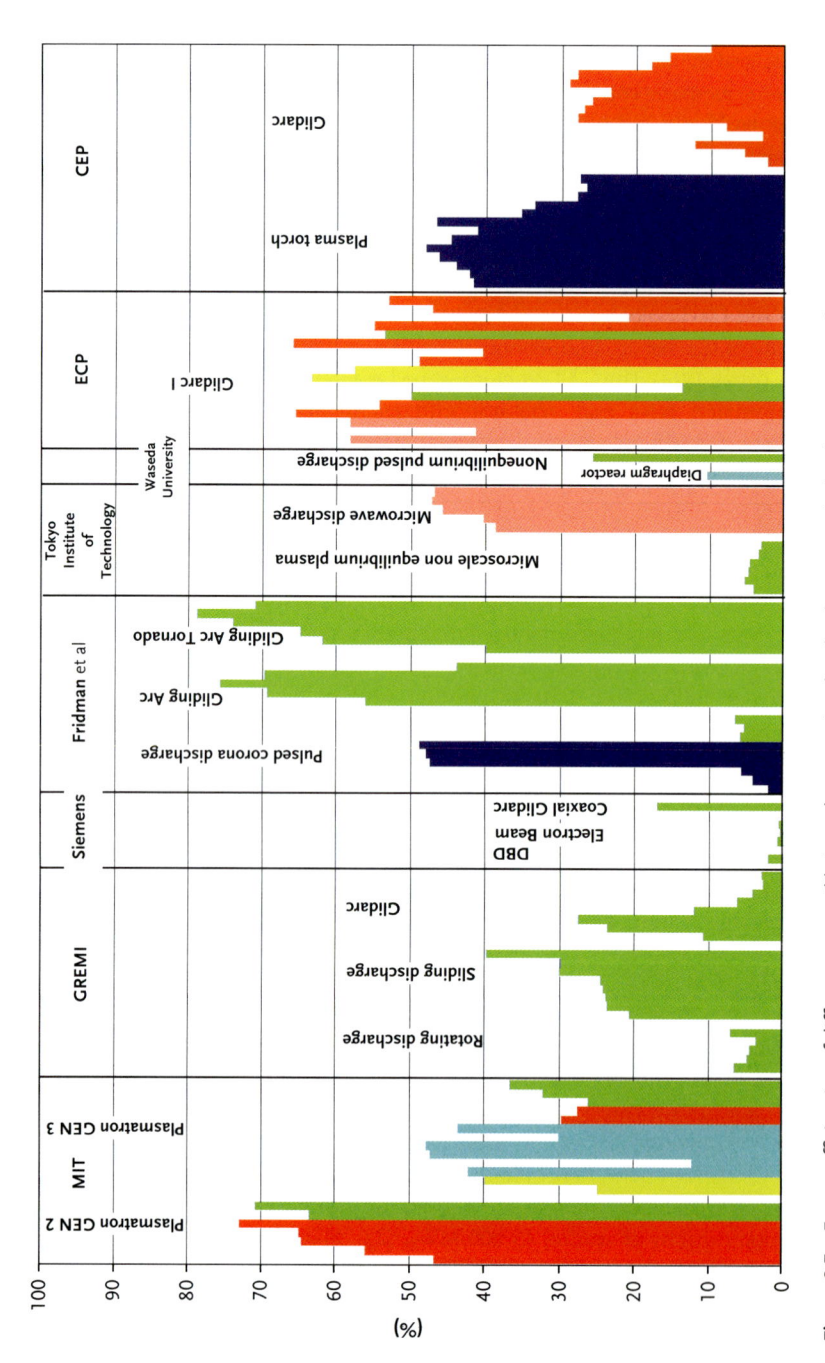

**Figure 2.7** Energy efficiencies of different nonequilibrium plasma-assisted technologies [23], by kind permission of MINES ParisTech – CEP, Plasma Group.

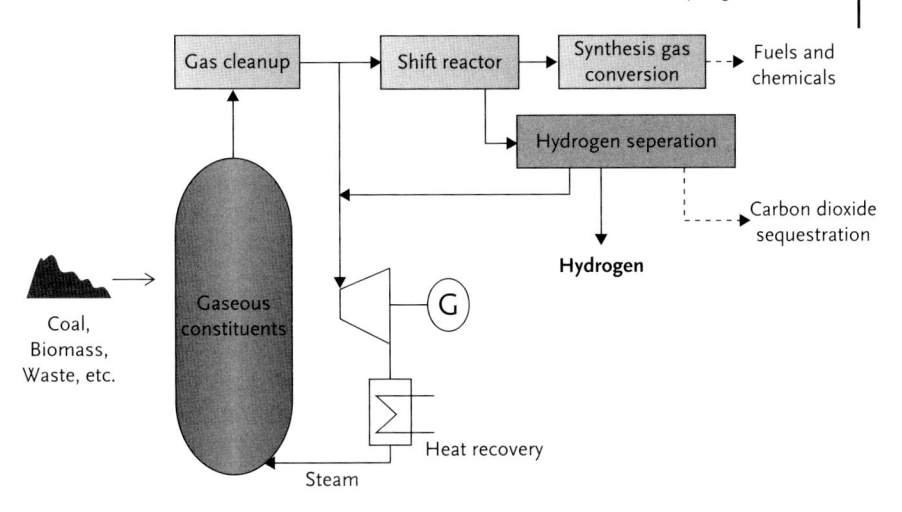

**Figure 2.8** Gasification-based energy-conversion options.

Figure 2.8 presents schematic process based on gasification. In this process, carbon-based feedstock is converted in the reactor (gasifier) to synthesis gas, being a mixture of hydrogen and CO. This process takes place in the presence of steam and oxygen at high temperatures and moderate pressure.

Stiegel and Ramezan [29] estimate that the availability of the gasifier must be greater than 97% for hydrogen production from coal. Department of Energy (DOE) Cost Targets in 2017 for coal gasification are estimated to be less than $1.10 for central production scale (with and without carbon capture and storage) [30].

## 2.3.2
### Water-Splitting Processes (Hydrogen from Water)

Hydrogen may be produced from water by water electrolysis in low-temperature process and from steam at elevated temperatures. Another way to generate hydrogen from water is the direct water-splitting process at high temperatures.

#### 2.3.2.1 Electrolysis of Water with Electricity from Renewable and Nonrenewable Energy Sources (Low-Temperature Water Splitting)

Electricity from renewable and nonrenewable energy sources may be converted to electricity through water electrolysis. Electrolysis has become a mature technology for both large- and small-scale hydrogen production and has been in the commercial use for more than 80 years. Electrolysis technology may be implemented at different scales wherever there is an electricity supply.

In water electrolysis, hydrogen is produced according to the following equation:

$$2H_2O_{liquid} + electricity \rightarrow 2H_2 + O_2 \qquad (2.11)$$

This reaction requires 39 kWh (140.4 MJ) of electricity to produce 1 kg of hydrogen at 25 °C at 1 atm.

$$\text{Reaction at the anode}: \quad H_2O \rightarrow \frac{1}{2}O_2 + 2H^+ + 2e^- \tag{2.12}$$

Equilibrium potential of the anode $E_a$ can be written as

$$E_a = E^0_{O_2} + \frac{RT}{2F}\ln\frac{a^{0.5}_{O_2}a_{H_2O}}{a^2_{OH^-}} \tag{2.13}$$

$$\text{Reaction at the cathode}: \quad 2H^+ + 2e^- \rightarrow H_2 \tag{2.14}$$

Equilibrium potential of the cathode $E_c$ can be written as

$$E_c = E^0_{H_2} + \frac{RT}{2F}\ln\frac{a^2_{H_2O}}{a^2_{H_2O} - a_{H_2}} \tag{2.15}$$

where $a$ is the activity. At 25 °C and pressure of 1 atm (0.1 MPa), the lowest voltage needed for water splitting is 1.229 V (reversible or equilibrium voltage). For this ideal case, the needed electricity is 83.7% of heating value of $H_2$, the rest is provided by heat from outside. Under isothermal conditions, it is possible to reach 100% conversion efficiency for producing hydrogen at voltage 1.47 V (known as the thermoneutral voltage) [31].

**Photovoltaic-Electrolysis System** Solar energy can be used to electrolyze water to produce hydrogen. Solar radiation is collected and converted into the useful form (heat or electricity) and then can power the electrolyzer. There are several solar methods of producing electricity from solar radiation like photovoltaics, solar power systems using central receiver thermal systems, parabolic troughs, and dish/stirling systems [31]. One of the methods to use solar energy is using PV cells. These cells produce direct currents; the electric power can be used directly for electrolysis of water to generate hydrogen. A schematic presenting photovoltaic-electrolysis hydrogen production system is shown in Figure 2.9.

The major components of such system are photovoltaic array being a set of photovoltaic modules (multiple interconnected solar cells), maximum power point tracker (MPPT), DC–DC converter supplying needed current for electrolysis process, and hydrogen storage system. Photovoltaic (PV) systems coupled to electrolyzers are commercially available. The systems offer some flexibility, as the output can be electricity from photovoltaic cells or hydrogen from the electrolyzer.

Direct photoelectrolysis is an advanced alternative to a photovoltaic-electrolysis system by combining both processes in a single apparatus. Light is used to directly split water into hydrogen and oxygen. Such systems offer a potential for cost reduction of electrolytic hydrogen compared with conventional two-step technologies. Present costs of electricity from a photovoltaic module is 6–10 times more than the electricity from coal or natural gas [32].

Photovoltaic solar cells and modules are produced for many applications, among others large-scale power generation (building integrated photovoltaics,

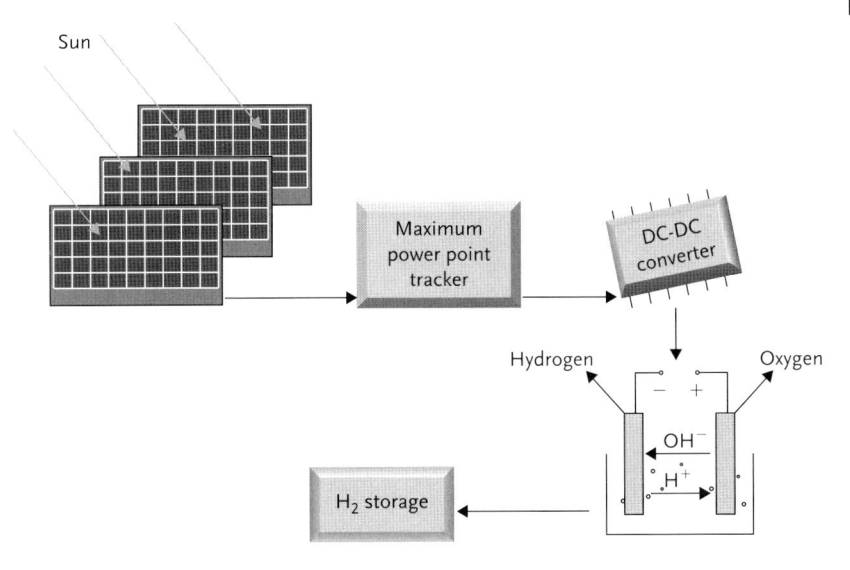

**Figure 2.9** Photovoltaic-electrolysis hydrogen production.

BIPV), supplying power in remote locations, satellites, and space vehicles [33]. Several types of PV technology can be distinguished: (i) wafer-based crystalline silicon, estimating that about 90% of current production uses wafer-based crystalline silicon technology [34], (ii) thin-film solar cells (TFSCs) using layers of semiconductors, and (iii) concentrating PV using lenses or mirrors. Solar cells are considered as a solution for space applications. Otte *et al.* [35] have reported that thin-film solar cells with cadmium, indium, gallium, selenium – $Cd(In,Ga)Se_2$ (CIGS) as absorber layer may become competitive option for future flexible thin-film space power generators when compared to traditional crystalline solar cells, mainly due to low mass and storage volume, high power/mass ratio (greater than 100 $W\,kg^{-1}$ at array level), high radiation resistance against proton and electron radiation, and lower production costs. Thin-film double-junction solar cells using amorphous silicon (a-Si:H) and microcrystalline silicon (μc-Si:H) are a further interesting candidate for use on satellites [36]. It is currently the most promising thin-film technology in terms of market impact. NCPV/NREL scientists have achieved world record, total-area efficiencies of 19.3% for a thin-film CIGS solar cell and 16.5% for thin-film cadmium telluride (CdTe) solar cell [37]. A schematic showing the structure of thin-film CIGS solar cell is presented in Figure 2.10.

Present and coming status of PV technologies and future development paths have been given by Kazmerski [38], highlighting the importance of cell and module performances. Figure 2.11 presents the research progress on PV market over the past three decades. International Energy Agency (IEA) has presented the PV Technology Roadmap targets including the essential role of R&D [39]. In case of crystalline silicon technologies, the main R&D aspects are new silicon materials and its processing and new device structures with novel concepts. Si consumption

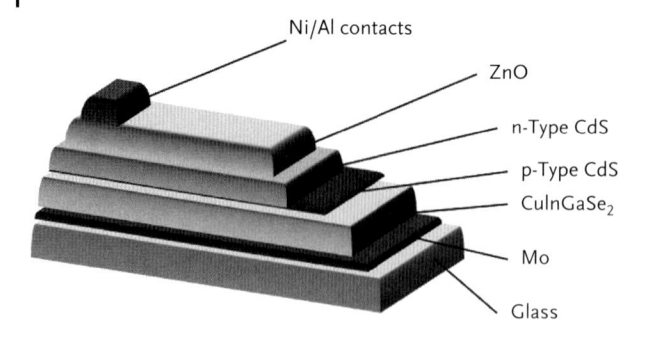

**Figure 2.10** Thin-film CIGS solar cell structure.

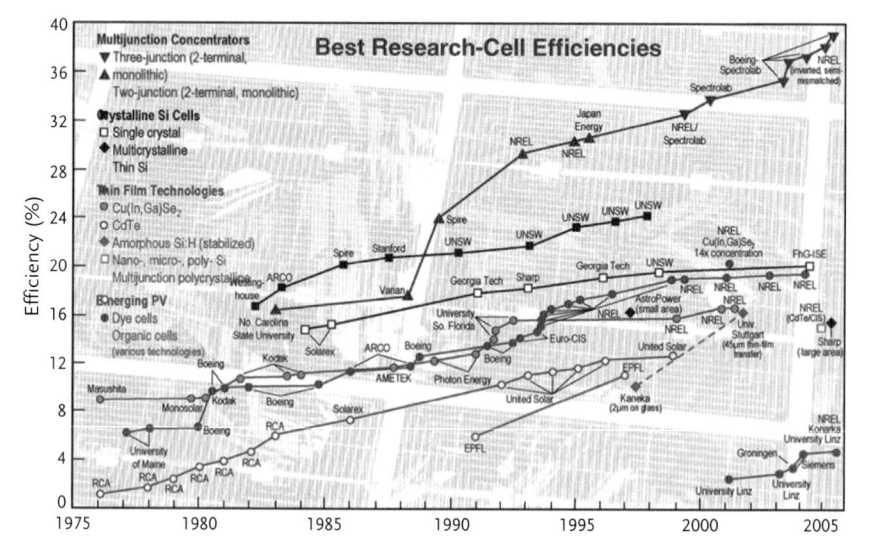

**Figure 2.11** Research progress of best research cells as a function of the technology type [38], reprinted by permission.

is expected to be less than 2 g W$^{-1}$ in 2020–2030/2050. Thin-film technologies goal for 2020–2030 is on advanced materials and concepts and simplified production process.

In photoelectrolytic processes, sunlight and semiconductors similar to photovoltaics are used. The basic photoelectrochemical (PEC) device consists of a semiconductor immersed in an aqueous electrolyte. Various materials have been investigated for use in photoelectrodes like thin film $WO_3$, $Fe_2O_3$, $TiO_2$; n-GaAs, n-GaN, CdS, ZnS for the photoanodes; and CIGS/Pt, p-InP/Pt, and p-SiC/Pt for the photocathodes [40]. Gallium arsenide (GaAs) solar panels are often used in the aerospace industry on satellites. They have efficiencies above 33%, giving them high-power densities. Current photoelectrodes used in photoelectrolysis that are stable in an aqueous electrolyte have low efficiency (1–2%) or low durability in a liquid electrolytic environment [41]. The target efficiency is greater than 16% solar energy to hydrogen. Photoelectrochemical hydrogen production is still in an early stage of development; the research and development is focusing on highly efficient, low-cost, and durable materials being able to reach 1.6–2.0 eV for single photoelectrode cells and 1.6–2.0 eV/0.8–1.2 eV for top/bottom cells in tandem configurations [40].

**Wind–Electrolysis System** Using the wind turbines as the source of electricity, pollution-free hydrogen can be produced by electrolysis (water is electrolyzed into hydrogen and oxygen). More specifically, the kinetic energy of the wind is converted into the mechanical energy by rotors of the turbines and then into electricity. Wind energy may be converted to hydrogen through water electrolysis by alkaline (KOH) electrolyzers or PEM electrolyzers. Hydrogen produced can be used, for example, as a fuel in fuel cells, internal combustion engines (ICEs), and turbines.

A principle of wind–electrolysis system is shown in Figure 2.12. In this case, wind turbine is connected to the electrolyzer to produce hydrogen. This may be useful if the distribution of hydrogen by pipeline system is more economically attractive than transmission of electricity to long distances (e.g., in remote areas).

Another option (Figure 2.13) is the integrated hydrogen–wind system, where produced and stored hydrogen can be used for transportation (fuel stations) or be sent to fuel cells to generate power again.

An innovative wind–hydrogen 3 MW system has been designed and constructed at the wind park of the Centre for Renewable Energy Sources (CRES), near Athens, Greece. The system consists of 500-kW wind turbine supplying power of 25 kW to a water electrolyzer with a nominal capacity of 0.45 kg h$^{-1}$ hydrogen, metal hydride tanks with a storage capacity of 3.78 kg hydrogen (42 Nm$^3$), a hydrogen compressor of 7.5 kW for filling high-pressure cylinders, nitrogen cylinder for inertization, water cooling system, and air compressor. Primary results showed that the overall efficiency of the system was approximately 50% [42]. A demonstration project combining wind power and hydrogen facility started to operate in western Norway on the island UTSIRA (*UTility Systems In Remote Areas*) in 2004. Two wind turbines (2 × 600 kW) combined with

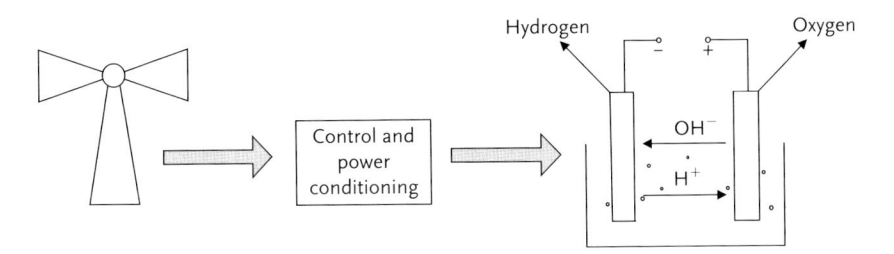

**Figure 2.12** Grid-independent wind–hydrogen generation system.

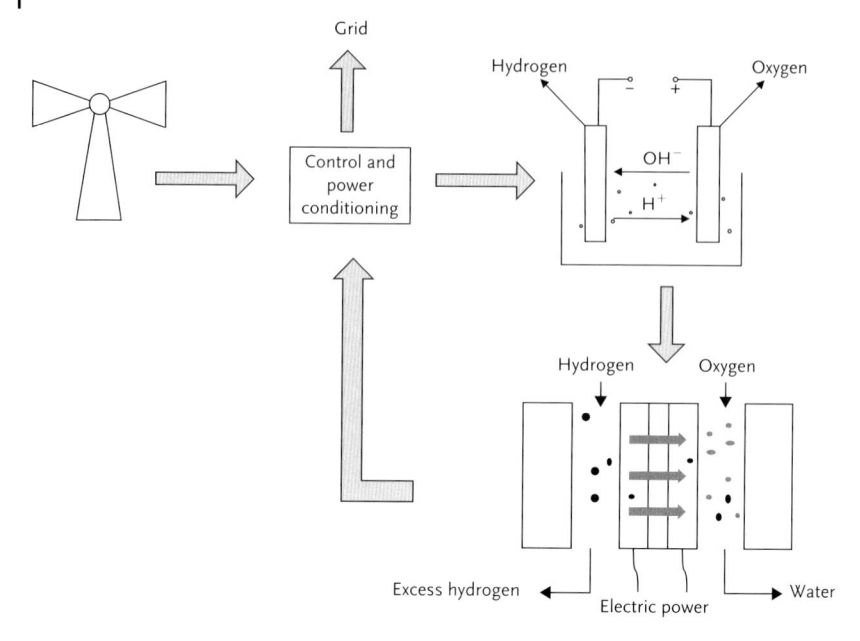

**Figure 2.13** General view of wind–hydrogen system.

**Table 2.5** New power capacity installed in the EU in 2008 (Total 23.851 MW) [45].

| Wind power | Gas | PV | Oil | Coal | Hydro | Nuclear |
| --- | --- | --- | --- | --- | --- | --- |
| 8.484 MW | 6.932 MW | 4.200 MW | 2.495 MW | 762 MW | 473 MW | 60 MW |
| 36% | 29% | 18% | 10% | 3% | 2% | 0.3% |

electrolyzer (10 000 Nm$^3$ h$^{-1}$, 48 kW) produce power for 10 private households, being part of the project. Surplus electricity is stored as chemical energy in the form of hydrogen (hydrogen storage capacity 2400 Nm$^3$, 20 MPa). When there is no wind, a hydrogen combustion engine (50 kW) and fuel cell (10 kW) convert the stored hydrogen back into electricity [43].

According to Sherif et al. [12], wind is the fastest growing renewable energy sector with annual growth of 27%. The World Wind Energy Association (WWEA) has published a forecast for worldwide wind installations in 2009 [44]. In 2008, more wind power was installed in the EU than any other generating technology, 36% of all capacity additions coming from wind, second to the natural gas as reported by the European Wind Energy Association (EWEA) [45]. It is said that a total of 23.851 MW of new power capacity was constructed in the EU in 2008. Table 2.5 shows the distribution of power capacity as related to the source.

Nearly 2.3% of total installed EU capacity is offshore. GE Wind was the number one manufacturer of wind turbines supplying the US market in 2008, with 43% of domestic turbine installations. Following GE were Vestas (13%), Siemens (9%), Suzlon (9%), Gamesa (7%), Clipper (7%), Mitsubishi (6%), Acciona (5%), and REpower (1%) [46].

Wind power can be used to generate hydrogen in grid connected or stand-alone applications. The economics of the wind hydrogen mainly depends on wind availability, costs of electrolyzers, hydrogen storage, transmission, and distribution availability. A key issue is whether hydrogen production systems shall be located near the wind farms or closer to the consumption sites. The future trend will be influenced by hydrogen transport and distribution costs as well as electricity market position [47]. A current trend, especially in Europe due to the high costs of ground, is to locate the wind turbines on platforms in the ocean, which is on the other hand expensive and complicated [12]. Hocevar and Summers [32] estimated that the costs in the future will be reduced by improvements in turbine design, more efficient power controls, optimization in rotor blades, and further improvements in materials. Wind electrolysis hydrogen generation is presently far from optimization, as the design of the turbine is to collect produced electricity and not hydrogen.

### 2.3.2.2 Different Types of Electrolyzers

Currently, three types of industrial electrolysis systems generating hydrogen are alkaline electrolyzers, proton exchange membrane (PEM) electrolyzers, or solid-oxide electrolysis cells (SOEC). Commercial low-temperature electrolyzers have system efficiencies of 57–75% (alkaline) and 55–70% (PEM) and are well established in industry and mature [41]. DOE 2017 cost targets set for water electrolysis are as following: alkaline and PEM electrolyzer less than $2.70 and less than $2.00 (distributed and semicentral/central production scale, respectively), and SOEC less than $2.00 (central production scale) [30]. Production of hydrogen by electrolysis is also expensive. According to Hocevar and Summers [32], the total cost of electrolytic hydrogen from currently available technology reaches a 14% capital cost recovery factor and has to include the total costs associated with the fueling facility.

**Alkaline Electrolyzers** Alkaline electrolyzers use approximately 30 wt% of alkaline electrolyte (usually KOH, potassium hydroxide) for transferring hydroxyl ($OH^-$) ions. Typically, alkaline electrolyzers can have the current density in the range of 100–300 mA cm$^{-2}$ and the efficiency (based on LHV of hydrogen) is 50–60% [40]. Alkaline electrolyzers are suitable for stationary applications and operate at pressures up to 25 bar (2.5 Pa).

The overall reactions at the electrodes are as following:

$$\text{Anode}: 4OH^- \rightarrow O_2 + 2H_2O + 4e^- \tag{2.16}$$

$$\text{Cathode}: 2H_2O + 2e^- \rightarrow H_2 + 2OH^- \tag{2.17}$$

$$\text{Overall reaction}: H_2O \rightarrow H_2 + \frac{1}{2}O_2 \tag{2.18}$$

Capacities of alkaline electrolyzers are in a wide range, from about 20 kW up to several mega watts. Several large-scale alkaline electrolyzers (above 100 MW) have been installed; currently, the largest unit is running in Egypt with capacity of 33 000 Nm$^3$h$^{-1}$ (156 MW). Among well-established manufacturers of alkaline electrolyzers are Norsk Hydro (Norway), Hydrogenics, DeNora (Italy), and Electrolyser Corporation Ltd. (Canada). At Juelich Research Center, the prototype of an alkaline high-pressure 120-bar (12 MPa) 5 kW electrolyzer has been developed and built [48]. Depending on temperature and current density, energy efficiencies of 75–90% were achieved. Energy efficiency was independent of the operating pressure (no additional energy was required for electrolysis at high pressures). Cross section of the FZ Juelich 5 kW high-pressure electrolyzer is shown in Figure 2.14.

Christodoulou *et al.* [49] reports about a "green" hydrogen production by alkaline water electrolysis combined with electricity produced by stand-alone photovoltaic (PV)–wind turbine (WT) hybrid system. Pilot unit, located inside the factory of Hystore Technologies Ltd. in Ergates Industrial Area, in Nicosia-Cyprus, consists of 940 W PV subsystem and 1300 W wind turbine subsystem, 3000 Wh battery bank, and an electric energy inverter (220 ACV/50 Hz). It was reported that total daily electricity production was of about 7313 Wh, enough to start-up electrolyzer to produce 1330 NL H$_2$ per day. The pilot unit was able to produce 4000 NL H$_2$ per month with a purity of more than 99.999%. The alkaline electrolyzer used in the pilot plant was a Model G2 from Erre Due (Italy). A schematic and the pictures of the subunits of the pilot system are shown in Figures 2.15 and 2.16, respectively.

**PEM Electrolyzers** PEM electrolyzers are used for production of gases for fuel cells having a high degree of purity (more than 99.99% for hydrogen) but also in analytical chemistry, welding processes, and metallurgy. Typically, precious metal catalysts (Pt, Pt/Rh, and iridium) are used together with solid polymer electrolyte (mainly Nafion®).

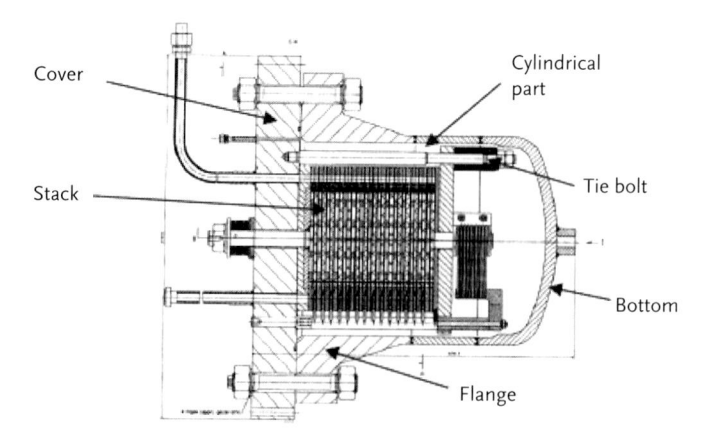

**Figure 2.14** Stack cross section of the FZ Juelich 5 kW high-pressure electrolyzer [48].

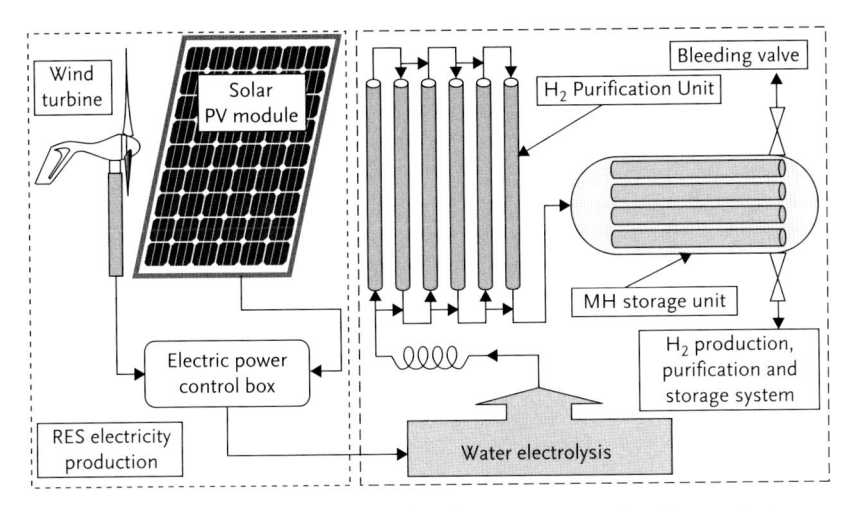

**Figure 2.15** Schematic of the subunits of the pilot system [49], reprinted by permission.

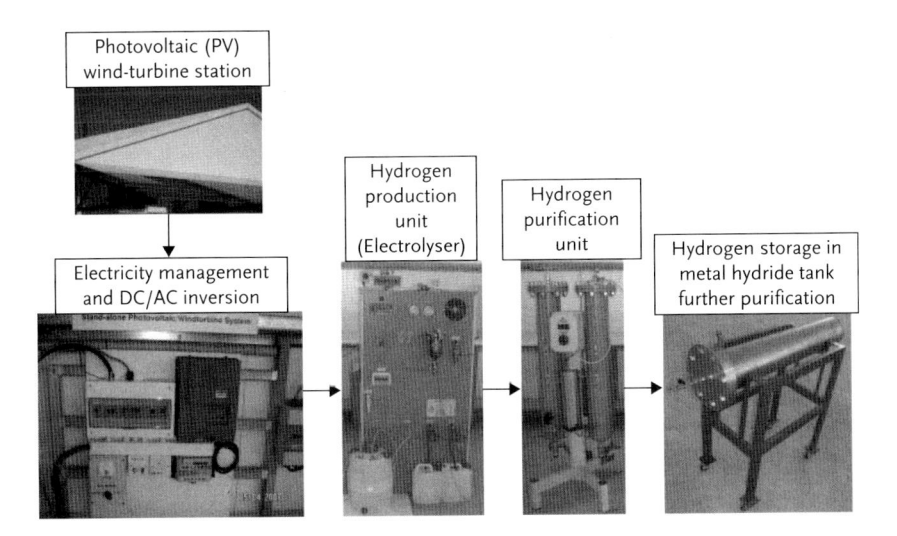

**Figure 2.16** Pictures of the subunits of the pilot system [49], reprinted by permission.

The overall reactions at the electrodes of PEM electrolyzers are as follows:

$$\text{Anode}: 2H_2O \rightarrow O_2 + 4H^+ + 4e^- \qquad (2.19)$$

$$\text{Cathode}: 4H^+ + 4e^- \rightarrow 2H_2 \qquad (2.20)$$

$$\text{Overall reaction}: H_2O \rightarrow H_2 + \frac{1}{2}O_2 \qquad (2.21)$$

PEM electrolyzers have low ionic resistances; and as a result, high current densities greater than 1600 mA cm$^{-2}$ can be achieved with high efficiencies of 55–70% of the process [40]. Recent achievements in PEM electrolyzer technology are presented by Millet *et al.* [50]. GenHy® is a trademark of PEM water electrolyzers developed and commercialized by the French Compagnie Européenne des Technologies de l'Hydrogène (CETH) in the course of EC-supported R&D program GenHyPEM (Sixth Framework Programme). It is reported that nearly 80% efficiencies of PEM electrolyzer with hydrogen production capacity up to 1 Nm$^3$ H$_2$ h$^{-1}$ operating at the pressure range from 1 to 50 bar have been reached with GenHy®1000. GenHy electrolyzers can be powered by the main power supply using appropriate AC/DC converters or directly from PV panels using high-efficiency DC/DC converters.

New generation of low-cost and efficient possible catalysts materials, such as cobalt clatrochelates or polyoxometalates, have been presented for the hydrogen evolution reaction at the cathodes. Figure 2.17 presents a GenHy®1000 filter-press stack with 12 membrane-electrode assemblies.

Photograph of GenHy®3000 PEM electrolyzer that is able to produce up to 3 Nm$^3$ H$_2$ h$^{-1}$ is presented in Figure 2.18 [50].

As can be seen from Figure 2.19, the stack efficiency is directly related to the operating temperature and current density. At 1 A cm$^{-2}$ and about 85 °C, a conversion efficiency $\varepsilon_{\Delta H}$ close to 80% and a mean cell voltage value of 1.88 V were obtained. During production, stable electrochemical performances were obtained after a few hours of continuous operation, when the temperature of the system was stable. To reach high operating temperature, water circuits have to be properly insulated to reduce heat transfers to the surroundings. Several hundred hours of

**Figure 2.17** GenHy®1000 PEM electrolyzer, reprinted by kind permission of P. Millet, Université Paris Sud 11 (UMR 8182) and CETH.

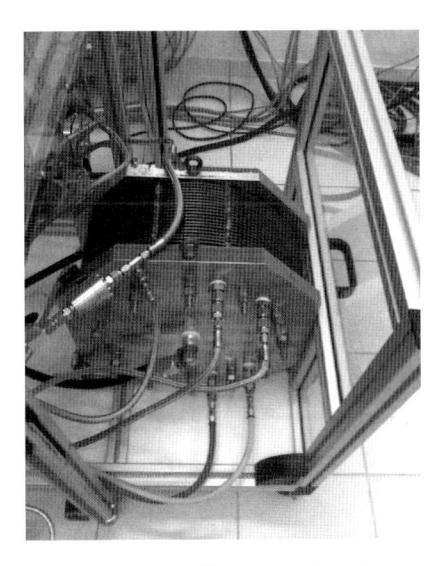

**Figure 2.18** GenHy®3000 PEM electrolyzer, reprinted by kind permission of P. Millet, Université Paris Sud 11 (UMR 8182) and CETH.

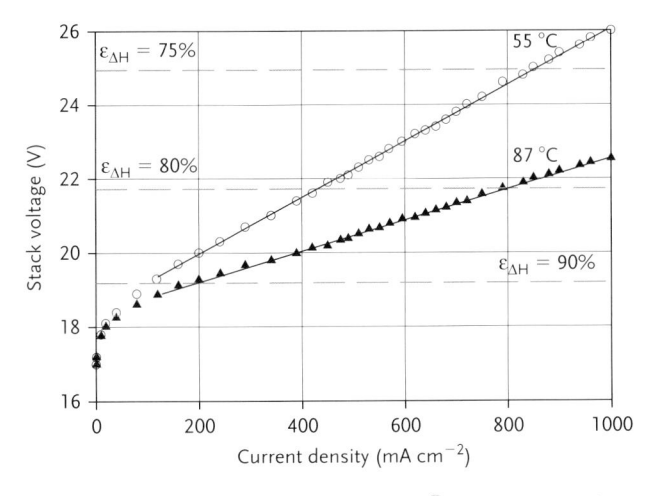

**Figure 2.19** Polarization curves of the GenHy®1000 PEM electrolyzer at different operating temperatures and pressures, reprinted by kind permission of P. Millet, Université Paris Sud 11 (UMR 8182) and CETH.

intermittent operation have been satisfactorily performed without noticeable degradation of electrochemical performances as presented in Figure 2.19.

In the course of the GenHyPEM project, some experiments have been made at high operating pressures under stationary conditions using a stack of cells made of PTFE-reinforced sulfonated polyethers (Nafion® 1100). Some results obtained in the 1–130 bar (0.1–13 MPa) pressure range are compiled in Table 2.6 [50].

**Table 2.6** Main test results reached with high-pressure PEM water electrolysis stack under stationary operating conditions, reprinted by kind permission of P. Millet, Université Paris Sud 11 (UMR 8182) and CETH.

| Parameter | Unit | Measured values at operating pressure (bar) | | | | | |
|---|---|---|---|---|---|---|---|
| | | 1 | 25 | 50 | 75 | 100 | 130 |
| Electric current | A | 123.5 | 126.5 | 125.5 | 124.5 | 125.0 | 124.0 |
| Operating temperature | °C | 86 | 87 | 5 | 89 | 84 | 88 |
| $H_2$ production rate | $Nm^3\,h^{-1}$ | 0.42 | 0.43 | 0.42 | 0.40 | 0.38 | 0.36 |
| $H_2$ purity before recombiner | % vol | 99.98 | 99.76 | 99.18 | 98.56 | 98.01 | 97.34 |
| $H_2$ purity after recombiner | % vol | 99.999 | 99.999 | 99.997 | 99.995 | 99.993 | 99.991 |
| Individual cell voltage | V | | | | | | |
| 1 | | 1.70 | 1.70 | 1.71 | 1.71 | 1.72 | 1.73 |
| 2 | | 1.68 | 1.69 | 1.70 | 1.71 | 1.71 | 1.72 |
| 3 | | 1.71 | 1.71 | 1.71 | 1.72 | 1.71 | 1.71 |
| 4 | | 1.70 | 1.71 | 1.71 | 1.73 | 1.74 | 1.74 |
| 5 | | 1.71 | 1.70 | 1.71 | 1.71 | 1.73 | 1.74 |
| 6 | | 1.69 | 1.69 | 1.70 | 1.71 | 1.73 | 1.73 |
| 7 | | 1.70 | 1.69 | 1.69 | 1.70 | 1.72 | 1.73 |
| 8 | | 1.68 | 1.70 | 1.70 | 1.73 | 1.74 | 1.74 |
| Stack voltage | V | 13.57 | 13.59 | 13.63 | 13.72 | 13.80 | 13.84 |
| Power consumption | kW | 1.676 | 1.719 | 1.711 | 1.708 | 1.725 | 1.716 |
| Specific energy consumption | $kWh\,N^{-1}\,m^{-3}$ | 3.99 | 4.00 | 4.07 | 4.27 | 4.54 | 4.77 |
| Current efficiency | % | 99.98 | 99.68 | 98.13 | 96.57 | 92.91 | 90.45 |
| Energy efficiency, $\varepsilon_{\Delta G}$ | % | 72.5 | 72.4 | 72.2 | 71.7 | 71.3 | 71.0 |

GenHy®1000 PEM electrolyzer systems are designed for the automated production of hydrogen using renewable sources of energy and can be used for the direct storage of hydrogen in metal hydride tank at low pressure or in cylinder under medium pressure. Larger electrolyzers (production capacity up to 5 $Nm^3$ $H_2\,h^{-1}$) have been developed at CETH in parallel with the GenHyPEM project, using 600 $cm^2$ MEAs (Figure 2.20). Dual units that can produce 10 $Nm^3\,H_2\,h^{-1}$ are now under testing. Modular units that can produce up to 100 $Nm^3\,H_2\,h^{-1}$ are planned for 2011.

**High-Temperature Electrolysis of Steam** Hydrogen can be produced by high-temperature electrolysis of water vapor offering significant thermodynamic and kinetic advantages and higher energy efficiency when compared with electrolysis performed at lower temperatures. Increasing the temperature to 1050 K results in about 35% reduction of the thermal and electrical demands of the electrolysis process [51]. As this process requires high temperatures for operation, it is coupled with advanced high-temperature nuclear reactors.

High-temperature advanced nuclear reactor combined with a high-efficiency high-temperature electrolyzer could achieve thermal to hydrogen conversion

**Figure 2.20** GenHy®5000 PEM electrolyzer, reprinted by kind permission of P. Millet, Université Paris Sud 11 (UMR 8182) and CETH.

efficiency of 45–55% [52]. Water is decomposing at temperatures around 2500 °C. High-temperature electrolysis of steam (HTES) concept has been proposed by INEEL [53, 54] using high-temperature gas-cooled reactor (HTGR) being a combination of thermal energy and electricity to decompose water in electrolyzer similar to the solid-oxide fuel cell (SOFC). High working temperature increases the efficiency of electrolyzer by decreasing the electrodes overpotentials, which result in power losses. In addition, the electricity demand for water splitting is reduced. As noncorrosive solid electrolytes are used, there are no liquid and flow distribution issues. The high operating temperature results in longer start-up times and mechanical/chemical compatibility issues. It is assumed that efficiencies up to 95% may be possible, but also estimated that it might take up to 10 years until solid oxide systems are available [32]. A concept of a very high temperature gas-cooled reactor (VHTR) coupled with high-temperature steam electrolysis for hydrogen production as a function of operating temperature 600–1000 °C has been evaluated by Korea Atomic Energy Research [55]. At 1000 °C, the maximal overall thermal efficiency was estimated about 48%. The overall process consisted of VHTR to produce high-temperature thermal energy, power cycle to generate electricity, AC/DC converter to supply direct current power to the electrolyzer, heat exchangers supplying thermal energy to superheated steam generator, high-temperature electrolyzer to produce hydrogen from steam, water–hydrogen separation unit,

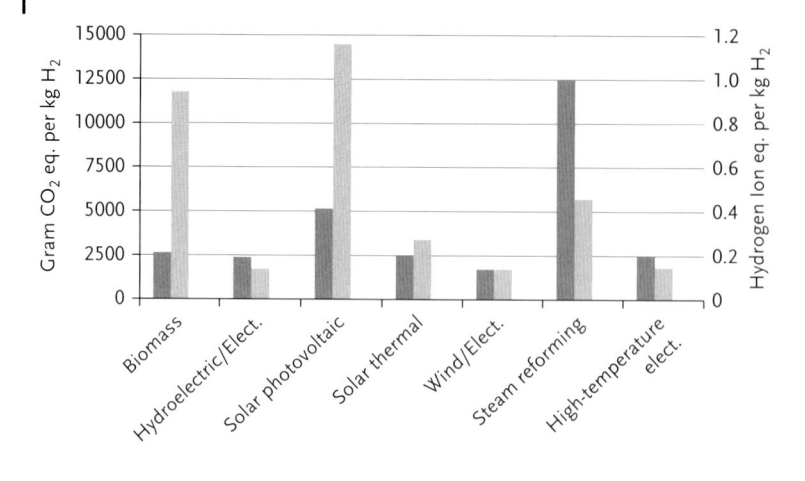

**Figure 2.21** Greenhouse emissions (dark gray columns) and acidification potential (light gray columns) of hydrogen production by several processes [51].

and dehumidifier to remove residual water from the hydrogen. Scenarios for large-scale hydrogen production by coupling high-temperature reactor with steam electrolyzers were presented and assessed by CEA [56]. Such combinations allow lowering the electrical consumption of electrolyzer being supplied by a high-temperature steam. A result of 55% efficiency has been reached. Life cycle assessment (LCA) of high-temperature electrolysis shown that environmental impacts of this process with respect to acidification potential and $H_2$ production greenhouse emissions are comparable with solar thermal processes and far lower than those of solar photovoltaic, biomass, and steam reforming [51], as presented in Figure 2.21.

A combination of solid-oxide fuel-assisted electrolysis cell (SOFEC) and SOFC hybrid stack for cogeneration of hydrogen and electricity has been proposed and demonstrated with efficiencies up to 69% using methane as a fuel [57]. Figure 2.22 presents the cogeneration idea, where SOFEC directly applies the energy of fuel to replace the external electrical energy required to produce hydrogen from water or water steam. SOFEC and SOFC are build in the same stack; hydrogen is produced by the SOFEC and electricity by the SOFC.

A 5-kW SOFEC–SOFC hybrid system design as developed by the Materials Science and Research, Inc., UT consists of following elements: three hybrid SOFEC–SOFC modules, three dedicated SOFC modules, steam generators, tail-gas combustors, process gas heaters, and Balance of Plant (BOP) zone. This is presented in Figure 2.23.

Estimated efficiency of such hybrid system is presented in Figure 2.24. Calculated values are based on sum of hydrogen produced (LHV) and net electrical output divided by consumed fuel (LHV of methane). Efficiency varies with output level, operating mode, and fuel utilization. Peak cogeneration efficiency is 54%.

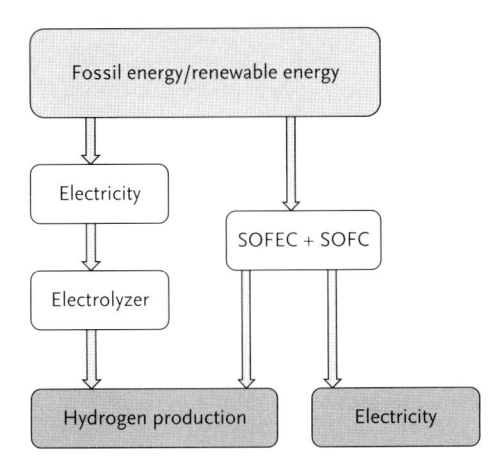

**Figure 2.22** SOFEC and SOFC hybrid stack, process schematic, adapted by permission of Materials and Systems Research, Inc.

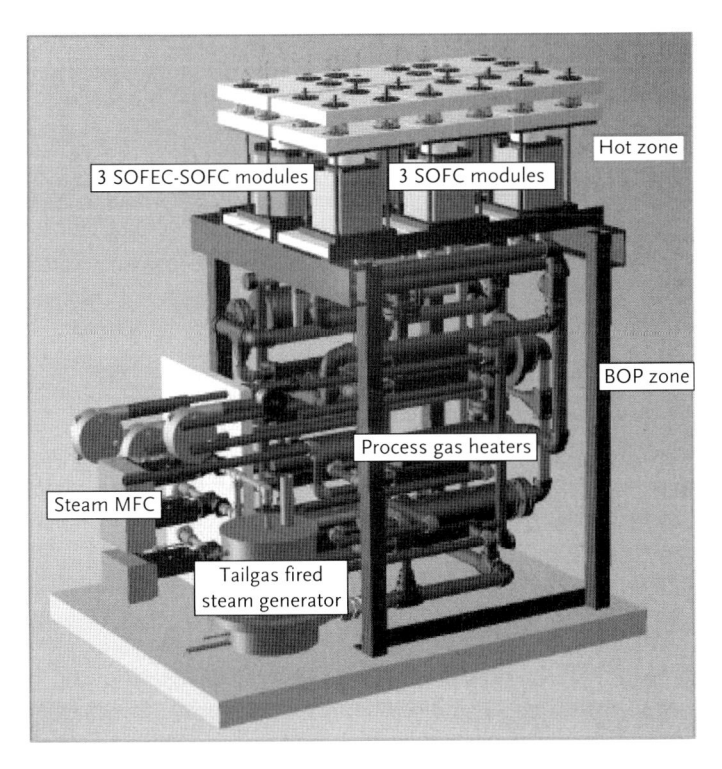

**Figure 2.23** Five kilowatt SOFEC–SOFC hybrid system design, by permission of Materials and Systems Research, Inc.

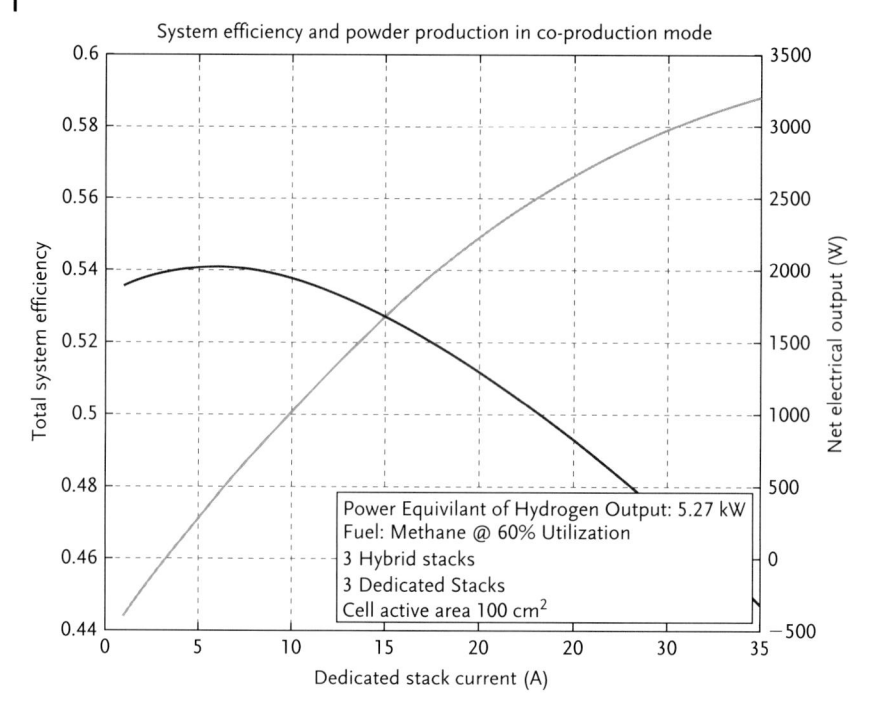

**Figure 2.24** Estimated efficiency of 5 kW SOFEC–SOFC hybrid system, by permission of Materials and Systems Research, Inc.

#### 2.3.2.3 High-Temperature Water Splitting in Combination with High-Temperature Nuclear Energy and Solar Energy

Hydrogen may be converted into electricity by means of endothermic water-splitting processes, supported by nuclear process heat. Nuclear power systems being relatively clean and abundant source of energy have a big potential to contribute to hydrogen economy. Hydrogen from nuclear reactor is produced by high-temperature water electrolysis (described above) or by thermochemical cycles, sulfur–iodine or UT-3, using nuclear heat or alternative energy sources for example wind, water, or sun. In electrolysis processes, energy needed for the process can be supplied by electricity generated from fossil fuels. In general, these processes are attractive because no carbon dioxide is released during the decomposition of the feedstock.

Direct solar hydrogen production used in thermal processes has been investigated in various research concepts through laboratory demonstrator to pilot scale over years.

The German Aerospace Center (DLR) is working since 1970s on the possibilities of direct use of solar energy, including photovoltaics and solar thermal and on iron-based thermochemical cycle processes [58].

Reactive material based on iron oxide is alternatively oxidized and reduced. Ceramic cylinder coated with metal oxide is heated in a solar receiver reactor with concentrated solar radiation to about 800 °C. In the first step, water vapor reacts with the coating liberating hydrogen. In the second step, the water vapor is turned off, reactor is heated up to 1200 °C, and oxygen is released from the reactive coating. and the cycle starts anew. The process is presented by the following reactions:

$$MO_{red} + H_2O \rightarrow H_2 + MO_{ox}; \quad T = 800\ °C \tag{2.22}$$

$$MO_{ox} \rightarrow MO_{red} + O_2; \quad T = 1200\ °C \tag{2.23}$$

After successful solar thermochemical hydrogen production in pilot tests (HYDROSOL-2) at the DLR, Cologne, the process has been introduced on the Spanish Platforma Solar de Almeria (PSA) in Tabernas (Figure 2.24). Two solar receiver reactors are located on 28 m small solar power system (SSPS) tower, in a modular mode designated for the output of 100 kW. Sun heliostats (Figure 2.25) have a thermal capacity of 2.7 MW, and assuring a direct solar radiation of 1900 kWh year$^{-1}$ m$^{-2}$ with more than 3000 h of sunshine a year, the location provides the best solar power in Europe [46] (Figure 2.26).

Hydrogen can also be generated using nuclear energy of thermochemical cycles (e.g., sulfur–iodine or UT-3) and hybrid cycles with electrolytic and thermochemical step. These processes need large amount of heat or electricity to generate hydrogen which could be supplied by the nuclear energy.

It is estimated that one of the most promising thermochemical cycles for massive scale hydrogen production is the sulfur–iodine (S–I) cycle (originally developed by the General Atomic) with thermal to hydrogen efficiency of 52% and adiabatic UT-3 cycle (calcium–bromine cycle) developed at the University of

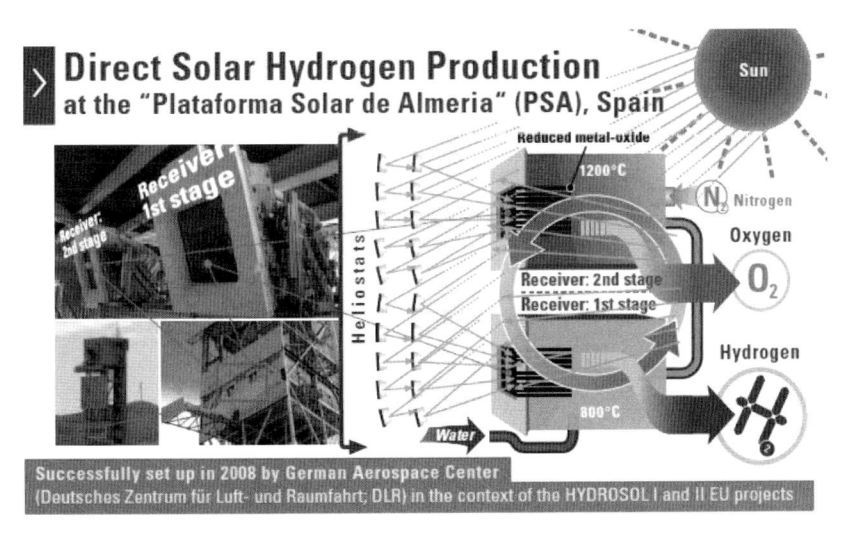

**Figure 2.25** Direct solar hydrogen production – process schematic at Platforma Solar de Almeria (PSA), [58] reprinted by permission of Arno A. Evers, FAIR-PR, Founder (1995) of the Group Exhibit Hydrogen and Fuel Cells at the Annual Hannover Fair.

**Figure 2.26** Direct solar radiation is collected by a field of 300 heliostats, each with a surface area of 40 m². The heliostats are operated to follow the sun by a computer-driven servo motors, reprinted by permission of Arno A. Evers, FAIR-PR, Founder (1995) of the Group Exhibit Hydrogen and Fuel Cells at the Annual Hannover Fair.

Tokyo with efficiency of around 50%. Sulfur–iodine cycle is a combination of three chemical reactions decomposing water using iodine and sulfur compounds in a closed-cycle by use of high-temperature nuclear heat supplied by HTGR, finally leading to water splitting:

Bunsen reaction (exothermic):
$$2H_2O + SO_2 + I_2 \rightarrow H_2S_4 + 2HI \ (T < 120\,°C) \quad \Delta H = -216 \text{ kJ mol}^{-1} \tag{2.24}$$

$H_2S_4$ decomposition (endothermic):
$$H_2S_4 \rightarrow H_2O + SO_2 + \frac{1}{2}O_2 (T > 800\,°C); \quad \Delta H = 371 \text{ kJ mol}^{-1} \tag{2.25}$$

HI decomposition: $\quad 2HI \rightarrow H_2 + I_2 (T > 300\,°C); \quad \Delta H = 12 \text{ kJ mol}^{-1}$
$$\tag{2.26}$$

As a result, hydrogen and oxygen are formed according to the reaction:

$$\text{Overall net result of iodine-sulfur cycle: } H_2O \rightarrow H_2 + \frac{1}{2}O_2 \tag{2.27}$$

Another thermochemical process may be UT-3 cycle using calcium–bromine. This process has been proposed in 1978 by Kameyama and Yoshida from University of Tokyo and since then has been intensively investigated. Under the contract of Japan Atomic Energy Research Institute (JAERI), the feasibility study of commercial UT-3 cycle plant showed hydrogen capacity of 20 000 Nm³ h$^{-1}$.

UT-3 process is a combination of following four gas–solid reactions [59]:

$$CaBr_{2(g)} + H_2O_{(g)} \rightarrow CaO_{(s)} + 2HBr_{(g)} \quad (897\,°C) \qquad (2.28)$$

$$CaO_{(g)} + Br_{2(g)} \rightarrow CaBr_{2(s)} + \frac{1}{2}O_{2(g)} \quad (427\,°C) \qquad (2.29)$$

$$Fe_3O_{4(s)} + 8HBr_{(g)} \rightarrow 3FeBr_{2(s)} + 4H_2O_{(g)} + H_{2(g)} \quad (-143\,°C) \qquad (2.30)$$

$$3FeBr_{2(s)} + 4H_2O_{(g)} \rightarrow Fe_3O_{4(s)} + 6HBr_{(g)} + H_{2(g)} \quad (537\,°C) \qquad (2.31)$$

In general, there are more than 300 water-splitting processes reported in the literature allowing reducing the temperature from 2500 °C, but they usually work at higher pressures. Some examples are Ispra Mark process, sulfur acid decomposition, and SynMet process of ZnO reduction and methane reforming within a solar reactor developed at Paul Scherrer Institute.

JAEARI aimed at connecting nuclear system with a chemical process producing hydrogen. High-temperature steam electrolyzers using nuclear energy are able to generate hydrogen without the corrosive conditions of thermochemical processes and greenhouse gas emissions associated with hydrocarbon processes. Using conventional steam reformers with 80% efficiency, from 1 mol of methane, 2.7 mol of hydrogen is produced. By using nuclear-heated reactors, from 1 mol of methane, 4 mol of hydrogen can be obtained, as no methane is consumed for combustion and the yield of hydrogen is nearly stoichiometric [60]. A comparison between methane combustion steam reforming and nuclear-heated steam reforming is shown in Figure 2.27.

Helium-cooled high-temperature engineering test reactor (HTTR) operating at 30-MWth output and 950 °C outlet coolant temperature was connected to a conventional steam methane/natural gas reforming (SMR) process. The goal was to produce 4200 Nm$^3$ h$^{-1}$ of hydrogen from 1800 Nm$^3$ h$^{-1}$ natural gas using 10 MW thermal heat supply at 880 °C from the HTTR. This HTTR–SMR plant was scheduled to produce hydrogen in 2008 [60].

## 2.3.3
### Hydrogen from Biomass

Biomass is one of the possibilities to produce hydrogen by different processes, however with low efficiency. The use of variety of biomass instead of fossil fuels for hydrogen generation may contribute to the reduction of $CO_2$ emissions to the atmosphere. It is estimated that about 12% of today's world energy supply comes from biomass, and the contribution from developing countries is significantly higher, of about 40–50% [61]. In general, four main types of biomass feedstock are defined: (i) energy crops: herbaceous, woody, industrial, agricultural, and water, (ii) agricultural residues and waste, (iii) forest waste and residues, and (iv) industrial and municipal wastes [61]. Hydrogen production from biomass can be classified into two categories: biological and thermochemical processes, which are shown in Figure 2.28.

**46** | *2 Hydrogen – Fundamentals*

Fossil-combustion steam reforming

0.83 mol $CH_4$ Steam Reform

0.17 mol $CH_4$ Combust

3.3 mol $H_2$ (803 kJ)

Nuclear-heated steam reforming

1 mol $CH_4$ Steam Reform

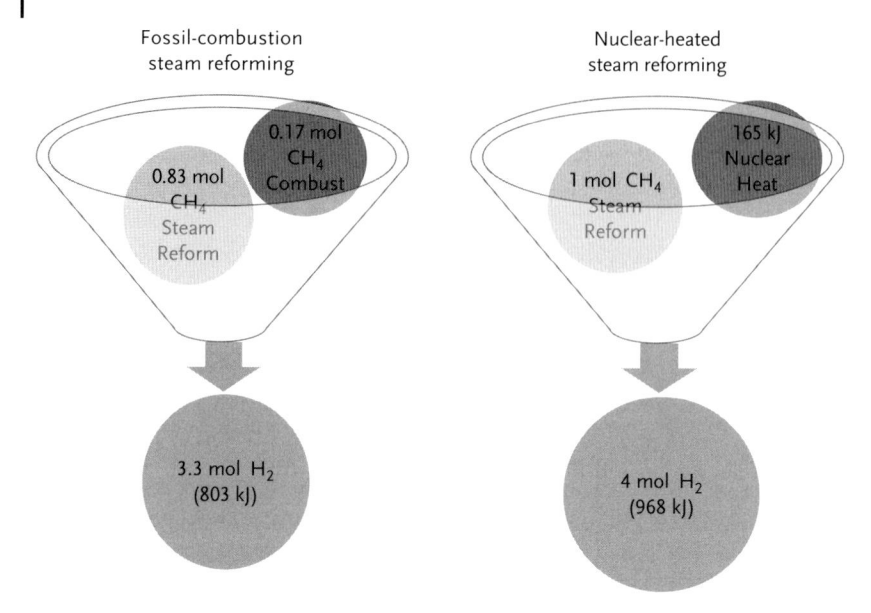

165 kJ Nuclear Heat

4 mol $H_2$ (968 kJ)

**Figure 2.27** Advantageous effect of nuclear heat supply to SMR [60].

Hydrogen production from biomass

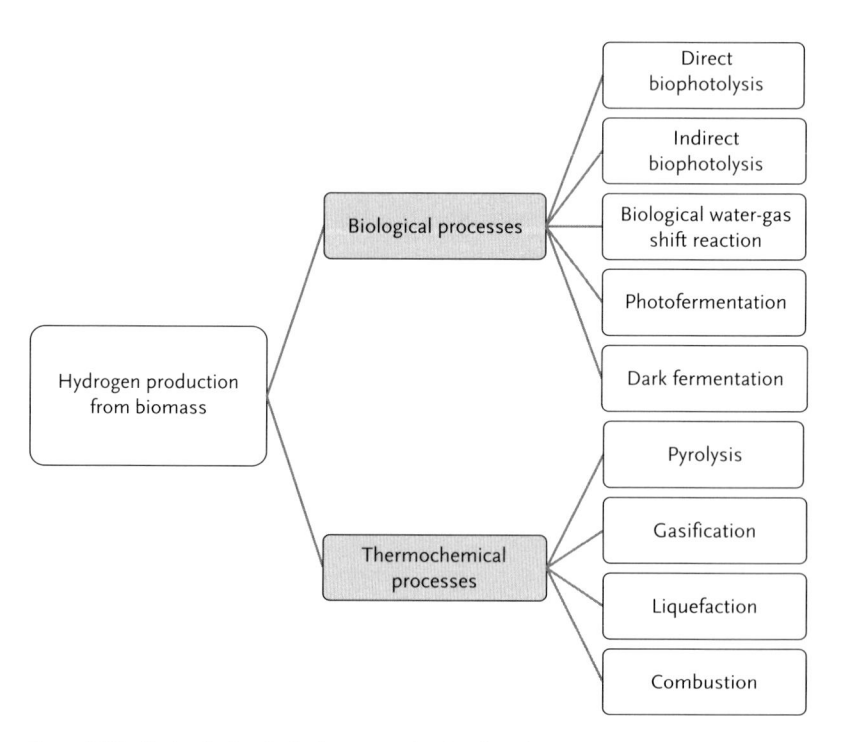

- Biological processes
  - Direct biophotolysis
  - Indirect biophotolysis
  - Biological water-gas shift reaction
  - Photofermentation
  - Dark fermentation
- Thermochemical processes
  - Pyrolysis
  - Gasification
  - Liquefaction
  - Combustion

**Figure 2.28** Technologies for hydrogen production from biomass.

### 2.3.3.1 Thermochemical Processes

During pyrolysis, the biomass is heated at 370–550 °C at 0.1–0.5 MPa in the absence of air to convert into liquid oils, solid charcoal, and gaseous compounds. A variety of different types of biomass have been tested for pyrolysis for hydrogen generation. Evans *et al.* [62] used pelletized peanut shells by pyrolysis integrated with catalytic steam reforming. Peanut shells have been investigated also by Yeboah *et al.* [63] and Abedi *et al.* [64]. Czernik *et al.* [65] have tested postconsumer wastes such as plastics, trap grease, mixed biomass, and synthetic polymers demonstrating that hydrogen could be efficiently produced from plastics; 34 g hydrogen was obtained from 100 g of polypropylene. Reforming of "trap grease" resulted in 25 g hydrogen per 100 g grease. During combustion process, biomass is burned in air. Generated chemical energy of biomass could be transferred into heat, mechanical power, or electricity. In biomass liquefaction, biomass is heated to 525–600 K in water under a pressure of 5–20 MPa in the absence of air. Liquefaction is not favorable for hydrogen production due to low hydrogen production and difficulties to achieve the operation conditions [61]. Biomass gasification is carried out in the presence of oxygen and is applicable in case of biomass having moisture content less than 35%. The direct biomass gasification process is similar to the coal gasification process. Bowen *et al.* have tested [66] three candidate biomass sources: bagasse, switchgrass, and a nutshell mix consisting of 40% almond nutshell, 40% almond prunings, and 20% walnut shell. The authors have compared the costs of hydrogen production and concluded that they are comparable to steam methane reforming (Figure 2.29).

### 2.3.3.2 Biological Processes

In general, biological processes may offer long-term potential for sustainable hydrogen production with rather low environmental impact, but they are in early

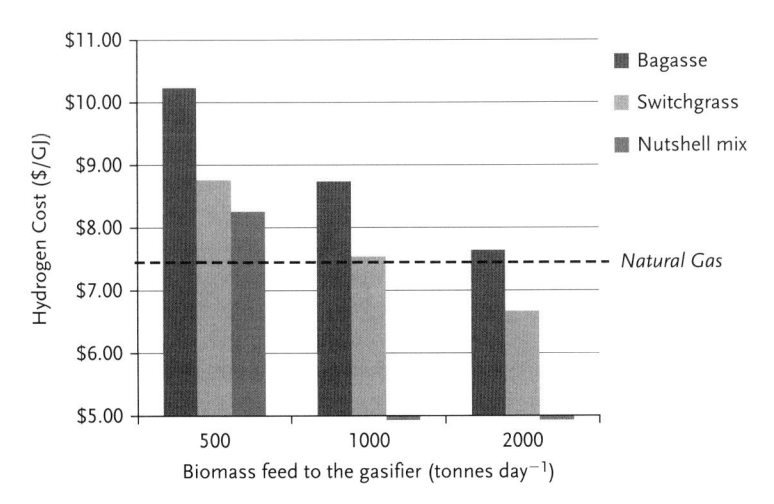

**Figure 2.29** Estimated cost comparison of hydrogen production by biomass gasification and natural gas steam reforming [66].

stages of development. Photolytic processes are based on the metabolism of green algae or cyanobacteria (blue-green algae) and Gram-positive bacteria, with sunlight to split the water, controlled by hydrogen-producing enzymes like hydrogenase and nitrogenase. In photosynthetic hydrogen production, sunlight is the source for photosynthetic bacteria to break down organics, releasing hydrogen. Opposite to that, dark-fermentative (anaerobic) bacteria grown on carbohydrate-rich substrates do not need sunlight for decomposition of microorganisms. Direct and indirect photolysis systems produce pure hydrogen, whereas in dark fermentation, a mixed biogas containing primarily hydrogen and carbon dioxide is produced and may contain some amounts of methane, carbon monoxide, and/or hydrogen sulfide. Bacteria known for hydrogen production under dark fermentation conditions include *Enterobacter*, *Bacillus*, and *Clostridium* [67]. Hydrogen production by these bacteria depends on process conditions such as pH, hydraulic retention time, and gas partial pressure. Microbial electrolysis uses microbes and an electricity for decomposing the biomass and releasing hydrogen. Photobiological hydrogen production of cyanobacteria do not last long, and energy-conversion efficiencies are generally low. Two steps for photobiological hydrogen production can be described by the following equations (direct biophotolysis):

$$\text{Photosynthesis}: 2H_2O \rightarrow 4H^+ + 4e^- + O_2 \tag{2.32}$$

$$\text{Hydrogen production}: 4H^+ + 4e^- \rightarrow 2H_2 \tag{2.33}$$

The basic principle of this process is shown in Figure 2.30.

Indirect biophotolysis is a cyclic process in which microalgae is raised in the atmosphere of $CO_2$ and light to produce biomass high in carbohydrates, which then are metabolized to produce hydrogen by first keeping algae in dark and then exposing it to light to complete the cycle.

Masukawa *et al.* [68] aimed at promoting and demonstrating evolution of hydrogen at high efficiencies (energy-conversion efficiency about 1%), lasting for a long time. Some of the hydrogenase mutants created from *Anabaena variabilis* were able to produce hydrogen at a rate 4−7 times that of wild type under optimal conditions. Troshina *et al.* [69] demonstrated that the cyanobacterium *Gleocapsa alpicola* (hydrogenase based system) grown in the presence of nitrate can intensify

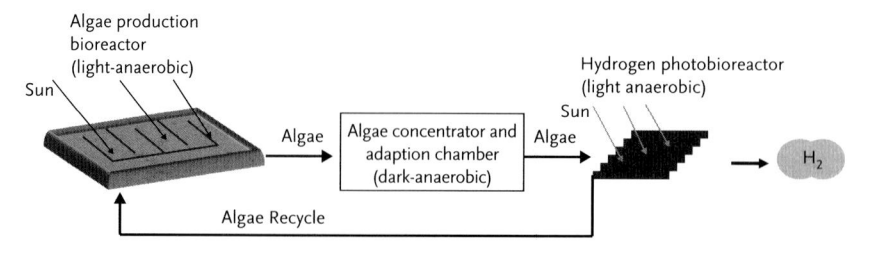

**Figure 2.30** Basics of biological production of hydrogen from algae.

hydrogen production under dark conditions. The optimal hydrogen production was obtained at alkaline conditions (pH 6.8–8.3) and the process temperature increase of 10 °C (up to 40 °C) doubled the hydrogen production. The authors claimed that so far *Gleocapsa alpicola* possesses a unique ability with respect to fermentative hydrogen production. Lin and Lay [70] demonstrated that the proper C/N ratio enhances biohydrogen production by shifting the microorganism metabolic pathway. Experimental study on the hydrogen production ability of the anaerobic microflora dominated by *Clostridium pasteurianum* showed that C/N ratio of 47 provided the optimal biohydrogen production rate. Many recent developments have been made in the research of green algae *Chlamydomonas reinhardtii* for hydrogen production at NREL and the University of California, Berkeley. The rate of hydrogen production by *C. reindardtii* reported by Kosourov et al. [71] was 7.95 mmol $H_2$ $L^{-1}$ of culture after 100 h, corresponding to about 0.07 mmol $H_2$ $(L \times h)^{-1}$.

Biological processes for hydrogen generation are still at early stage of development, mainly on a laboratory scale; however, there is an intensive research going on to find ways to improve hydrogen yield and hydrogen production rates. Advantages and disadvantages of different biological processes for hydrogen production are presented in Table 2.7 [72].

Table 2.7 Merits and demerits of different biological processes for hydrogen production [72].

| Type of microorganism | Advantages | Disadvantages |
| --- | --- | --- |
| **Green algae** | Can produce $H_2$ from water | Require light for $H_2$ production |
| *Chlamydomonas reinhardtii, Chlamydomonas moevussi, Scenedesmus obliquus* | Solar conversion energy is increased by 10-fold as compared to trees, crops | $O_2$ can be dangerous for the system |
| **Cyanobacteria** | Can produce $H_2$ from water | Uptake hydrogenase enzymes are to be removed to stop degradation of $H_2$ |
| *Anabaena azollae, Anabaena CA, Anabaena cylindrica, Anabaena carlabilis, Nostoc muscorum, Nostoc sponglaeforme, Westiellopsis prolific* | Nitrogenase enzyme mainly produces $H_2$<br>Has the ability to fix $N_2$ from atmosphere | Require sun light<br>About 30% of $O_2$ is present in gas mixture with $H_2$<br>$CO_2$ is present in the gas<br>$O_2$ has inhibitory effect on nitrogenase enzyme |
| **Photosynthetic bacteria** | Can use different materials like whey, distillery effluents, and so on. | Require light for $H_2$ production |

(Continued)

**Table 2.7** (Continued)

| Type of microorganism | Advantages | Disadvantages |
|---|---|---|
| *Rhodobacter sphaeroides, Rhodobacter capsulatus, Rhodobacter sulidophilus, Rhodopseumodomonas sphaeroides, Rhodopseumodomonas paluxtris, Rhodopseumodomonas capsulate, Rhodospilillum rubnum, Holabacterium halobium, Chlorobium limicola, Chloroflexu aurantiacus, Thiocapsa roseopersicina* | Can use wide spectrum of light | Fermented broth will cause water pollution problem |
| **Fermentative bacteria** | Can produce $H_2$ all day long without light | Fermented broth is required to undergo further treatment before disposal, otherwise it will create water pollution problem |
| *Enterobacter aerogenes, Enterobacter cloacae, Clostridium butyricum, Clostridium pasteurianum, Desulfovibrio vulgaris, Magashaera elsdenii, Citrobacter intermedius, Escherichia coli* | Can utilize different carbon sources like starch, cellobiose, sucrose, xylose, and so on<br><br>Different types of raw materials can be used<br><br>Produce valuable metabolites such as butyric acid, lactic acid, and so on as by-products<br><br>Anaerobic process, therefore no oxygen limitation problems | $CO_2$ is present in the gas |

## 2.3.4
### Hydrogen from Aluminum

When considering significant amount of hydrogen produced per gram of aluminum, it seems that it is a promising and cheap material for onboard pure hydrogen generation for the use in microfuel cells. Aluminum offers a hydrogen source with high power density (1 g of aluminum can produce 1.245 L of hydrogen, close to 95% of the theoretical maximum) [73]. Hydrogen is generated through aluminum oxidation reaction according to the following formula:

$$Al + 3H_2O \rightarrow Al(OH)_3 + \frac{3}{2}H_2 \quad (2.34)$$

It is claimed [73, 74] that sodium hydroxide can catalyze the reaction (2.25) and can be regenerated using Eq. (2.26), as shown in the equations:

$$2Al + 6H_2O + 2NaOH \rightarrow 2NaAl(OH)_4 + 3H_2 \quad (2.35)$$

$$NaAl(OH)_4 \rightarrow NaOH + Al(OH)_3 \quad (2.36)$$

10 W class PEM fuel cell was developed, running on the hydrogen produced from aluminum powder and water and could operate laptop PC by Hitachi Maxell [75]. By exchanging aluminum and water cartridges, the continuous operation of the laptop was obtained. Total amount of generated hydrogen was 1.36 L per 1 g of aluminum (25 °C, 1 atm). Wang et al. [76] proposed a safe and simple hydrogen generator based on hydrogen produced by the reaction of aluminum and NaOH solution for feeding PEM fuel cell. This generator operated stably a single cell under 500 mA for nearly 5 h, with about 77 wt% $H_2$ utilization ratio. About 38 mL min$^{-1}$ of hydrogen rate was obtained with 25% concentration and 0.01 mL s$^{-1}$ dropping rate of NaOH solution. Schematic diagram and the laboratory setup for hydrogen generation from aluminum is shown in Figure 2.31.

Performance of the single cell operated under 500 mA is shown in Figure 2.32. The cell is operated until the aluminum was totally consumed. Stable operation for about 5 h was proven.

For generation of hydrogen from aluminum, Jung et al. [73] used two flow-type reactors, one allowing the direct feed of liquid water in channels containing aluminum pellets and the second utilizing heat from the reaction to vaporize liquid water before going to the reactor. Adding calcium oxide and sodium hydroxide enhanced the maximum conversion of aluminum. In case of the feed of liquid water, the conversion rate was 74.4% (reactor 1) and 78.6% when water was partially vaporized.

It seems that aluminum could provide alternative source of hydrogen for onboard use. Further investigations in this direction are ongoing.

## 2.3.5
### Outlook

Fossil fuels, electricity, nuclear energy, biomass, and solar energy are the potential sources to be used for wide variety of ways for hydrogen production. Each of them has advantages and disadvantages. It is obvious that current hydrogen processes are still depending mainly on fossil fuels contributing significantly to the $CO_2$ emission to the atmosphere and climate change. For example, the most common way of producing industrial hydrogen is by steam reforming generating high

## 2 Hydrogen – Fundamentals

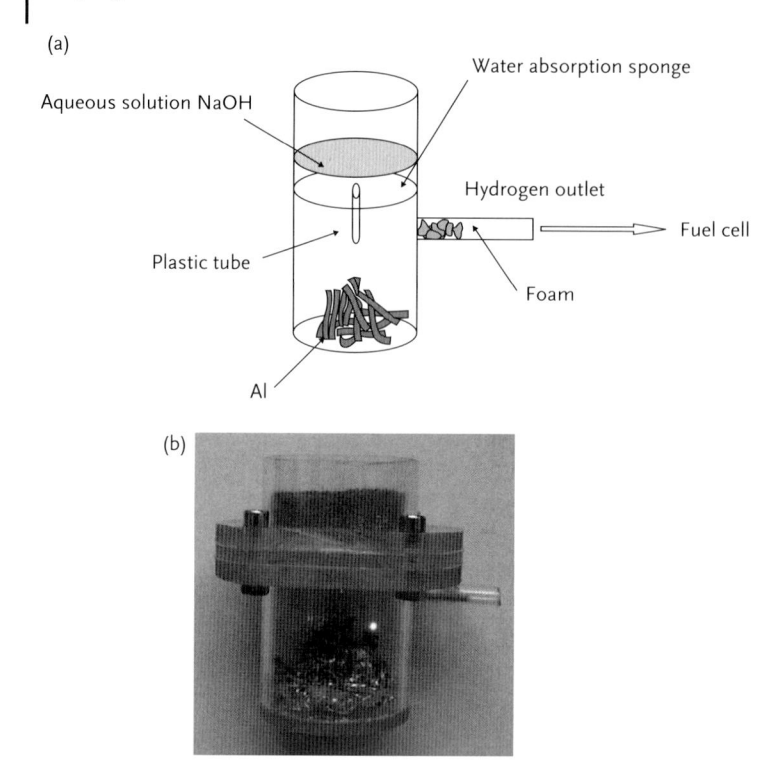

(a)

Aqueous solution NaOH  
Water absorption sponge  
Hydrogen outlet  
Fuel cell  
Plastic tube  
Foam  
Al

(b)

**Figure 2.31** Schematic diagram (a) and laboratory prototype (b) for hydrogen production from aluminum [76], reprinted by permission.

amount of carbon dioxide as a by-product. Water electrolysis is expensive and dependent on the electricity costs but has a high potential to reduce $CO_2$ emission. Life cycle emission analysis of high-temperature electrolysis by nuclear energy showed that this process is "far superior to the conventional steam-reforming process with respect to the global warming potential and acidification potential" [51]. Nuclear technologies might be a viable option for the future hydrogen economy but still a lot of research has to be performed. High-temperature processes need material development. In addition, public acceptance of nuclear reactors is still a big issue in many countries. There is wide field for development of alternative hydrogen production methods like biomass-to-hydrogen processes. It is important especially that a lot of biomass is available everywhere and is inexpensive. The photoelectrolysis is at an early stage of development and material cost and practical issues have to be solved. Similarly, the photobiological processes are at a very early stage of development with so far low conversion efficiencies obtained. It is assumed that solar energy-based hydrogen production might be a viable option for the future considering that hydrogen is a good storage medium for solar energy [31].

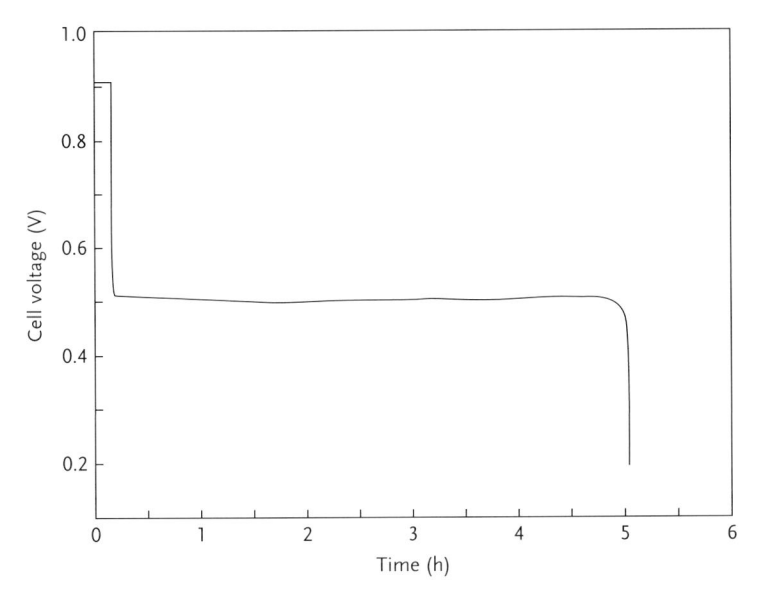

**Figure 2.32** Performance of the single cell under 500 mA [76], reprinted by permission.

Analysis of economic and environmental aspects of various hydrogen production methods [77] indicated that in spite the renewable energy cannot completely fulfill the energy demands, electrolyses coupled with solar, wind, hydropower, and biomass are available sources for hydrogen production for the future. Especially wind and hydroenergy are favorable sources of electricity for pollution-free hydrogen generation.

Based on the work performed for the project roads2hycom, the efficiency projections for 2030 based on estimated advances in the state-of-the-art in hydrogen production methods have been presented. They are shown in Figure 2.33.

It can be seen that hydrogen production from renewable sources may become comparable with fossil fuel. The most mature technologies are reforming and gasification. Hydrogen obtained by gasification of biomass might be competitive with hydrogen coal gasification.

## 2.4
## Hydrogen Storage Safety Aspects

The use of hydrogen must address a proper and detailed assessment of safety aspects, taking into account every imaginable hazardous situation with hydrogen from the transportation to storage and safe use. Due to physical and chemical properties of hydrogen, safety aspects are especially critical in storage systems, which depend on application, either mobile or stationary. In general, storage

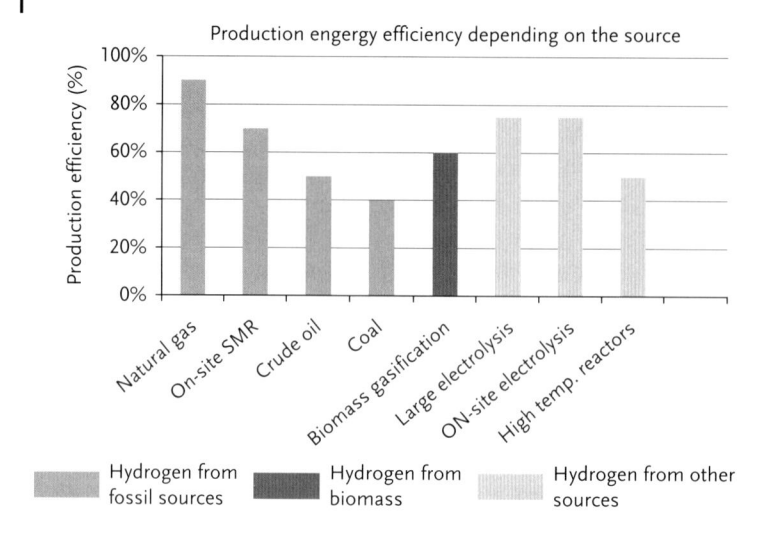

**Figure 2.33** Energy chain efficiencies for 2030, primary source to hydrogen fuel (source: roads2hycom).

systems have to fulfill criteria such as high hydrogen content, fast hydrogen supply, low-energy consumption during storage and supply, as well as economically viability. Currently, the most common way of hydrogen storage is in a compressed form. Compressed gas storage in pressurized cylinders is the best option for onboard hydrogen storage. The safety of these cylinders is a concern, especially in highly populated areas. The relatively low hydrogen density together with the very high gas pressures in the system are important drawbacks of high-pressure storage method. In case of liquefied hydrogen, there are additional safety concerns, namely low heat of vaporization (0.904 kJ mol$^{-1}$ H$_2$) and large volume of expansion from liquid to room temperatures. Liquid hydrogen is stored in cryogenic cylinders at 21.2 K at ambient pressure. The critical temperature of hydrogen is 33 K, above which hydrogen is a noncondensable gas. The pressure required to maintain liquid density at 300 K is 202 MPa; therefore, it requires the use of pressure relief systems on any confined cryogen space. The pressure relief valves should be of sufficient size to accommodate, at safe pressure levels, the maximum gas evolution rate that could occur [78]. The challenges of liquid hydrogen storage are the energy-efficient liquefaction process and the thermal insulation of the cryogenic storage vessel in order to reduce the boil-off of hydrogen. The boil-off effect could have severe implications in automotive industry if vehicles are left in a closed space (garage).

A lot of interest is spent on the development of new solid-state hydrogen storage materials and their reversibility in order to understand and reduce risks dealing with its synthesis, handling, and utilization. Presently, many different hydrides are under development. Usually, they need heating up to in excess of about 300 °C to release hydrogen, which could create severe problems, especially for mobile applications [79]. For practical applications, rapid hydrogen desorption requires around 100 °C and 1 bar, which corresponds to an enthalpy of formation around 15–25 kJ mol$^{-1}$ H$_2$. Detailed critical challenges for the development of solid-state hydrogen storage materials have been addressed [80]. The International Standard (ISO 16111:2008 (E) describes details on service conditions, design criteria, batch, and routine tests for transportable hydride-based hydrogen storage systems, so-called metal hydride assemblies [81]. Landucci *et al.* [82] defined and evaluated safety performance and reliability aspects of alternative hydrogen storage technologies (compressed gas, liquefied hydrogen, and hydrogen storage materials) by applying process flow diagrams available for each technology. Based on the results, strategies for the prevention of incidents and to reduce damages have been defined. In this study, following criteria were taken into consideration: technical specifications of the equipment, definition of substances and operating parameters, characterization of operative conditions of process lines and piping, and inventories of main equipment. It was concluded that metal hydrides and chemical hydrides storage systems present less-severe operating conditions when compared with conventional technologies, despite of a need of more complex system and higher number of auxiliary units.

For commercial exploitation of solid materials, international standards are required in the same manner like for compressed or liquid systems. Many of solid-state storage systems are sensitive toward air, water, and water vapor. In order to evaluate these materials, a special program within the frames of the International Partnership for Hydrogen and Fuel Cells in the Economy (IPHE) has started in 2006, involving numerous laboratories from Europe, United States, and Japan. Within this program involved laboratories work on: (i) the determination of the materials behavior under internationally recognized and standardized testing conditions, (ii) the calculation of thermodynamic energy released due to anticipated environmental reactions and comparison to measurement of chemical kinetics and experimental evaluation of energy release rates, (iii) the mitigation of observed risks through the development and testing of methods and solutions which will hinder or stop high-energy release rate reactions through engineering solutions, and (iv) prototype system testing of risk reduction strategies in subscale hydrogen storage components [83].

The first IPHE project, led by the Russian Academy of Sciences focused on the integration of solid-state hydrogen storage systems with a hydrogen fuel cell. Other large-scale projects include StorHy, with 13 countries involved in both physical as well as materials-based hydrogen storage, and NessHy, with 22 partners from 12 countries with primary focus on solid hydrides.

### 2.4.1
### Hydrogen Properties Related to Safety

According to Material Safety Data Sheet (MSDS), hydrogen is easily ignited with low-ignition energy, including static electricity. Hydrogen is lighter than air and can accumulate in the upper sections of enclosed spaces [84]. Hydrogen is a very reactive fuel. Its flame is very sensitive to instabilities and acceleration by

**Table 2.8** Limits of flammability of selected combustible vapors in air and oxygen at 25 °C and 1 atm (selected data for the table are based on [87]).

| Combustibles | Air LFL (25 °C, 1 atm) | Air UFL (25 °C, 1 atm) | Oxygen LFL (25 °C, 1 atm) | Oxygen UFL (25 °C, 1 atm) |
|---|---|---|---|---|
| Methane | 5.0 | 15.0 | 5.0 | 61.0 |
| Ethane | 3.0 | 12.4 | 3.0 | 66.0 |
| Propane | 2.1 | 9.5 | 2.3 | 55.0 |
| Propylene | 2.4 | 11.0 | 2.1 | 53.0 |
| Cyclopropane | 2.4 | 10.4 | 2.5 | 60.0 |
| **Hydrogen** | **4.0** | **75.0** | **4.0** | **95.0** |
| Methanol | 6.7 | 36.0 | <6.7 | 93.0 |
| Ammonia | 15.0 | 28.0 | ~15.0 | ~79.0 |

turbulence, as a result producing transition of deflagration into detonation over relatively short run-up distances. The major safety concerns for hydrogen are its combustion and detonation properties. When compared with most hydrocarbons (Table 2.8), it can be seen that hydrogen has much wider flammability range, from 4% (lower flammability limit, LFL) to 75 vol% (upper flammability limit, UFL) in air (4–95 vol% in oxygen) and detonability limits of 11–59 vol% in air [85, 86]. Limits of flammability of hydrogen–air mixtures with added nitrogen and carbon dioxide at 25 °C and 1 bar are presented in Figure 2.34.

It can be seen that for most homologous series of organic combustibles, flammability limits both in air and oxygen decrease with increasing molar weight or number of carbon atoms. The upper values are significantly higher in oxygen than in air. Increased hazards of hydrogen and other combustibles with oxygen are mainly due to upper limit values. Flammability limits of hydrogen increase with the temperature. The lower limit drops from 4 vol% at NTP to 3% at 100 °C; detonability limits expand with the scale of a mixture [88]. Hydrogen has very low minimum ignition energy (MIE) of 0.017 MJ in air and 0.0012 MJ in oxygen at 25 °C and 1 bar [87]. For comparison, MIE values for most combustibles are in the range of 0.1–0.3 MJ. For example, in air methanol, it is 0.14 MJ, ethane 0.26 MJ, methane 0.3 MJ, and benzene 0.22 MJ. MEI values in oxygen are at least an order of magnitude lower [87].

Minimum autoignition temperature (AIT) of hydrogen in air is 520 °C and in oxygen the value is lower, about 400 °C. Hydrogen flame velocity is high, 2.7 m s$^{-1}$ [89].

Schroeder and Holtappels [90] summarized the most frequently used standard test methods for atmospheric pressure and the experimental parameters used for the determination of the hydrogen explosion limits. These are EN 1839 (T) tube method, EN 1839 (B) bomb method, ASTM E 681-01 (USA), and DIN 51649, part 1 (Germany). In addition, they described the pressure and temperature influence on the explosion limits of hydrogen–air mixtures and hydrogen–oxygen mixtures up to an initial pressure of 200 bar. The temperature influence on the explosion limits of hydrogen–air mixtures has shown that the explosion range becomes

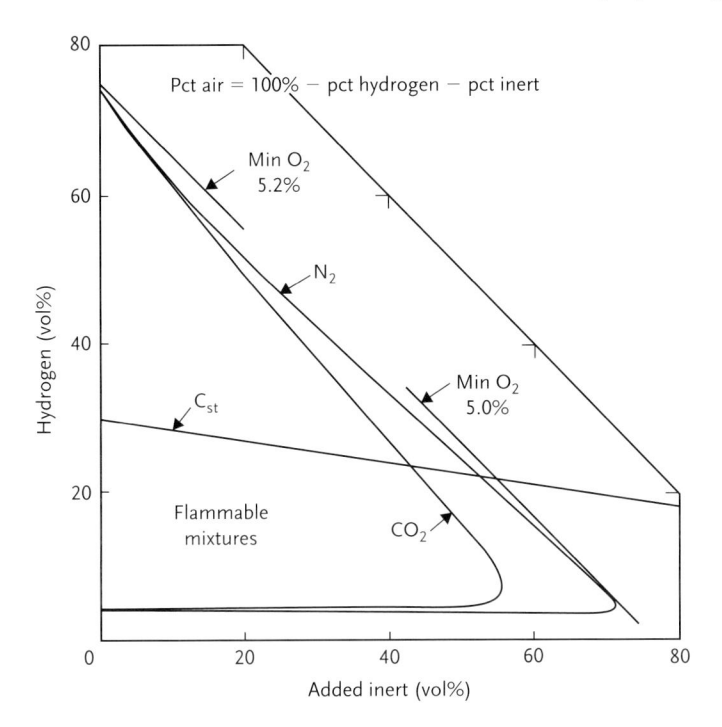

**Figure 2.34** Limits of flammability of hydrogen–air mixtures with added nitrogen and $CO_2$ at 25 °C and 1 atm [87].

wider with increasing temperature, as shown in Figure 2.35. It was observed that the lower was the temperature, the higher was the lower explosion limit in mol% $H_2$ (LEL), but lower the upper explosion limit in mol% $H_2$ (UEL). For example, LEL at 20 °C was 3.9 mol% $H_2$, at 200 °C 2.9 mol% $H_2$, and at 400 °C the LEL was 1.5 mol% $H_2$. UEL at 20 °C was 75.2 mol% $H_2$, at 200 °C 81.3 mol% $H_2$ and at 400 °C the LEL was 87.6 mol% $H_2$ [91].

Pressure influence on explosion limits of hydrogen–air mixtures, as presented in Figure 2.36, shows that LEL increases with increasing initial pressure; hydrogen behavior is opposite to most flammable hydrocarbon gases. UEL after decreasing to initial pressure of 20 bar increases again [90].

Pressure influence on the explosion limits of hydrogen–oxygen mixtures at room temperatures and 80 °C is shown in Figure 2.37.

A special aspect of hydrogen has been highlighted by Houf and Schefer [91], namely that the lower flammability limit is different for differently propagating flames due to low density of hydrogen in comparison with air. It has been shown that that the lower flammability limit in air can rise to 7.2% for horizontal propagating flames and to 9.5% for downward propagating flames. For conventional hydrocarbons, such limits are independent on the direction of flame propagation. Based on the simulation plots, it was shown that while the decay distance to the lower flammability limit for upward-propagating hydrogen flames (0.04 mole

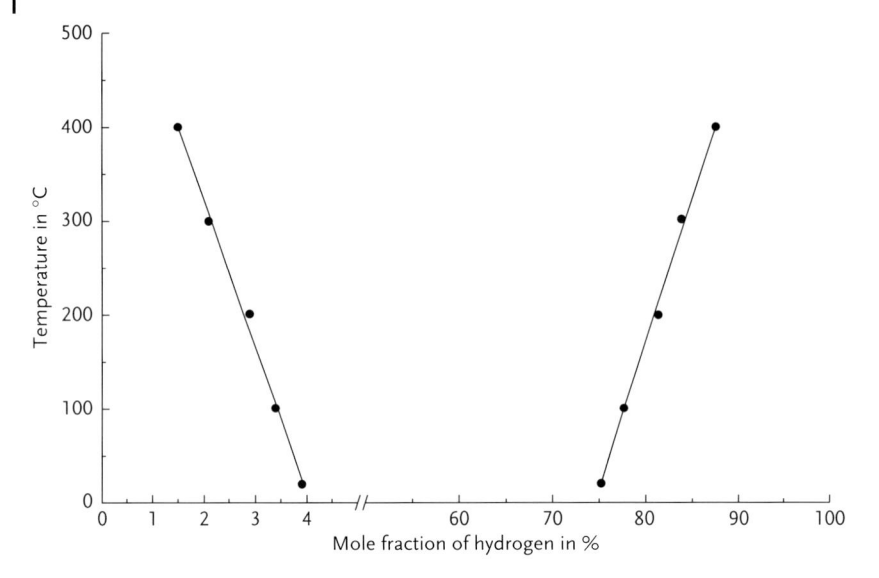

**Figure 2.35** Influence of the temperature on the explosion limits of hydrogen–air mixtures measured at atmospheric pressure according to DIN 51649. On the left values for LEL, on the right for UEL [90].

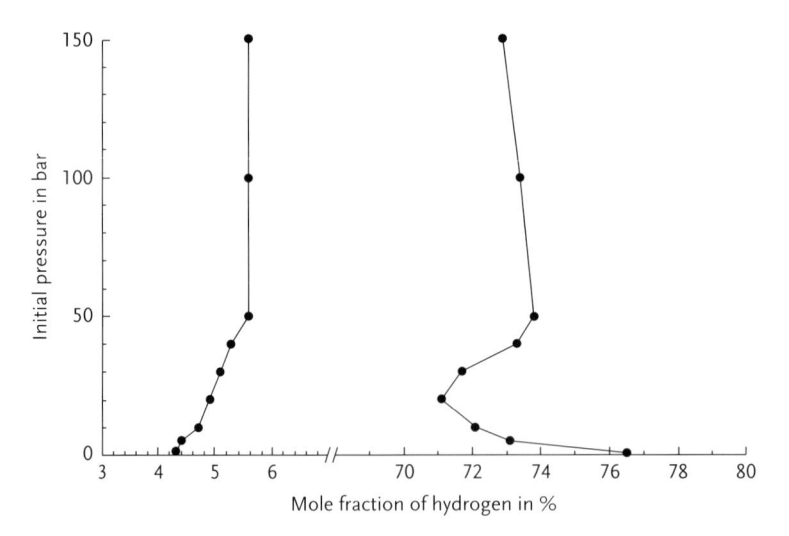

**Figure 2.36** Influence of the initial pressure on explosion limits of hydrogen–air mixtures at room temperatures [90].

fraction) was approximately 3.5 times greater than that for natural gas, the hydrogen decay distance to the lower flammability limit for downward-propagating flames (8.5–9.5%) was only a factor of 1.5 greater [91]. Schefer *et al.* characterized dimensional and radiative properties of an ignited hydrogen jet,

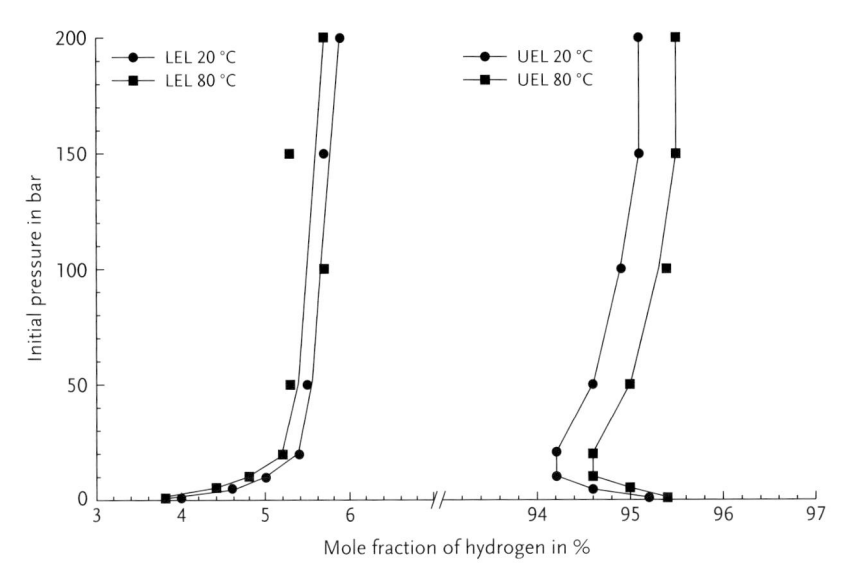

**Figure 2.37** Influence on the explosion limits of hydrogen–oxygen mixtures at 20 °C and 80 °C [90].

finding that such data are relevant to a safety scenario of sudden leak in a high-pressure hydrogen vessel [92]. The results showed that flame length increases with total jet mass flow rate and jet nozzle diameter. Assuming turbulent flames, a correlation of flame length as a function of hydrogen concentration in mixture with commercially available natural gas has been shown by Choudhuri and Gollahalli [93]. With the increase of hydrogen concentration, the reactivity of fuel mixture increases, thus shortening combustion time, and as a consequence reduces the overall flame length. In addition, it was observed that (i) the overall flame temperature increased with hydrogen content due to higher energy output and lower flame radiation, (ii) CO emission index decreases, (iii) $NO_x$ emission index increases, and (iv) volumetric soot concentration decreases. Furthermore, it was noted that radiative heat loss decreases as hydrogen content increases in the fuel mixture. This observation is of high value for designing combustion chamber because the flame temperature indicates how cold or hot are the flames in relation to the surrounding [93].

Wu et al. [94] addressed a problem of the liftoff and blowup stability of hydrogen jet flames with addition of methane and propane, and compared it with mixture of $H_2/Ar$ and $H_2/CO_2$. Experiments have shown that the liftoff velocity measured for pure hydrogen flame was $730\,m\,s^{-1}$ and its height increased linearly with jet velocity (Figure 2.38).

As can be seen in Figure 2.38, the addition of the methane, $CO_2$, and propane always increased the liftoff height of the initially lifted hydrogen diffusion flames. Addition of propane produced the highest liftoff height. After adding propane to already lifted hydrogen jet flames, the liftoff height was about 2.6 times higher

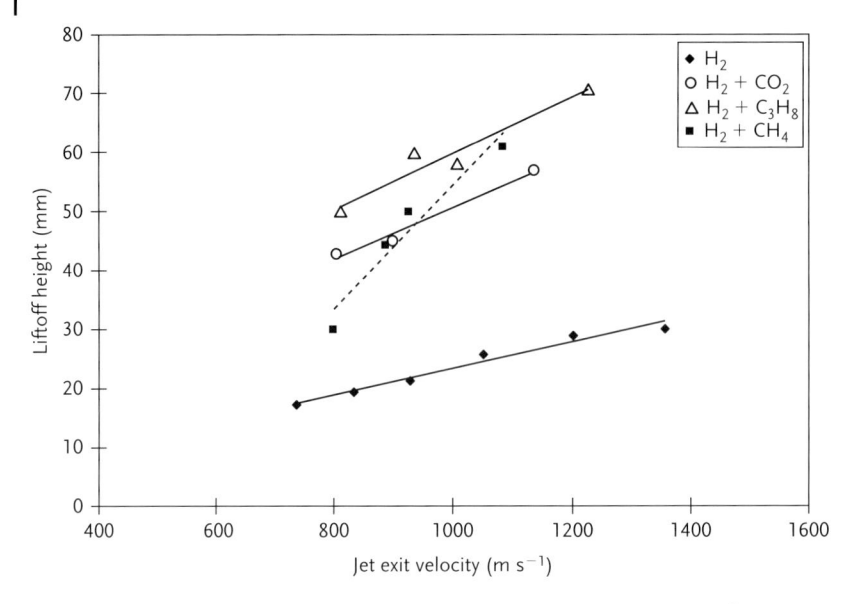

**Figure 2.38** Comparison of the liftoff height of $H_2/C_3H_8$, $H_2/CH_4$, and $H_2/CO_2$ flames [94].

before the blowout was observed; this was when the propane concentration was of 4–5%.

Two flame stability regimes were identified for $CO_2$ addition to attached hydrogen flames. After adding $CO_2$, it was noted that if the $CO_2$ concentration is higher than 6.4%, a stable flame was produced; when the concentration was lower that 6.4%, $CO_2$ addition resulted in a lifted flames, blown up at higher velocities. After adding $CO_2$ to lifted hydrogen flames, the liftoff height was nearly twice of the initial value for pure hydrogen flames. The authors assume that hydrocarbons may be the dominant factor in determining burning velocities of hydrogen mixed with hydrocarbons, probably influenced by chemical kinetics mechanism of the hydrocarbon–air reaction [94].

Important problem of the ignition of hydrogen–air mixtures by thermal sources was highlighted by Morreale *et al.* [95]. Tests conducted in dynamic conditions with the flammable mixture flowing over a carbon steel sheet or wire placed in a tubular reactor showed that carbon steel provokes ignition at 820 °C ($H_2$ concentration 5.7% in the mixture). An increase in hydrogen content lowers the ignition temperature to 760 °C ($H_2$ concentration 6.5% in the mixture) and further decrease in the ignition temperature to 720 °C resulted from higher residence time of the gaseous mixture [95]. Hydrogen can hardly be detected with human senses being odorless. When it is outside in an open space, hydrogen will easily disperse, because it is much lighter than air. Leaking in semiconfined or confined space could lead to fill up the space, and given the abundance of potential ignition sources, a flash fire or an explosion is possible. This property requires a lot of attention and put on leak-free connection techniques [96].

As recently stated: "From a safety point of view, care has to be taken because of its explosion and fire proneness when mixed with air. Although much is known about its properties, there are still aspects not well explained or not completely mapped despite recent research efforts in different countries" [96].

## 2.4.2
### Selected Incidents with Hydrogen

On March 19, 1936, the airship "Hindenburg" was rolled out. The airship had a total length of 246.7 m and a maximum diameter of 44.7 m; these dimensions are close to the dimensions of the titanic (269.4 m length and 66.4 m height and 28 m width). The total mass of the airship was 242 tons, and the empty weight was of 118 tons. About 200 000 m$^3$ of hydrogen in 16 cells have been on board. Helium was not available at that time in Germany. After a tour through Germany, the airship was in Lakehurst (United States) and 7 times in Rio de Janeiro. In total, 1600 passengers have been carried and the airship had 3000 flying hours. On May 6, the accident of the Hindenburg occurred. Thirty five of 97 people on board died. The causation of the accident was deeply investigated. The most probable reason is the electrostatic caused fire of the fabric of the covering and subsequent burning of the hydrogen. The covering was coated with flammable materials. This could be interpreted that such an accident is possible even with the helium. Nowadays, new materials are available for the covering and in addition the preferred helium is available. The Hindenburg accident is nowadays is synonym for the danger of hydrogen, but that is not what we can learn from this accident.

Another accident which was worldwide noticed was the accident of the American space shuttle: "Challenger" on January 28, 1986 (Figure 2.39). The space shuttle broke apart 73 s into its flight, leading to the deaths of its seven crew members. After an explosion, the spacecraft fell into the ocean, The shuttle had no escape system; therefore, the crew members did not have any chance to survive. The accident was deeply investigated and on September 29, 1988 (2.5 year later), the "Discovery" was the next shuttle to start into the space. For the causation of the accident, a failure of a small O-ring was found. The O-ring failure caused a breach in the solid rocket booster joint, allowing pressurized hot gas from within the solid rocket motor to reach the outside and affect the adjacent solid rocket booster attachment hardware and external fuel tank. This led to the separation of the right-hand solid rocket booster back attachment and the structural failure of the external tank. Aerodynamic forces promptly broke up the orbiter. A study reported that 85% of Americans had heard the news within an hour of the accident. This accident is also not a typical accident, although hydrogen was used as fuel, but with all other fuels an accident with the same impacts would happen. The Challenger disaster has been used as a case study in many discussions of engineering safety and workplace ethics.

In order to minimize any potential risk and to assure safe operation, handling and usage of hydrogen and hydrogen-related systems, careful and comprehensive

**Figure 2.39** Challenger disaster in 1986 [97] (Credit: NASA).

safety management is needed. Sharing the details on hydrogen safety issues is the best way to prevent them. The Hydrogen Incident Reporting Tool [98] contains records of safety events when using and working with hydrogen and hydrogen systems. The information are publicly available. Typical record consists of the description of the accident, contributing factors, involved equipment, and some information on the consequences and steps implemented to avoid similar cases in the future. The status of methodology of best practice tool for hydrogen safety was discussed by Weiner *et al.* [99]. The definition of best practice is: "Methods and techniques that have consistently shown results superior than those achieved with other means, and which are used as benchmarks to strive for." There is, however, no practice that is best for everyone or in every situation, and no best practice remains best for very long as people keep on finding better ways of doing things. A plenty of experiences regarding the safe handling and use of hydrogen exists. Hydrogen Safety Best Practices [100] captures this knowledge base and makes it available to the people working already or starting with hydrogen and hydrogen-related system. It is hoped that the mentioned online safety knowledge tools can provide a mechanism for sharing, discussing, and learning from the experience of others.

### 2.4.3
### Human Health Impact

Two different way for contact with hydrogen are possible, hydrogen as gas or as liquid. Gaseous hydrogen is asphyxiating. After inhalation, high concentrations of hydrogen so as to exclude an adequate supply of oxygen to the lungs causes headache, drowsiness, dizziness, excitation, excess salivation, vomiting, and unconsciousness. Lack of oxygen can result in death. Hydrogen is biologically inactive and essentially nontoxic. The same is valid for liquid hydrogen because at the low boiling point hydrogen gases will always be present together with the liquid hydrogen. The cold liquid can cause severe frostbite. After skin contact, the cold gas or liquid can cause severe frostbite. Swallowing is an unlike route of exposure. After eye contact, cold gas or liquid may cause severe frostbite.

### 2.4.4
### Sensors

Of high importance is the use of very sensitive hydrogen safety sensors and the requirements. Currently, there are numerous hydrogen safety sensors available commercially. New hydrogen sensor technologies are still being developed and can be expected to be commercialized successfully. US DOE target specifications for R&D on hydrogen safety sensors are presented in Table 2.9.

Recently, a detailed overview on hydrogen safety sensors is given by Buttner et al. [102]. At present, most common hydrogen sensors can be classified into following groups: electrochemical (EC), metal oxide (MOX), pellistor-type combustible gas sensors (CGS), thermal conductivity (TC), optical, Pd-film, and Pd-alloy films (Pd). General performance of each type of sensor group is discussed in detail and results are presented in the numerical performance metrics. The ranking is based on general performance parameters and values could change with each application. The authors are referencing to several national and international organizations developing standards and codes addressing the performance of hydrogen detectors. It is observed that the requirements recommended by these standards are covering broader range of specification that are set by US DOE [102] (Table 2.10).

### 2.4.5
### Regulations, Codes, and Standards (RCS)

The main task of the European Integrated Hydrogen Project (EIHP) was to enhance the safety of hydrogen vehicles and to facilitate the approval of hydrogen

**Table 2.9** US Department of Energy Targets for Hydrogen Safety Sensors R&D, Multi-Year Research and Development Plan, table 3.8.2, p. 3.8−7, [101].

| US DOE Targets for Hydrogen Safety Sensor R&D | |
|---|---|
| Measurement range | 0.1−10% |
| Operating temperature | −30 to +80 °C |
| Response time | Under 1 s |
| Accuracy | 5% of full scale |
| Gas environment | Ambient air, 10−98% relative humidity range |
| Lifetime | 10 years |
| Interference resistant | Hydrocarbons |

**Table 2.10** Generalized ranking of various hydrogen sensor platforms to selected performance metrics (5 = ideal, 0 = poor) [102].

| Metric | EC | MOX | CGS | TC | Optical | Pd |
|---|---|---|---|---|---|---|
| Analytical metrics | | | | | | |
| Detection limit | 4.0 | 4.0 | 4.0 | 4.0 | 4.0 | 4.0 |
| Selectivity | 3.5 | 3.0 | 3.5 | 3.5 | 4.5 | 4.9 |
| Linear range | 4.9 | 3.0 | 4.0 | 4.0 | 3.0 | 4.0 |
| Response time | 4.0 | 3.0 | 4.0 | 4.5 | 3.0 | 3.5 |
| Repeatability | 3.5 | 4.0 | 4.0 | 4.0 | 3.0 | 4.0 |
| Environmental effect | 3.5 | 3.5 | 4.0 | 3.0 | 4.0 | 4.0 |
| Logistic metrics | | | | | | |
| Level of maturity | 5.0 | 5.0 | 5.0 | 4.5 | 3.5 | 4.5 |
| Size | 4.0 | 4.0 | 4.0 | 4.5 | 4.0 | 3.5 |
| Power | 4.0 | 3.0 | 3.0 | 4.5 | 4.0 | 3.5 |
| Maintenance | 3.5 | 3.5 | 4.0 | 4.0 | 5.0 | 4.0 |
| Lifetime | 3.5 | 4.0 | 3.5 | 4.0 | 4.0 | 3.8 |
| Matrix | 4.0 | 3.0 | 3.0 | 4.5 | 4.0 | 3.5 |

vehicles by the development of draft regulations for the use of hydrogen as a vehicle fuel. The final report was coordinated and compiled by Wurster [103]. These documents were forewarded to the WP29 of the Economic Comission for Europe (ECE), a United Nations (UN) Organization. Actually, two proposals for draft regulation have been developed:

Uniform provisions concerning the approval of:

1. Specific components of motor vehicles using compressed gaseous hydrogen.
2. Vehicles with regard to the installation of specific components for the use of compressed gaseous hydrogen.

Uniform provisions concerning the approval of:

1. Specific components of motor vehicles using liquid hydrogen.
2. Vehicles with regard to the installation of specific components for the use of liquid hydrogen.

These harmonized ECE regulations will replace the national legislation and regulations. Members of the ECE are not only European countries. The ECE is free for the accession of countries from other continents. About 60 countries are members of the ECE. The ECE is considered to be the platform for the future global harmonization for legal requirements for vehicles.

The actual legal situation in Germany is shown in Figure 2.40.

In case of hydrogen vehicles, there is a link in the StVZO to the law for pressure vessels which refers to many other bylaws, codes of practice standards, and so on. But these additional requirements were not developed for vehicles applications and often they do not meet the requirements for vehicle applications.

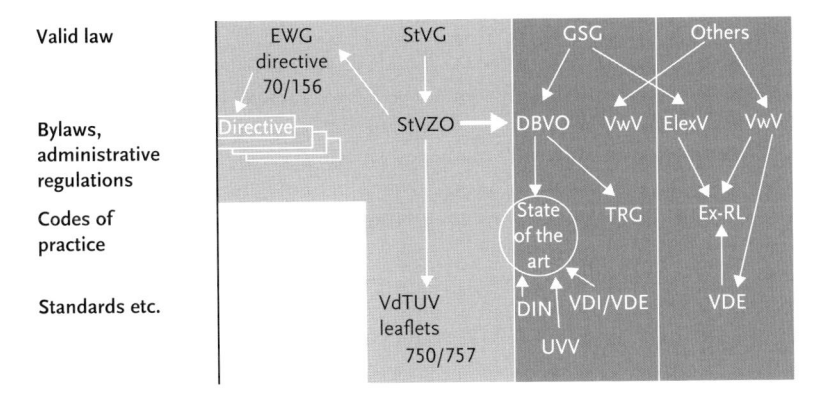

**Figure 2.40** Legal situation in Germany for approving a hydrogen road vehicle [103].

In the second phase, EIHP2, the draft regulation documents for Europe shall be improved step by step (revised draft versions) and developed to a global regulation for hydrogen-powered road vehicles. Procedures for periodic vehicle inspections will be developed. Requirements for new draft standards, refueling procedures, and periodic inspections for hydrogen refueling infrastructure components and systems will be developed and inputs will be forwarded to regulatory bodies (e.g., for vehicles to UN ECE in Geneva) [104].

EIHP2 will furthermore undertake actions to validate the EIHP1 draft regulations by developing and realizing hydrogen components and vehicles according to these drafts and getting them approved by selected authorities. In parallel, the EIHP1 draft proposal will be monitored during its discussion process at ECE in order to assist in achieving a valid regulation [105].

Main results expected from EIHP2 are:

- Development of a worldwide harmonized regulation for hydrogen-powered road vehicles.
- Development of procedures for periodic vehicle inspections (roadworthiness).
- As far as possible, development of a worldwide standard or regulation and of periodic inspection procedures for the relevant refueling infrastructure, subsystems, or components.

Final acceptance of the ECE regulations into EU-law and into national law of the member states has to be undertaken. In order to develop the draft regulations into global regulations, the support by the signatory states of ECEWP29 is needed.

### 2.4.6
### Safety Aspects in the Hydrogen Chain from Production to the User

By using hydrogen as an energy carrier, for example, transportation purpose, the situation is totally different to the traditional use of hydrogen. The use of hydrogen in industry can be characterized as following. Hydrogen is produced mainly in a few big centralized plants distributed by the gas companies and is used mainly in chemical plants, which means that only a clear defined group of people are handling the hydrogen and these people are well educated and instructed to deal with hydrogen. This situation will be different when hydrogen is becoming an important energy carrier. This will change the situation for hydrogen production, the distribution, the storage, and the use by the private persons. The following requirements are coming up with the broad use of hydrogen:

- fast filling and deflating of storage units
- high number of items
- higher dynamic
- difficult controlling
- consumer instead of professional
- new materials, lack of experience.

The production of hydrogen would be more decentralized; it might be that hydrogen will be produced even in households. Hydrogen will be distributed in thousands of filling stations. Hydrogen will be stored in high amount of small storage tanks in cars and probably for other application like forklifts boats and so on. This is a challenge for all the people who are working on the maintenance or repair of such hydrogen system, for example, in garages for repairing cars. All these garages should be well equipped to handle hydrogen systems and all the people dealing with these systems need special education and instructions. All these measures will finally end up in a safe use of hydrogen in broad way.

### 2.4.6.1 Hydrogen Production

One important route for producing hydrogen from renewable energy is the water electrolysis. Coupling these electrolyzers with electric power from wind energy or solar energy requires a high dynamic of electrolyzers. An efficient way for compressing the hydrogen is compressing direct in the high-pressure electrolyzer. Electrolyzers at high pressures running at low power have to solve the problems coming up with the affected gas purity. The higher is the pressure, the higher is the driving force for the hydrogen and oxygen permeability. Not only safety aspects show the need to reduce this cross permeation but also the negative effect on the current efficiency shows the requirement to find efficient countermeasures. The bigger problem occurs at the oxygen side, while hydrogen has the higher permeability compared to oxygen. But in principle on both cathode and anode side, an explosive can be generated. For example, for PEM electrolysis, two different countermeasures are reported by Fateev *et al.* [106]. In principle, there are two strategies to minimize the risk of an explosive gas mixture in the electrolyzer. To reduce the cross-permeation phenomena, the modification of the solid polymer electrolyte was invetsigated. The second investigated path was the use of catalytic $H_2/O_2$ recombiners to maintain $H_2$ levels in the $O_2$ and $O_2$ levels in the $H_2$ at valus compatible with the requirements. The easiest way to reduce the permeation rate and finally the concentration levels is to increase the membrane thickness, but this costs an increase in efficiency due to the increase of membrane resistance. Fateev *et al.* finally showed that PEM–water electrolysis up to 70 bar is possible. To avoid critical concentartions, there are two possiblities: a surface modification of the solid electrolyte or adding gas recombiners direct in the electrolyzer cell or along the production line. Very promising results were reported from the GenHyPEM project by Fateev *et al.* [107]. In this project, for application above 50 bar, thick Nafion-reinforced membranes were used resulting in a slightly reduced efficiency. Experiments with gas recombiners in the gas circuits have been performed up to 130 bar. With the combination of these countermeasures, high gas purities could be obtained. At 100 bar, the hydrogen concentration before the recombiner was 98.01% and after the recombiner was 99.991. The test was performed at a current density of 500 mA cm$^{-2}$ and at a mean temperature of 88 °C.

The same countermeasures are applicable for high-pressure electrolyzers with an alkaline electrolyte. A proper selection of the diaphragm and a reasonable thickness of the diaphragm together with the use of recombiners allows the same operational pressures compared to the PEM systems (see Chapter 4.2.2.2).

### 2.4.6.2 Hydrogen Refuelling Stations

To introduce hydrogen in transportation application, a safe refuelling must be granted. The requirements for the filling station are different for different storage systems.

In case of liquid hydrogen, most requirements are due to the low temperatures. For the gaskets of the connection unit, this results in a strong load, also caused by the number of filling procedures.

In case of high-pressure gas storage, the filling of the high-pressure storage tank has different risks. These risks are reported by Faudou *et al.* [108]. Four risks can be defined for filling of, for example, 350 bar vessels. Composite pressure vessels either with a aluminum liner or (Type 3) with a polyethylene liner (Type 4) are typically certified for a temperature range of −40 °C to 85 °C.

- **Low-temperature risk**
  This risk occurs in the low-temperatures domain (below −40 °C). This risk may only occur in cold filling cases.
- **Overpressure domain**
  This overpressure domain is quite easy to handle as the pressure in the filling house is the same as in the vessel. This parameter is therefore easy to control.
- **Overfilling risk**
  This risk can occur when the ambient temperature is low. In this case, the filling can be stopped at high pressure but the temperature of the gas at the end of the filling coil is not so high. After an exposure of the vessel to higher temperatures, the pressure may reach unexpected values.
- **Overheating risk**
  The overheating risk occurs when the remaining pressure in vessel is low and when the ambient temperature is high.

Several measures were proposed to remain in the operational window for the filling. First, a modeling tool for the prediction of the gas temperature evolution during filling process was developed. Further, a filling management tool was developed. This tool is able to predict the safest way of refuelling.

The aim of the EU project HyApproval [109] (under European Commission FP6 contract No. 019813) was to elaborate a "Handbook for Approval of Hydrogen Refuelling Stations," which could be used to certify public hydrogen filling stations in Europe. The handbook with appendices and executive summary (in various languages) was published in 2008 [110]. In Appendix I, "safety data sheets for hydrogen and refrigerated hydrogen" are given. Appendix II provides "approval requirements in five EU countries (Netherlands, Italy, France, Germany, and Spain) and the USA." Appendix III describes "Emergency Response Plan." In Appendix IV, "Quantitative Risk Assessment of Hydrogen Refuelling Station with on-site production" is described, Appendix V provides a "Consequence Assessment Summary Report." Finally, Appendix VI is dealing with "Vehicle Description and Requirements." Very detailed guidelines for design, operation, and maintenance of hydrogen refueling stations (HRS) are presented taking into account design, construction, and operating requirements and recommendations. This include among others general recommendations for hydrogen plants, safety distances, hazardous areas, as well as emergency safety plan. Flow chart of HRS approval process in Germany is presented in Figure 2.41 [110]. "Handbook for Approval of Hydrogen Refuelling Stations" was provided to hydrogen refuelling station operators and local authorities. It has to be highlighted that some interviewee answered that the handbook will not be used because it has no legal status. It can therefore only be referred to as best practice information.

As can be expected from the response, the use of the handbook will depend on the relevance of the information in the book to the problems that the various stakeholders are confronted with. A point to be noted is that the (legal) status of the handbook will also affect its use.

### 2.4.6.3 Storage/Transportation (Compressed/Liquid/Metal Hydride)

The three main possibilities for storing hydrogen for transportation are compressed hydrogen, liquid hydrogen, or as metal hydride. For each possibility, different safety aspects are valid. In case of compressed hydrogen, the lifetime cycle stability is of main interest. In this case, the judgment of the fatigue caused reliability of composite pressure vessels plays an important role [111]. A different approach is required for carbon fiber composite vessels. They have a different behavior compared to the behavior of standard metal vessels. While the material-specific variations for metals are of the same dimensions in the Wöhler diagram, for composites there are significant differences especially between the composites and the metals. While for metals, the lifetime determining stress is a function of the filling pressure, for composite material the situation is different because of the relaxation processes. Taking this into account, adapted proof procedures are necessary for the safety assessment for this composite pressure vessels.

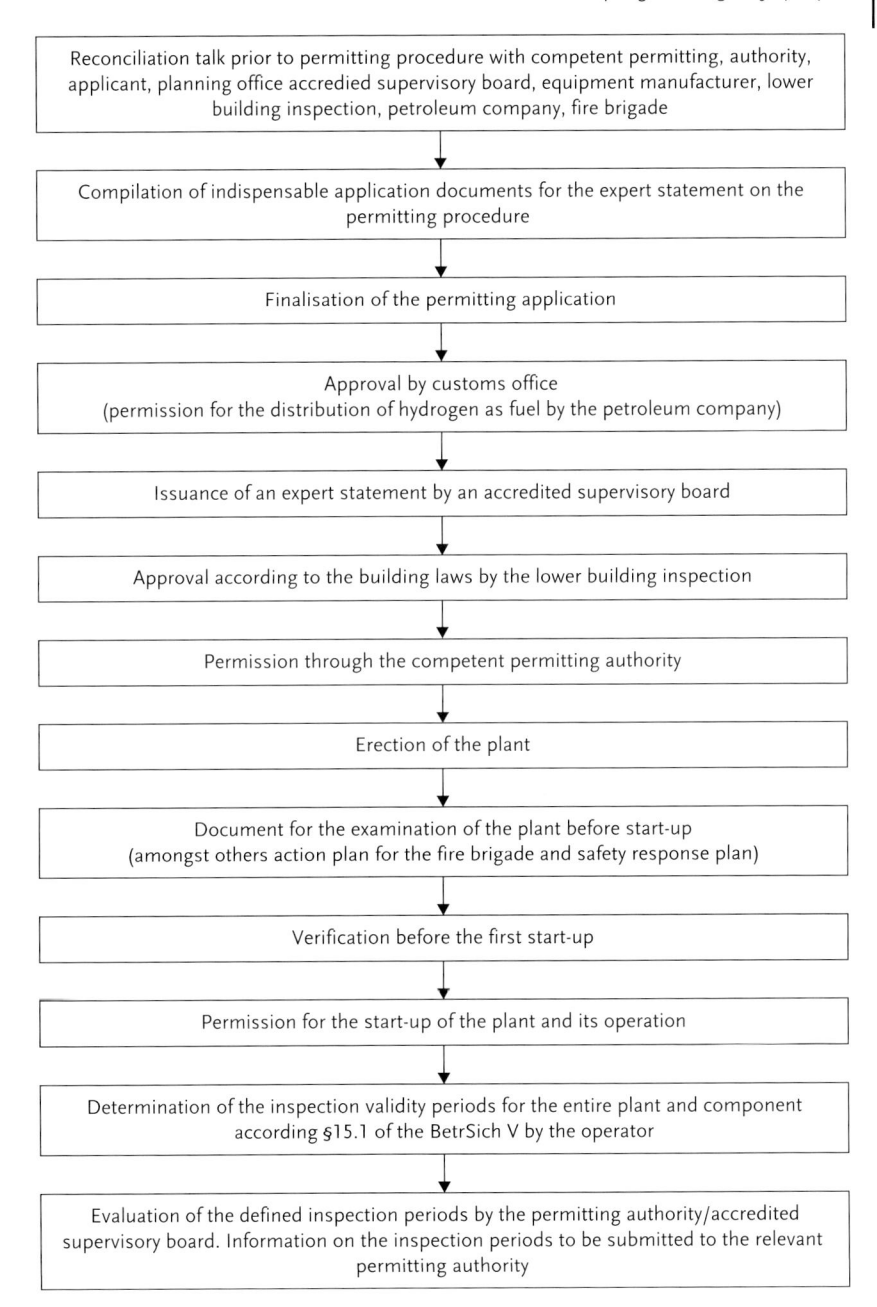

```
┌─────────────────────────────────────────────────────────────────────────┐
│ Reconciliation talk prior to permitting procedure with competent        │
│ permitting, authority, applicant, planning office accredied supervisory │
│ board, equipment manufacturer, lower building inspection, petroleum     │
│ company, fire brigade                                                   │
└─────────────────────────────────────────────────────────────────────────┘
                                    ↓
┌─────────────────────────────────────────────────────────────────────────┐
│ Compilation of indispensable application documents for the expert       │
│ statement on the permitting procedure                                   │
└─────────────────────────────────────────────────────────────────────────┘
                                    ↓
┌─────────────────────────────────────────────────────────────────────────┐
│ Finalisation of the permitting application                              │
└─────────────────────────────────────────────────────────────────────────┘
                                    ↓
┌─────────────────────────────────────────────────────────────────────────┐
│ Approval by customs office                                              │
│ (permission for the distribution of hydrogen as fuel by the petroleum   │
│ company)                                                                │
└─────────────────────────────────────────────────────────────────────────┘
                                    ↓
┌─────────────────────────────────────────────────────────────────────────┐
│ Issuance of an expert statement by an accredited supervisory board      │
└─────────────────────────────────────────────────────────────────────────┘
                                    ↓
┌─────────────────────────────────────────────────────────────────────────┐
│ Approval according to the building laws by the lower building inspection│
└─────────────────────────────────────────────────────────────────────────┘
                                    ↓
┌─────────────────────────────────────────────────────────────────────────┐
│ Permission through the competent permitting authority                   │
└─────────────────────────────────────────────────────────────────────────┘
                                    ↓
┌─────────────────────────────────────────────────────────────────────────┐
│ Erection of the plant                                                   │
└─────────────────────────────────────────────────────────────────────────┘
                                    ↓
┌─────────────────────────────────────────────────────────────────────────┐
│ Document for the examination of the plant before start-up               │
│ (amongst others action plan for the fire brigade and safety response    │
│ plan)                                                                   │
└─────────────────────────────────────────────────────────────────────────┘
                                    ↓
┌─────────────────────────────────────────────────────────────────────────┐
│ Verification before the first start-up                                  │
└─────────────────────────────────────────────────────────────────────────┘
                                    ↓
┌─────────────────────────────────────────────────────────────────────────┐
│ Permission for the start-up of the plant and its operation              │
└─────────────────────────────────────────────────────────────────────────┘
                                    ↓
┌─────────────────────────────────────────────────────────────────────────┐
│ Determination of the inspection validity periods for the entire plant   │
│ and component according §15.1 of the BetrSich V by the operator         │
└─────────────────────────────────────────────────────────────────────────┘
                                    ↓
┌─────────────────────────────────────────────────────────────────────────┐
│ Evaluation of the defined inspection periods by the permitting          │
│ authority/accredited supervisory board. Information on the inspection   │
│ periods to be submitted to the relevant permitting authority            │
└─────────────────────────────────────────────────────────────────────────┘
```

**Figure 2.41** Flow chart of Hydrogen Refuelling Stations (HRS) approval process in Germany [110].

The influence of the temperature on the fatigue strength of compressed hydrogen tanks is important too, and has to be considered. Not only the mechanical properties have influence on the safety but also the hydrogen permeability has to be taken into account. Especially at high pressures, the hydrogen permeability of the used materials has to be considered. Dangerous concentrations of hydrogen in closed rooms, for example, garages, have to be avoided.

For a safe use of pressure vessels, the following safety issues have to be considered:

- Pressure vessels have to be provided with a shut-off valve to close the line when requested.
- A flow-limiting device prohibits the abrupt depletion of the vessel in case of damaged system.
- Implementation of a pressure relief valve for save depletion of the tank in case of a heat charge of a fire.

In case of storage of liquid hydrogen, the focus is the permanent release of hydrogen due boil-off. Dangerous hydrogen concentrations in closed rooms have to be avoided. During refuelling, the amount of filling has to be limited to about 85% to have space for thermal expansion.

Storing hydrogen in metal hydrides looks very safe in a first view because the pressure is relatively low and the hydrogen release is low due to the fact that energy is needed. But it is important to assess the impact on the environment in case that the metal hydride leaves the vessel.

### 2.4.6.4 Garage for Repairing Cars

The use of hydrogen in motor vehicles is a new technology, which has considerable differences to conventional cars. Because of the complexity of the hydrogen-based power trains, especially educated and trained professionals in the garages for repairing such cars are required. The need of additional general training for handling with hydrogen and vehicle-type-specific instructions is obvious.

For the professionals in the garage, an emergency plan has to be elaborated. And for the professionals itself, a personal protective equipment has to be available. But not only the education of the professionals and instructions are important measures but also the garage itself has to be especially equipped with, for example, additional vent lines. The BGI 5108 – Berufsgenossenschaftlicher Informationen und Grundsätze für Sicherheit und Gesundheit am Arbeitsplatz [112], for example, is valid in Germany. In the BGI 5108, the requirements and countermeasures in garages are described.

### 2.4.7
### Safety Aspects of Hydrogen Vehicles

In recent years, significant research and developmental efforts were put on hydrogen storage technologies with the goal of realizing a breakthrough for fuel cell vehicle applications. Automotive industry in collaboration with US DOE established specific technical targets for onboard hydrogen storage systems [113, 114]. Three main barriers to be overcome in order to enable a commercialization of hydrogen fuel cells vehicles have been summarized by Satyapal *et al.* [113]. Onboard hydrogen storage systems has to assure a vehicle driving range higher than 500 km and meeting vehicle packaging, cost, and performance requirements. Fuel cell system cost must be lowered to $30 per kilowatt by 2015 in parallel meeting performance and durability requirements, and the cost of safe and efficient hydrogen production and delivery must be lowered to be competitive with gasoline. In addition to the above-mentioned requirements for commercialization of hydrogen fuel cell vehicles, they must meet certain safety requirements associated with storing and dealing with hydrogen onboard.

With respect to hydrogen safety, fuel cell vehicles must fulfill the applicable regulations, especially the Federal Motor Vehicle Safety Standards and Regulations FMVSS 208 (Occupant crash protection) and FMVSS 301 (Fuel system integrity) for the United States [115]. Safety issues related to OpelHydroGen4 equipped with 70 MPa (700 bar) compressed hydrogen systems and designed in order to meet these standards were presented recently [116] (Figure 2.42). As any leak of hydrogen could be a source of ignition, seven hydrogen sensors at four locations for hydrogen detection have been mounted, as can be seen in Figure 2.36.

Hydrogen warning system is activated when hydrogen concentration exceeds the threshold values. The propulsion in the hydrogen warning mode is provided by OpelHydroGen4 high voltage traction battery. In order to validate hydrogen vessel performance, several tests under harsh conditions have been performed including permeation testing, drop testing, vibration, extreme temperature, bonfire, and gunfire testing. It was reported that after testing of several vessels at various pressures up to 70 MPa and extreme impact conditions, all of them remained leak tight and no pressure drop was noted. Vehicle run test with operating fuel cell and hydrogen supply onboard including a crash to a barrier with mass of 1368 kg at a speed of 80.5 km s$^{-1}$ showed that OpelHydroGen4 sensor system worked properly, closing main shut-off valves automatically thus showing no external hydrogen leakage [116].

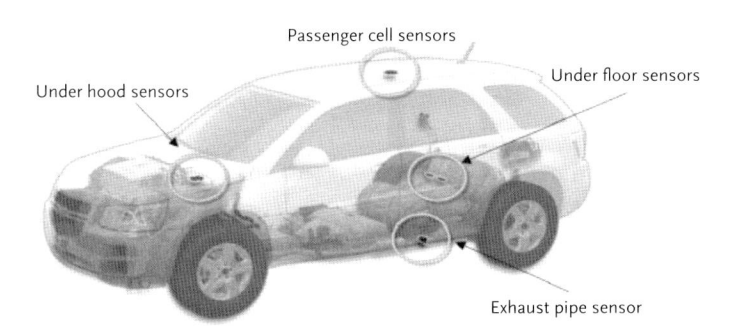

**Figure 2.42** Location of hydrogen sensors onboard the HydroGen4 [116].

## 2 Hydrogen – Fundamentals

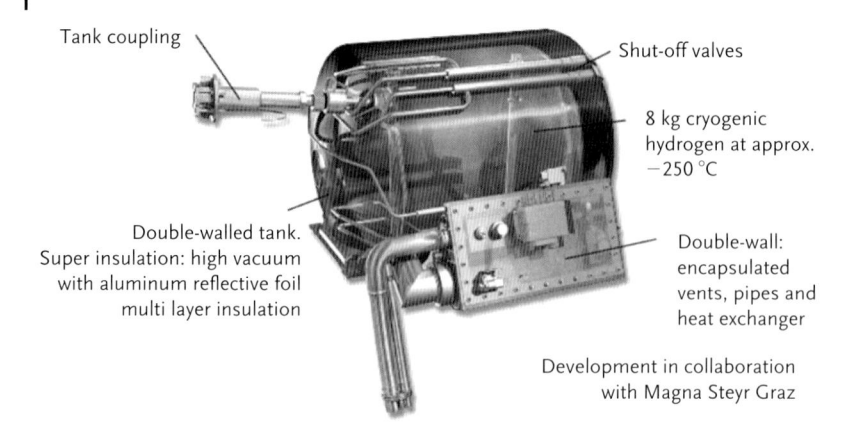

**Figure 2.43** LH$_2$ fuel storage with filling pipe [117].

Tank coupling
Shut-off valves
8 kg cryogenic hydrogen at approx. −250 °C
Double-walled tank. Super insulation: high vacuum with aluminum reflective foil multi layer insulation
Double-wall: encapsulated vents, pipes and heat exchanger
Development in collaboration with Magna Steyr Graz

BMW Hydrogen 7 with bifuelled internal combustion engine concept (H$_2$-ICE) has been developed with the engine, tank system, and car electronics being able to work on hydrogen [117]. The engine with power output of 191 kW was designed to switch from hydrogen to petrol without significant delay. BMW used in addition to a 74 L petrol tank a LH$_2$ vessel able to hold about 8 kg of liquid hydrogen for hydrogen supply. (Figure 2.43).

Special attention was put to safety requirements associated with liquid hydrogen properties during development process, especially boil-off gas. Four-level safety concept (design, fail, safe, safety identification, and warning) has been evaluated and confirmed by the internal and external authorities. Five hydrogen sensors have monitored the entire vehicle. Component and system tests have been carried out. After exposing the hydrogen tank to a temperature greater than 500 °C for at least 5 min (test in flames), the tank stayed closed off. BMW Hydrogen 7 was exposed to crash program including front crash with a speed of: 50 km h$^{-1}$ against a fixed barrier (US New Car Assessment Program, NCAP Front-crash), 64 km h$^{-1}$ against a deformable barrier (EU NCAP) and rear end-crash at 80 km h$^{-1}$ against a mobile barrier. It was proven that safety system reacted closing the hydrogen vessel valves and the vessel remained intact despite of its distortion [117].

A transition to pure hydrogen cars will require dealing with the problem of parking in enclosed areas or driving through channel. Explosion risk associated with hydrogen vehicles in tunnels was examined [118]. As the examples for the modeling, releases from hydrogen cars (with 700 bar H$_2$ tanks) and buses (with 350 bar H$_2$ tanks) have been investigated with two different tunnel layouts and a range of longitudinal ventilation condition. The simulations were based on the assumption that during incidents the hydrogen tanks were full. It was observed that pressures resulting from hydrogen explosions could be very high, with the exception of small clouds. From the study, it was found that pressure loads around 0.1–0.2 bar can be expected for a hydrogen incident. The results are obtained based on a set of assumptions about incident/ignition probabilities and need further studies based on experience, accident data, and experimental studies [118].

## 2.4.8
### Safe Removal of Hydrogen

A removal of hydrogen in enclosed spaces by a passive catalytic system is widely established and has shown its reliability based on the simplicity of the process. Such a passive system has the big advantage that when it is once well designed and built the system works independent on lack of power supply or other failure cases. The process is the catalytic recombination of hydrogen and oxygen with water as a product. Using catalysts active at room temperature allows the passive reaction. Typically, a hydrophobic platinum catalyst is used for this purpose. The reaction is exothermic and therefore, temperature of the catalyst will increase. The catalyst system should be designed as ignited in such a way that the ignition temperature is avoided. Reinecke *et al.* [119] investigated such passive autocatalytic recombiners (PAR). The work resulted in specifically designed catalysts combined with passive cooling devices; the realization of a combined concept in a prototype is planned.

## References

1 Bossel, U. (2000) The birth of the fuel cell 1835–1845. *European Fuel Cell Forum*, ISBN 3-905592-09-6.
2 Wilson, J., Wilson, W., and Wilson, J.M. (2007) *William Robert Grove. The Lawyer Who Invented Fuel Cell*, Metolius Ltd., United Kingdom, ISBN 978-0-9557193-0-1.
3 Rifkin, J. (2003) *The Hydrogen Economy*, Jeremy P Tarcher/Penguin Group, Inc., New York, ISBN 1-58542-254-1.
4 Ciba, J., Trojanowska, J., and Zolotajkin, M. (1996) *Mala Encyklopedia Pierwiastkow*, PWN, Warszawa, ISBN 83-204-2019-9.
5 Van Nostrand, R. (2005) *Encyclopedia of Chemistry*, 5th edn, John Wiley & Sons, Hoboken, NJ, ISBN 0-471-61525-0.
6 Atkins, P.W. (1998) *Physical Chemistry*, 6th edn, Oxford University Press, Oxford, Melbourne, Tokyo.
7 Zuettel, A. (2004) *Hydrogen Storage Methods*, Springer Verlag, Naturwissenschaften **91**, 157–172.
8 Nuckols, M.L., Wood-Putnam, J.L., and Van Zandt, K.W. (2009) New development in thermal technologies for hot and cold exposures. Special Operations Forces Industry Conference (SOFIC), Tampa, FL, 1–4 June SOFIC Paper No. 9702.
9 Adamson, K.-A. and Pearsons, P. (2000) Hydrogen and methanol: A comparison of safety, economics, efficiencies and emissions. *J. Power Sources*, **86**, 548–555.
10 Hake, J.-F., Linssen, J., and Walbeck, M. (2006) Prospects for hydrogen in the german energy system. *Energy Policy*, **34**, 1271–1283.
11 Muradov, N.Z. and Veziroglu, T.N. (2005) From hydrocarbon to hydrogen-carbon to hydrogen economy. *Int. J. Hydrogen Energy*, **30** (3), 225–237.
12 Sherif, S.A., Barbir, F., and Veziroglu, T.N. (2005) Wind energy and the hydrogen economy – review of the technology. *Solar Energy*, **78**, 647–660.
13 Ghenziu, A.F. (2002) Review of fuel processing for hydrogen production in PEM fuel cell systems. *Curr. Opin. Soild State Mater. Sci.*, **6**, 389–399.

14 Rostrup-Nielsen, T. (2005) Manufacture of hydrogen. *Catalysis Today*, **106**, 293–296.
15 Winter-Madsen, S. and Olsson, H. (2007) Steam reforming solutions. *Hydrocarbon Eng.*, July.
16 Wieland, S., Melin, T., and Lamm, A. (2002) Membrane reactors for hydrogen production. *Chem. Eng. Sci.*, **57**, 1571–1578.
17 Önsan, Z.I. (2007) Catalytic processes for clean hydrogen production from hydrocarbons. *Turk J. Chem.*, **31**, 531–550.
18 Kolb, G. (2008) *Fuel Processing for Fuel Cells*, Wiley-VCH Verlag GmbH & Co., Weinheim, ISGN 978-3-527-31581-9.
19 Hagh, B.F. (2004) Stoichiometric analysis of autothermal fuel processing. *J. Power Sources*, **130**, 85–94.
20 Ersoz, A., Olgun, H., Ozdogan, S., Gungor, C., Akgun, F., and Tiris, M. (2003) Autothermal reforming as a hydrocarbon fuel processing option for PEM fuel cell. *J. Power Sources*, **118**, 384–392.
21 Lin, S.-Y., Suzuki, Y., Hatano, H., and Harada, M. (2002) Developing an innovative method, HyPr-RING, to produce hydrogen from hydrocarbons. *Energy Convers. Manage.*, **43**, 1283–1290.
22 Lin, S., Suzuki, Y., Hatano, H., Oya, M., and Harada, M.(2001) Innovative hydrogen production by reaction integrated novel gasification process (HyPr-RING). *J. S. Afr. Inst. Min. Metall.*, January/February, 53–59.
23 Petitpas, G., Gonzalez-Aguilar, J., Darmon, A., Metkemeijer, R., and Fulcheri, L. (2007) Prague: non-equilibrium plasma assisted hydrogen production: state-of-the-art. 28th International Conference on Phenomena in Ionized Gases, ICPIG, Prague, Czech Republic, July 15–20.
24 Deminsky, M., Jivotov, V., Potapkin, B., and Rusanov, V. (2002) Plasma-assisted production of hydrogen from hydrocarbons. *Pure Appl. Chem.*, **74** (3), 413–418.
25 Sarmiento, B., Brey, J.J., Viera, I.G., Gonzalez-Elipe, A.R., Cotrino, J., and Rico, V.J. (2007) Hydrogen production by reforming of hydrocarbons and alcohols in a dielectric barrier discharge. *J. Power Sources*, **169**, 140–143.
26 Ouni, F., Ahmar, E.E., Khacef, A., and Cormier, J.M. (2006) Novel sliding discharge reactor for hydrogen production through hydrocarbons. WHEC 16, Lyon France, June 13–16.
27 Shoko, E., McLellan,B., Dicks, A.L., and da Costa, J.C.D. (2006) Hydrogen from coal; production and utilisation technologies. *Int. J. Coal Geol.*, **65**, 213–222.
28 Childress, J. (2005) The World Gasification Industry 2004–2010, Major Factors & Trends Driving Growth. Second International Freiberg Conference on IGCC & XtLTechnologies, Freiberg, Germany, June.
29 Stiegel, G.J. and Ramezan, M. (2006) Hydrogen from coal gasification: an economical pathway to a sustainable energy future. *Int. J. Coal Geol.*, **65**, 173–190.
30 DOE Hydrogen Production Roadmap. http://www1.eere.energy.gov/hydrogenandfuelcells/pdfs/h2_production_roadmap.pdf.
31 Goel, N., Mirabal, S.T., Ingley, H.A., and Goswami, D.Y. (2003) Hydrogen production in *Advances in Solar Energy, An Annual Review of Research and Development*, Vol **15** (ed. D.Y. Goswami), American Solar Energy Society, Inc., Boulder, CO, ISBN 0-89553-258-1.
32 Hocevar, S. and Summers, W. (2008) *Hydrogen production in Hydrogen Technology, Mobile and Portable Applications*, (ed. A. Leon), Springer-Verlag, Berlin, Heilderberg, pp. 15–79.
33 Miles, R.W. (2006) Photovoltaic solar cells: choice of materials and production methods. *Vacuum*, **80** (10), 1090–1097.
34 Jäger-Waldau, A. (2005) *PV Status Report 2005, Research, Solar Cell Production and Market Implementation of Photovoltaics*, Office for Official Publications of the European Communities, Luxembourg, ISBN 92-79-00174-4.
35 Otte, K., Makhova, L., Braun, A., and Konovalov, I. (2006) Flexible Cu(In,Ga)Se$_2$

thin-film solar cells for space application. *Thin Solid Films*, **511–512**, 613–622.
36 Kuendig, J., Goetz, M., Shah, A., Gerlach, L., and Fernandez, E. (2003) Thin film silicon solar cells for space applications: study of proton irradiation and thermal annealing effects on the characteristics of solar cells and individual layers. *Solar Energy Mater. Solar Cells*, **79**, 425–438.
37 Ullal, H.S. (2004) Polycrystalline thin-film photovoltaic technologies: progress and technical issues, Prepared for the 19th European PV Solar Energy Conference and Exhibition, Paris, France, June 7–11. NREL/CP-520-36241. www.nrel.gov/ncpv/thin_film/docs/nrel_partnership_ullal_pv_sec_2004_status.doc.
38 Kazmerski, L.L. (2006) Solar photovoltaics R&D at the tipping point: a 2005 technology overview. *J. Electron. Spectrosc. Relat. Phenom.*, **150** (2–3), 105–135.
39 http://www.iea.org/papers/2009/PV_roadmap_targets_printing.pdf. [Online] 2009.
40 Holladay, J.D., Hu, J., King, D.L., and Wang, Y. (2009) An overview of hydrogen production technologies. *Catalysis Today*, **139**, 244–260.
41 Singliar, M. (2007) Solar energy using for hydrogen production. *Pet. Coal*, **49** (2), 40–47. ISSN 1337-7027.
42 Varkaraki, E., Lymberopoulos, N., Zoulias, E., Kalyvas, E., Christodoulou, C., Vionis, P., and Chaviaropoulos, P. (2006) Integrated wind–hydrogen systems for wind parks. EWEC, Athens, February 27–March 2.
43 Riis, T. (2005) The Utsira Wind–Hydrogen Project. IPHE ILC Meeting, Rio de Janeiro, Brazil, March 22.
44 World Wind Energy Association (2009) Wind Energy Market worldwide continues strong growth, WWEA Press Release, June 2009.
45 European Wind Energy Association (EWEA) (2009) Wind Now Leads EU Power Sector, EWEA Press Release, February 2.
46 Wiser, R., Bolinger, M., Barbose, G., Mills, A., Rosa, A., Porter, K., Fink, S., Tegen, S., Musial, W., Oteri, F., Heimiller, D., Roberts, B., Belyeu, K., and Stimmel, R. (2008) US DOE Energy Efficiency and Renewable Energy, 2008 Wind Technologies Market Report.
47 Gonzalez, A., McKeogh, E., and Gallachoir, B.O. (2003) The role of hydrogen in high wind energy penetration electricity systems: the Irish case. *Renewable Energy*, **29**, 471–489.
48 Janssen, H., Bringmann, J.C., Emonts, B., and Schroeder, V. (2004) Safety-related studies on hydrogen production in high-pressure electrolysers. *Int. J. Hydrogen Energy*, **29**, 759–770.
49 Christodoulou, C.N., Karagiorgis, G.N., and Poullikkas, A.K. (2008) Renewable energy sources (RES) electricity storage in the form of "green hydrogen". Proceedings of Deregulated Electricity Markets Issues in South-Eastern Europe (DEMSEE 2008), Hilton, Nicosia-Cyprus, September 22–23.
50 Millet, P., Ngameni, R., Grigoriev, S.A., Mbemba, N., Brisset, F., Ranjbari, A., and Etievant, C. (2009) PEM water electrolyzers: from electrocatalysis to stack development. *Int. J. Hydrogen Energy*, doi 10.1016/j.ijhydene.2009.09.015.
51 Utgikar, V. and Thiesen, T. (2006) Life cycle assessment of high temperature electrolysis for hydrogen production via nuclear energy. *Int. J. Hydrogen Energy*, **31**, 936–944.
52 O'Brien, J.E., Stoots, C.M., Herring, J.S., and Hartvigsen, J. (2005) Hydrogen production performance of a 10-cell planar solid oxide electrolysis stack. Proceedings of FUELCELL2005, Third International Conference on Fuel Cell Science, Engineering and Technology, Ypsilanti, MI, USA, May 23–25.
53 O'Brien, J.E., Stoots, C.M., Herring, J.S., and Lessing, P.A. (2004) Performance measurements of solid-oxide electrolysis cells for hydrogen production from nuclear energy. The 12th ICONE Meeting, Arlington, VA, USA, April 25–29.
54 Hawkes, G.L., O'Brien, J.E., Stoots, C.M., and Herring, J.S. (2005) CFD model of a planar solid oxide electrolysis cell for

hydrogen production from nuclear energy. The 11th International Topical Meeting on Nuclear Reactor Thermal-Hydraulics (NURETH-11), Avignon, France, October 2–6.
55 Shin, Y., Park, W., Chang, J., and Park, J. (2007) Evaluation of the high temperature electrolysis of steam to produce hydrogen. *Int. J. Hydrogen Energy*, **32**, 1486–1491.
56 Rodriguez, G. and Pinteaux, T. (2003) Studies and design of several scenarios for large production of hydrogen by coupling a high temperature reactor with steam electrolysers. 1st European Hydrogen Energy Conference, Grenoble, France, September 2–5.
57 Tao, G., Butler, B., Homel, M., and Virkar, A. (2009) Development of a novel efficient solid-oxide hybrid for co-generation of hydrogen and electricity using nearby resources for local application. 2009 DOE Hydrogen Program Annual Review, May 22.
58 Evers, A.A. (2009) A new approach to a flexible power system. 8th European Fuel Cell Forum, Luzerne, USA. C0007-Abstract 001.
59 Lemort, F., Lafon, C., Dedryvère, R., and Gonbeau, D. (2006) Physicochemical and thermodynamic investigation of the UT-3 hydrogen production cycle: a new technological assessment. *Int. J. Hydrogen Energy*, **31** (7), 906–918.
60 Marchetti, C., Spitalnik, J., Hori, M., Herring, J.S., O'Brien, J.E., Stoots, C.M., Lessing, P.A., Anderson, R.P., Hartvigsen, J.J., Elangovan, S., Vitart, X., Martinez Val, J.M., Talavera, J., Alonso, A., Miller, A.I., and Wade, D.C. (2004) *Nuclear Production of Hydrogen – Technologies and Perspectives for Global Deployment*. International Nuclear Societies Council., LaGrange Park, IL, USA, ISBN 0-89448-570-9.
61 Ni, M., Leung, D.Y.C., Leung, M.K.H., and Sumathy, K. (2006) An overview of hydrogen production from biomass. *Fuel Process. Technol.*, **87**, 461–472.
62 Evans, R., Boyd, L., Elam, C., Czernik, S., French, R., Feik, C., Philips, S., Chaornet, E., and Patern, Y. (2003) Hydrogen from biomass-catalytic reforming of pyrolysis vapors. *FY 3003, Hydrogen, Fuel Cells, and Infrastructure Technologies, Progress Report*, National Renewable Energy Laboratory, Golden, CO.
63 Yeboah, Y.D., Bota, K.B., and Wang, Z. (2003) Hydrogen from biomass for urban transportation. *FY 2003, Hydrogen, Fuel Cells, and Infrastructure Technologies, Progress Report*, National Renewable Energy Laboratory, Golden, CO.
64 Abedi, J., Yeboah, Y.D., Realff, M., McGee, D., Howard, J., and Bota, K.B., (2001) An integrated approach to hydrogen production from agricultural residues for use in urban transportation. Proceedings of the 2001 DOE Hydrogen Program Review, NREL/CO-570-30535, National Renewable Energy Laboratory, Golden, CO.
65 Czernik, S., French, R., Evans, R., and Chornet, E. (2003) Hydrogen from post-consumer residues. *FY 2003, Hydrogen, Fuel Cells, and Infrastructure Technologies*, Progress Report, National Renewable Energy Laboratory, Golden, CO.
66 Bowen, D.A., Lau, F., Zabransky, R., Remick, R., Slimane, R., and Doong., S. (2003) Techno-economic analysis of hydrogen production by gasification of biomass. *FY 2003 Progress Report, Hydrogen, Fuel Cells, and Infrastructure Technologies*, National Renewable Energy Laboratory, Golden, CO.
67 Levin, D.B., Pitt, L., and Love, M. (2004) Biohydrogen production: prospects and limitations to practical application. *Int. J. Hydrogen Energy*, **29**, 173–185.
68 Masukawa, H., Mochimaru, M., and Sakurai, H. (2002) Hydrogenases and photobiological hydrogen production utilizing nitrogenase system in cyanobacteria. *Int. J. Hydrogen Energy*, **27**, 1471–1474.
69 Troshina, O., Serebryakova, L., Sheremetieva, M., and Lindblad, P. (2002) Production of $H_2$ by the unicellular cyanobacterium *Gloeocapsa alpicola* CALU 743 during fermentation. *Int. J. Hydrogen Energy*, **27**, 1283–1289.
70 Lin, C.Y. and Lay, C.H. (2004) Carbon/nitrogen-ratio effect on fermentative hydrogen production by mixed

microflora. *Int. J. Hydrogen Energy*, **29**, 41–45.

71. Kosourov, S.N., Tsygankov, A., Seibert, M., and Ghirardi, M.L. (2002) Sustained hydrogen photoproduction by *Chlamydomonas reinhardtii*: effects of culture parameters. *Biotechnol. Bioeng.*, **78**, 731–740.

72. Das, D. and Veziroglu, T.N. (2001) Hydrogen production by biological processes: a survey of literature. *Int. J. Hydrogen Energy*, **26**, 13–28.

73. Jung, C.R., Kundu, A., Ku, B., Gil, J.H., Lee, H.R., and Jang, J.H. (2008) Hydrogen from aluminium in a flow reactor for fuel cell applications. *J. Power Sources*, **175**, 490–494.

74. Kundu, A., Jang, J.H., Gil, J.H., Jung, C. R., Lee, H.R., Kim, S.h., Ku, B., and Oh, Y.S. (2007) Micro-fuel cells – current development and applications. *J. Power Sources*, **170**, 67–78.

75. NIKKEI ELECTRONICS (2006) Using Aluminum and Water to Make Pure Hydrogen as a Fuel Source for Fuel Cells, Etc. November 20, vol. **939**. 2009, http://techon.nikkeibp.co.jp/english/NE/2006/1120.html.

76. Wang, E.-D., Shia, P.-F., Du, C.-Y., and Wang, X.-R. (2008) A mini-type hydrogen generator from aluminum for proton exchange membrane fuel cells. *J. Power Sources*, **181** (1), 144–148.

77. Kothari, R., Buddhi, D., and Sawhney, R.L. (2008) Comparison of environmental and economic aspects of various hydrogen production methods. *Renewable Sustainable Energy Rev.*, **12**, 553–563.

78. Edeskuty, F.J. and Reider, R. (1979) Hydrogen safety problems. *Int J. Hydrogen Energy*, **4**, 41–45.

79. Pargeter, R.J. and Hammond, R.I. (2007) A temperature controlled mechanical test facility to ensure safe materials performance in hydrogen at 1000 bar. 2nd International Conference on Hydrogen Safety, San Sebastian, Spain, September 11–13.

80. Guo, Z.X., Shang, C., and Aguey-Zinsou, K.F. (2008) Materials challenges for hydrogen storage. *J. Eur. Ceramic Soc.*, **28**, 1467–1473.

81. ISO 16111:2008 (E) *Transportable Gas Storage Devices – Hydrogen Absorbed in Reversible Metal Hydride*.

82. Landucci, G., Tugnoli, A., Nicolella, C., and Cozzani, V. (2007) Assessment of inherently safer technologies for hydrogen storage. IChemE Symposium Series No. 153.

83. Anton, D., Mosher, D., Fichtner, M., Dedrick, D., Chahine, R., Akiba, E., and Kuriyama, N. (2007) Fundamental safety testing and analysis of hydrogen storage materials & systems. Proceedings of the 2nd International Conference on Hydrogen Safety, San Sebastian, Spain, September 11–13.

84. MSDS number: 1009, product information: 1-800-752-1597. Air Products and Chemicals, Inc.

85. Zabetakis, M.G. (1965) *Flammability Characteristics of Combustible Gases and Vapors, Bulletin 627*. U.S. Dept of the Interior, Bureau of Mines, USA.

86. Biennial report on hydrogen safety. http://www.hysafe.org/BRHS

87. Kuchta, J.M. (1985) *Investigation of Fire and Explosion Accidents in the Chemical, Mining and Fuel Related Industries, Bulletin 680*. U.S Dept of the Interior, Bureau of Mines, USA.

88. Molkov, V. (2007) Hydrogen safety research: state-of-the-art. Proceedings of the 5th International Seminar on Fire and Explosion Hazards, Edinburgh, UK, 28–43.

89. Reider, R. and Edeskuty, F.J. (1979) Hydrogen safety problems. *Int. J. Hydrogen Energy*, **4**, 41–45.

90. Schroeder, V. and Holtappels, K. (2005) Explosion characteristics of hydrogen–air and hydrogen–oxygen limits at elevated pressures. International Conference on Hydrogen Safety, Pisa, Italy, September 8–10.

91. Houf, W. and Schefer, R. (2007) Predicting radiative heat fluxes and flammability envelopes from unintended releases of hydrogen. *Int. J. Hydrogen Energy*, **32**, 136–151.

92. Schefer, R.W., Houf, W.G., Bourne, B., and Colton, J. (2006) Spatial and radiative properties of an open-flame hydrogen plume. Int. J. Hydrogen Energy, **31**, 1332–1340.

93 Choudhuri, A.R. and Gollahalli, S.R. (2003) Characteristics of hydrogen–hydrocarbon composite fuel turbulent jet flames. *Int. J. Hydrogen Energy*, **28**, 445–454.

94 Wu, Y., Lu, Y., Al-Rahbi, I.S., and Kalghatgi, G.T. (2009) Prediction of the liftoff, blowout and blowoff stability limits of the pure hydrogen and hydrogen/hydrocarbon mixture jet flames. *Int. J. Hydrogen Energy*, **34** (14), 5940–5945.

95 Morreale, C., Marengo, S., Migliavacca, G., and Maggioni, A. (2010) $H_2$ ignition by hot surfaces: safety issues and test methods.18th World Hydrogen Energy Conference, Essen, Germany.

96 Pasman, H.J. and Rogers, W.J. (2010) Safety challenges in view of the upcoming hydrogen economy: an overview. *J. Loss Prev. Process Ind.*, **23**, 697–704.

97 http://news.bbc.co.uk/2/shared/spl/hi/pop_ups/06/sci_nat_1986_challenger_disaster/html/1.stm (21 may 2012).

98 www.h2incidents.org (21 may 2012).

99 Weiner, S.C., Fassbender, L.L., and Quick, K.A. (2009) International Conference on Hydrogen Safety, *Using Hydrogen Safety Best Practices and Learning from Safety Events*, Ajaccio, France, September 16–18.

100 www.h2bestpractices.org (21 may 2012).

101 http://www1.eere.energy.gov/hydrogenandfuelcells/mypp/pdfs/safety.pdf (21 may 2012).

102 Buttner, W.J., Post, M.B., Burgess, R., and Rivkin, C. (2011) An overview of hydrogen safety sensors and requirements. *Int. J. Hydrogen Energy*, **36**, 2462–2470.

103 Wurster, R. (2000) European Integrated Hydrogen Project [EIHP] Publishable Final Report, Joule Contract No. JOE3-CT97-0088.

104 http://www.unece.org/trans/main/wp29/wp29wgs/wp29grpe/grpeage.html (21 may 2012).

105 http://www.eihp.org (27 February 2011).

106 Fateev, V.N., Grigoriev, S.A., Millet, P., Koroptsev, S.V., Porembskiy, V.I., Aupretre, F., Pepic, M., Etievant, C., and Puyenchet, C. (2007) Hydrogen safety aspects related to high pressure PEM water electrolyse. International Conference on Hydrogen Safety, San Sebastian, Spain, September 11–13.

107 Fateev, V.N., Grigoriev, S.A., Millet, P., Koroptsev, S.V., Porembskiy, V.I., Pepic, M., Etievant, C., and Puyenchet, C. (2009) High pressure PEM electrolysis and corresponding safety issues. International Conference on Hydrogen Safety, Aiaccio, France, September 16–18.

108 Faudou, J.-Y., Lehman, J.-Y., and Pregassame, S. (2005) Hydrogen refueling stations: safe filling procedures. International Conference on Hydrogen Safety, Pisa, Italy, September 8–10.

109 http://www.hyapproval.org/ (21 may 2012).

110 www.hyapproval.org/publications.html (21 may 2012).

111 Mair G.W. and Lau, M. (2008) Beurteilung der ermüdungsbeingten Ausfallsicherheit von Composite-Druckgefäßen, TÜ Bd. 49 Nr. 11/12-Nov./Dez.

112 BGI 5108 – Wasserstoffsicherheit in Werkstätten (2009) Berufsgenossenschaftlicher Informationen und Grundsätze für Sicherheit und Gesundheit am Arbeitsplatz.

113 Satyapal, S., Petrovic, J., Read, C., Thomas, G., and Ordaz, G. (2007) The U.S. Department of Energy's National Hydrogen Storage Project: progress towards meeting hydrogen-powered vehicle requirements. *Catalysis Today*, **120**, 246–256.

114 http://www1.eere.energy.gov/hydrogenandfuelcells/pdfs/freedomcar_targets_explanations.pdf (21 may 2012).

115 http://www.carsafetylawyer.com/resources/federal-motor-vehicle-safety-standards-fmvss (21 may 2012).

116 Sachs, Ch. (2010) Safety aspects of hydrogen fuel cell vehicles. 18th World

Hydrogen Energy Conference, 16–21 May 2010, Essen, Germany.

117 Müller, C., Fürst, S., and von Klitzing, W. (2007) Hydrogen safety: new challenges based on BMW hydrogen 7. International Conference on Hydrogen Safety, San Sebastian, Spain, September 11–13.

118 Hansen, O.R. and Middha, P. (2009) CFD simulation study to investigate the risk from hydrogen vehicles in tunnels. *Int. J. Hydrogen Energy*, **34** (14), 5875–5886.

119 Reinecke, E.-A., Kelm, S., Struth, S., Granzow, Ch., and Schwarz, U. (2007) Design of catalytic recombiners for safe removal of hydrogen from flammable gas mixtures. International Conference on Hydrogen Safety, San Sebastian, Spain, September 11–13.

# 3
# Hydrogen Application: Infrastructural Requirements

## 3.1
### Transportation

Hydrogen application for transportation has a very long history. The earliest application was the Rivaz-hydrogen car with an atmospherical piston engine, patented in January 1807. Figure 3.1 explains the main principle of this vehicle which can be named as the first technological step to hydrogen application for transportation.

The main usage of hydrogen in transportation in twentieth century was first observed for nuclear submarines, airships, and launching systems from the 1960s onward.

Airship history is strongly linked with the family of aerostatic planes such as the Zeppelin-type family; the first Zeppelin-type airships were flown by Hugo Eckener in the early days of the twentieth century, [1].

The beginning of the commercial passenger transportation aviation in the 1920s and 1930s has improved the design of large vessels, especially the Hindenburg family. The membrane design of the hull of aerostatic flight vehicles in combination with accessories and the vicinity of engines has always had a strong influence on the construction of such vehicles.

In addition to that, demands of very low permeation rates lead to specific woven cloth fabrics such as liners, which were also used on the early Zeppelin hydrogen-lifted airships. It can be clearly stated that the unfortunate catastrophic failure of structural interfaces on the Hindenburg Airship has initiated a major drawback on the aviation use of hydrogen.

Without outlining the current investigation results of the Lakehurst accident in 1937, a technical implication of future storage systems could be always linked with the philosophy of this specific airship.

The development of fuel cells (F/C) as a major keystone to the success of hydrogen and transportation in the twentieth century has offered huge possibilities especially for ships, submarines, and large commercial vessels. The specific need of silent motion for an underwater vessel, submarine (as experienced during World War II (WWII)) had brought up the use of F/C for submarines in combination with nuclear power plant propulsion system. Hydrogen storage for the use

*Hydrogen Storage Technologies: New Materials, Transport, and Infrastructure*, First Edition.
Agata Godula-Jopek, Walter Jehle, and Jörg Wellnitz.
© 2012 Wiley-VCH Verlag GmbH & Co. KGaA.
Published 2012 by Wiley-VCH Verlag GmbH & Co. KGaA.

**1807:**

H$_2$ storage:
COH$_2$ ballon

H$_2$ combustion engine

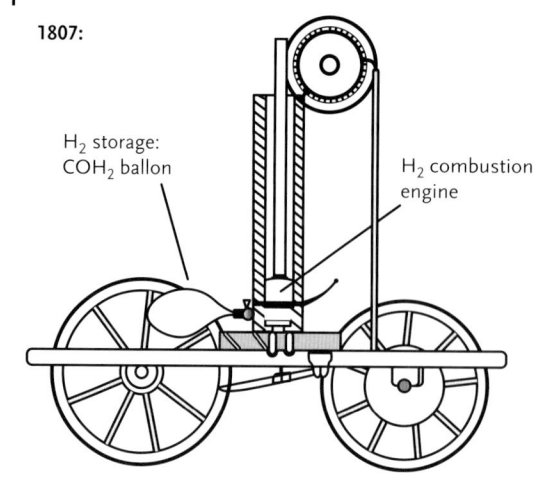

**Figure 3.1** Rivaz car patented as earliest vehicle ever running with hydrogen.

in this system is performed mainly with pressurized tanks of large sizes and/or cryogenic subsystems. Limitations due to cost and weight and availability of material play a minor role in comparison to the strong demands for silent diving.

The high-cost implications for the use of CFRP and F/C play a minor role; cost estimation per kilowatt is 8–12,000 Euro for an F/C.

The usage of hydrogen for ground transportation issues is strongly linked with the storage of hydrogen in gaseous, liquid, or solid state. The main principle is derived by the fact that pair vehicle specific amount of hydrogen can be filled from a filling station which will give the vehicle a substantial driving range. So, the capability of storing hydrogen in tanks plays the key role for the feasibility of individual or public ground transportation. This fact leads to the core problem to be solved for the hydrogen community: to ensure reliable hydrogen tank systems for the use for everybody.

In recent history, we have seen different developments for public ground transportation and individual transportation.

The ground transportation issue is mainly linked to buses for urban usage; this was mainly performed with CH$_2$ in cylindrical tanks sorted in racks, for example, mounted on the roof fuselage of the space frame body of buses.

The combination of CH$_2$ in high-strength cylindrical tanks with internal combustion engine (ICE) is a very powerful method to ensure a zero-emission environment in urban areas, airports, fair grounds, and so on. The storage of 1 kg of hydrogen can ensure a range of about 100 km, which is a sufficient distance in a city environment (1 kg per 100 km).

Application of this kinds can be seen on several applications around the world. Original equipment manufacturers (OEMs) like Honda, General Motors, and Mitsubishi have introduced bus fleets of this kind since the early 1980s. The availability of nearby filling stations can be maintained either by individual refill or by the installation of hydrogen-filling stations, such as at Berlin International Airport (BBI) or the Linde site in Unterschleißheim.

The installation of hydrogen transportation in buses with the combination of ICE and $CH_2$ is a very powerful method for the installation of zero-emission public transportation and allows a low-cost retrofit to convert a conventional combustion engine for the use of hydrogen (Figure 3.2).

For the use of hydrogen in individual ground transportation, the main concept would rely on the fact that the vehicle is refilled at a commercial public filling station with the refilling time below 3 min. Keeping in mind the necessity of providing a substantial range to the vehicle user, also the tank system plays the most important role for the feasibility for hydrogen transportation. In order to establish a range above 500 km per vehicle, mass of more than 5 kg hydrogen has to be carried per filling that requires enormous effort on the structural demands of the hydrogen tanks as state-of-the-art [2].

Figure 3.3 shows recent application of F/C vehicle from Daimler with a $CH_2$ tank system that appears to be a typical installation package for space frame-type vehicles.

The combination of fuel cell with a hydrogen tank system is one main branch (AST) of the car hydrogen development and can be seen on various vehicles from OEMs all over the world.

The F/C usage provides substantial advantages such as energy-efficient drivetrain, silent-mode operation behavior, and a very high efficiency in well-to-wheel

**Figure 3.2** Picture of a typical hydrogen-filling station at Linde site nearby Munich, Germany.

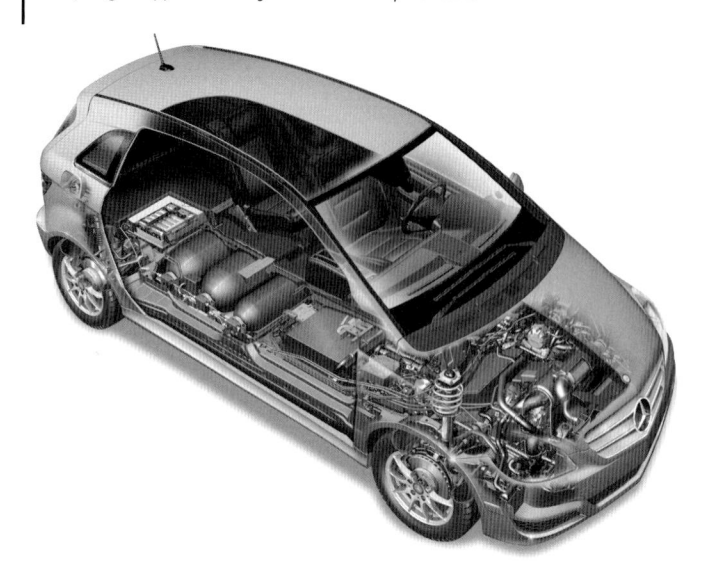

**Figure 3.3** B-class F/C car with $CH_2$ tank system (with permission from Daimler AG).

assessment. In [3], a brief overlook was given on the current worldwide F/C car fleets.

The second branch of personal vehicles would rely on the usage of ICE, again with different kinds of hydrogen storage means.

ICEs have a much lower efficiency rate than F/C but on the other hand would provide a high-power output related to low-mass and low-package needs, which give substantial advantages for the easy-to-use retrofit.

In particular, the combined usage of crude-oil-based fuels, biofuels, and hydrogen gives very good chances to achieve the range targets and the cost demands from the customer.

A famous example for this is the BMW fleet of seven-series cars that shows an experience of more than 35 years of hydrogen usage. The recent seven-series BMW hydrogen car was presented to the world public in 2007. The fleet was consequently established all over the world.

The combined usage of liquid conventional fuel and hydrogen would allow a dramatic cost reduction in F/C power trains. The ICE is recognized to show up with a much lower cost level than F/C vehicles, caused by the current cost range of platinum in the reacting cell of F/C.

The retrofit of a conventional ICE for the use of hydrogen is strongly linked to the endurance life limit and the environment in which the engine is run. The usage of $H_2$ would lead to an instant power drop of about 20–30% because of the extremely different combustion behavior of gaseous hydrogen. This power drop can be retrieved by different means. One option is direct injection and previous cooling of the LP supply system. Another one is to introduce a compressor system that can be run with an individual power supply. Other options are conventional turbo charger (with high temperatures and fatigue problems) and the setup of a special common rail injections system after the compressor circuit.

Such systems were mainly investigated in the so-called HyICE project, supported by the European Union, ref. HyICE.

The results from HyICE were directly used for the seven-series application and lead to an increase in power by about 20%.

To run an ICE engine a different type of oil supply system is used to run an ICE, and an increase in cooling power will in most cases lead to an increase of liquid-cooling fluid flux. The HyRacer project of the University of Ingolstadt (2005–2008) has given a clear insight into the retrofit of small car engines and the necessity of higher cooling power. The HyRacer was built in a mini-series production for BMW and different customers in Japan and Australia (Figure 3.4).

Derived from that, the Formula H project was launched in 2007 that showed the easy to use retrofit of a conventional bike engine (Rotax) for a conventional Le Mans racing car. This vehicle project showed the feasibility of a low cost and easy to use hydrogen application for an individual ground transportation.

The Formula H car used a conventional $CH_2$ tank with the pressure of 250 bar and carbon fiber reinforced the outer skin with an internal aluminum liner. These tank systems are most common and the base of the high-pressure tank family which ends up with an application of 750 bar tanks using high tensile strength carbon fibers.

**Figure 3.4** HyRacer.

## 3.2
## Filling Stations

The distribution of hydrogen for usage at filling stations is always a most important point of discussion when it comes to the feasibility of hydrogen ground transportation for the mass usage. Several examples of families of filling stations in approximately last 30 years were given by Air Liquid and Linde as well as by major oil companies like Shell, Aral, Esso, BP, Total, and so on. From all this experiences, it can be stated that filling stations are a state-of-the-art technique and the establishment of a net of stations is not a matter of time but is a matter of the presence of hydrogen car fleet.

The main key for the operation of filling stations is the technique of storing the hydrogen at the filling stations as well as an option for producing the hydrogen at the station which may apply to techniques like solar energy, wind energy, biomass, etc. (Figure 3.5).

Filling stations are designed that combine the storage of hydrogen in the gaseous and the liquid state. The hydrogen can be easily transformed from the liquid to the gaseous state that gives the possibility of high-pressure storage and filling of cars and vehicles.

The transportation of hydrogen to filling stations can be either maintained by pipelines/distribution systems or simply by trucks, whereas the truck transportation in a liquid state provides much higher efficiency and hydrogen capabilities than the gaseous storage tanks (see Figure 3.2).

**Figure 3.5** Filling station concept at Berlin Brandenburg International Airport (BBI) (photo courtesy: Pierre Adenis).

To provide a reliable operation of the filling station, a minimum hydrogen amount of between 200 and 500 kg for a 2-day operation is required. This can only be maintained by liquified hydrogen trucking or gaseous pipeline supply. In addition to that, production of hydrogen from solar power and wind energy with reverse-electrolyses would give additional capabilities for the production of $H_2$ which would also be available in individual filling stations.

It is a clear statement from the community that a filling station net can be provided countrywide in Europe, Japan, United States, and so on in a very fast manner. This is fully dependent on the presence of a hydrogen fleet or a forecast of hydrogen fleet installation per country.

In other words, filling stations and distribution of hydrogen are not the bottleneck of the hydrogen usage for mass-productive ground vehicles.

## 3.3
## Distribution

The main emphasis of distribution of hydrogen is linked up with the use of hydrogen in conventional individual passenger cars. Note that all techniques for storing of hydrogen, as well as for propulsion system and for distribution, have to cope with the high demand of vehicles worldwide, especially in new markets such as India, China, and South America.

Currently, the vehicle production worldwide is in a range between 65 and 70 million cars/year more than 16 million cars/year only in Europe (5.5 million cars/year in Germany).

This means that for a daily production rate of, for instance, 2500 cars, structural materials and components have to be supplied in an accordant manner.

To ensure an average range between 400 and 500 km per vehicle in a hybrid mode, at least 3 kg of hydrogen/vehicle has to be carried in a storage system.

Despite the demands of structural components such as carbon fiber, the demand of hydrogen for this kind of fleet is enormous.

It can be assumed, regarding the forecast in Figure 3.6, that until 2050 the fleet of hydrogen cars will be in a range of 20–30% worldwide. This would also include public transportation as well as other vehicles such as marine vessels or railroad. Even with this forecast, the demand of hydrogen would be in a range of 8–11 million tons per year. This leads to much higher efforts that need to be taken from now on to ensure mobility with hydrogen vehicles using hydrogen tank systems.

The world production of hydrogen today would allow the supply of about 110 million cars at a glance; of course, this amount is nowadays mainly used by the chemical industry.

In addition to that, hydrogen production as state-of-the-art is mainly driven by the usage of natural gas which is a direct derivative of crude oil.

To fulfill the high demands of hydrogen as one of the main propulsion sources of the near future, an enormous additional effort is necessary to establish a sufficient supply.

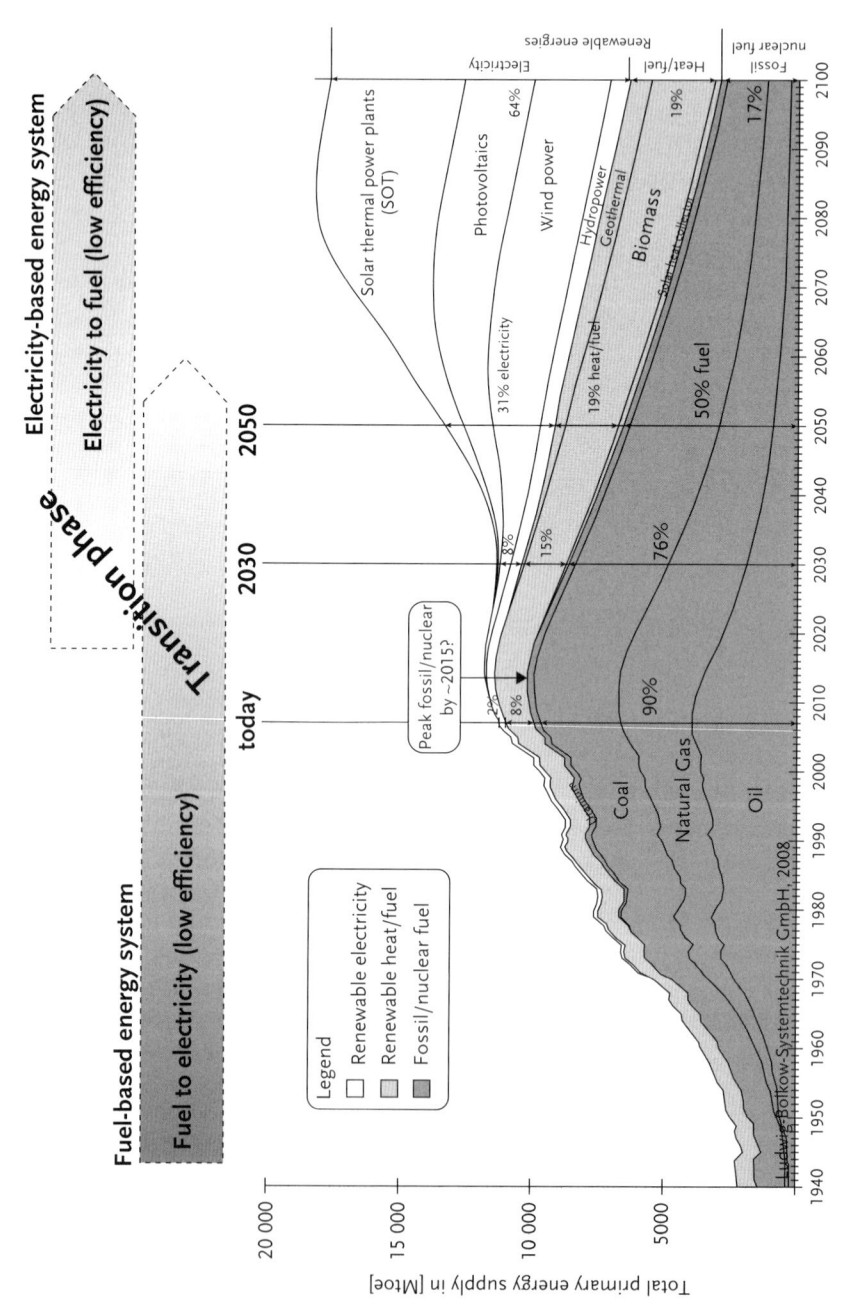

**Figure 3.6** Availability of sources for propulsion of vehicles in future.

This effort can be made by the stringent use and the introduction of electrolyses systems powered by solar energy as well as by wind energy (offshore wind parks and inland wind-energy parks) and by hydropower.

The consequent use of these techniques would immediately lead to establishing these demands and would give a real zero-emission down-line of future transportation. In Europe, more than 5000 windmill stations are planned to be build in the next 10 years, mainly in offshore parks in North Sea and East Sea.

Note that the use of nuclear power would lead to a major pathway to the production of hydrogen in future.

Despite local politics differences with regard to the usage of nuclear power plants, more than 140 new nuclear power plant applications are planned worldwide which gives enormous potential for the production of clean hydrogen.

As far as the distribution of hydrogen is concerned, the gas has to be filled in large tanks at the production side and transported to major distribution centers, which will be the same procedure as with crude oil.

For the production of hydrogen in off-regions, the usage of pipelines and a pipeline-distribution system is mandatory and is recognized as being one of the key infrastructure problems of the distribution philosophy in the next 30 years.

## 3.4
## Military

Hydrogen use for military vessels has been one of the main application fields for F/C over the past 50 years. Big advantages on the efficiency rate of F/C and the high-energy output in addition to "silent-mode" operations have given a major impulse for the necessity of hydrogen equipment.

In addition to that, cost and weight issues play a minor role compared to the extraordinary advantages of a combined operation. The main example for fuel cell operation and hydrogen storage application can be found in the development of submarines and nuclear power submarines.

From the early days of submarine operation, for example, CSS-Hunley (civil war 1861–1865) to WWI application in German and The British navy to WWII "mass-production" of German submarine fleets, the silent-mode operations of submarines played a major role for the survivability for maritime combat operation.

After WWII, the use of conventional electric battery-driven motors (charged by diesel engines) was replaced by a nuclear-power-plant power train sets in combination with F/C chargers. This technique was the major key to zero-noise emission for underwater navy operations.

Despite high-cost implications of the implementation of power packs from 100 kW and even more, the necessity of F/C is driven mainly by the design of naval vessels. Storage of hydrogen is performed by using compressed hydrogen tanks as well as cryogenic storage systems for long-range operations. The usage of solid

**Figure 3.7** Nuclear power plant driven on a submarine (U9) (with kind permission from the Auto & Technik Museum Sinsheim e. V).

storage hydrogen is of course limited due to high-weight implications and the low storage density within the selected agent such as magnesium-alanate (see Section 3.3; Figure 3.7).

F/C-operated vessels can be found in all big navy forces, and also in a high majority on aircraft carriers.

As far as army operations are concerned, portables play an important role as a subsidiary power plant for stationary infantristic applications. Production of hydrogen with solar pearl race with diesel engines can allow a silent-mode operation in combat for infantry as well as for mobile headquarters. This also allows operations with a low-infrared scanning pattern, because the hydrogen-powered cells would produce a low heat output and also low radiation emissions. This would give an excellent electronic camouflage effect. Smaller hydrogen devices such as portable batteries and portable F/C can be found in reconnaissance balloons and especially unmanned flying vehicles. This field of military operation belongs mainly to the air force combat portfolio (Figure 3.8).

Unmanned aircraft vehicle (UAV) as well as drones are recognized as the future flying combat vehicles in fighter and in bomber missions. This is the major field for the use of hydrogen-powered F/C, auxiliary power unit (APU), and in the form of onboard aggregate power supply.

A major role of the hydrogen-powered system is also recognized in the field of guided missiles, especially medium- and long-range cruise missiles. The demand of independent traveling flying bombs fits excellently into the properties of highly efficient F/C, whereby a low amount of hydrogen is needed to fulfill long-range missions. From the early days of F/C application, guided flying systems have played a very important role in the development of F/C, especially the "hot F/C," providing much higher energy and efficiency output (Figure 3.9).

**Figure 3.8** USS John F. Kennedy aircraft carrier.

**Figure 3.9** Fighter plane firing a missile.

## 3.5
## Portables

The application for portable hydrogen application is mainly driven by the requirements of the user. This section discusses the application of such means for use in households, outdoor, and also for the improvement and enhancement of daily life.

These requirements and the development of portable F/C would go with the demand of hydrogen much less than 1 kg, otherwise such storage systems would be unaffordable or be too big to be carried by hand or on a small vehicle.

Therefore, F/C operations are limited to a very small energy level and output; however, storage capacity can be designed for a "handheld" device.

One interesting example is the use of F/C-powered propellers for civil-maritime operations such as the "high-speed" diving aid for rescue operations in deep sea (also used for fun and scuba diving).

This is a good example of combat-package of an F/C system in which a small amount of hydrogen is stored in a compressed tank.

The market of small F/C (energy output much lower than 500 W) is very high and leads to various examples of portable hydrogen tools. Mainly hydrogen-operated F/C would add additional power to batteries or would give a good option to avoid costly battery operations. This can be seen also in the toy market for model planes, model ships, and mini-robots.

In Figure 3.10, a mini-self-teaching robot from the Graupner company is shown with the use of a small F/C pack as a power plant. This example shows, on one

**Figure 3.10** Graupner's F700 "Blue robot" with F/C pack carried by the robot himself.

hand, high package demands but on the other hand, the feasibility of the combination of hydrogen-powered tools within a limited cost environment.

## 3.6
## Infrastructure Requirements

Due to the limited range of hydrogen vehicles in the "pure hydrogen mode," which is in a range between 250 and maximum 350 km for a conventional passenger car, hydrogen infrastructure as far as fuelling stations are required is strictly mandatory for the use of hydrogen as "future gasoline" of the car in practice.

Hydrogen-filling stations are technically a feasible challenge which was realized in several applications worldwide, for example, Japan, Germany, California, and so on. The typical hydrogen-filling station should provide gaseous and liquid hydrogen supply, whereas options for "onside production" as well as for hydrogen delivery by truck should be concerned. In the research field of hydrogen infrastructure, several promising projects and pathways had been supplied. In Europe, the HyWays project was the main flagship of the demands and predictions of hydrogen-filling stations and fleet targets.

Imagine between 2010 and 2015 a limited number of around 400 small hydrogen stations are planned or demanded, serving round 10 000 hydrogen cars. This would mean 25 cars per station on an average. For corridors, another 500–600 small fillings stations would be required to fulfill travel demands from individual transportation (Figure 3.11).

The current development and car fleet predicted for electric mobility as well as hybrid cars would give a limitation to this number by estimated about 30%. This means that the car fleet powered by hydrogen between 2020 and 2035 would be in the neighborhood of around 15 million units for Europe. In addition to that the costs for the infrastructural setup of this huge effort are very high per country, so that we can recognize a substantial "chicken–egg" problem for introducing car fleets from hydrogen power source. The cumulative investment costs for hydrogen road transportation systems in Europe are estimated between 12 and 14.5 billion Euros. Between 2025 and 2030, the cost target would be around 60 billion Euros; of this, almost 50% will be covered by production investment, 20% by refueling investment, and the rest by transport, distribution, and liquefaction. This shows that more than 60% of the invest cost has to be brought up for the conventional part of the vehicle. So that car development for fuel cells and internal combustion engines describes the main action; this has to be resolved and longed before huge investments on hydrogen-filling stations can be made (Figure 3.12).

In Germany, the hydrogen demand and supply would be between 50 and 600 filling stations, covering an average supply of 300–1000 kg per day. It is estimated that the utilization of hydrogen refueling stations on an average would reach up to 50–80% from 2015 to 2020. Major influence of the strategy of the supply of hydrogen is the realization of electrolyte production as planned on site, which is a very powerful local method to produce hydrogen. As far as transportation issues

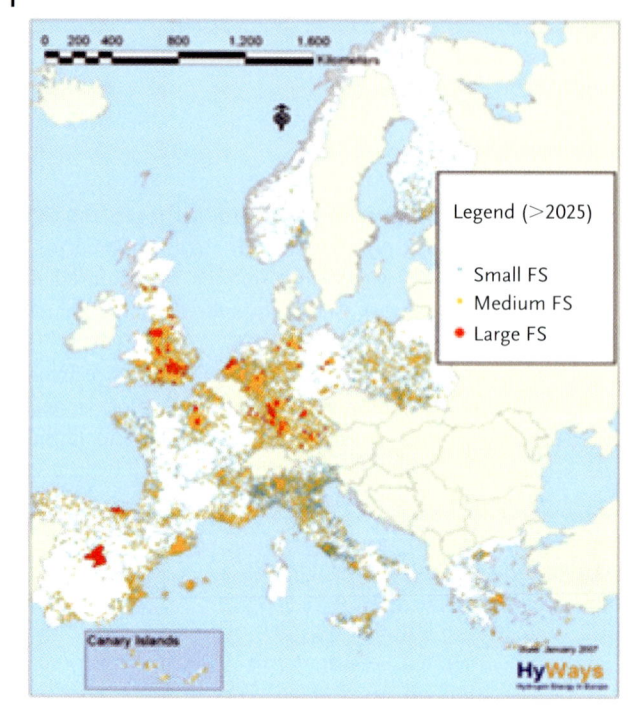

**94** | *3 Hydrogen Application: Infrastructural Requirements*

Example: fuelling stations spatial coverage for 8% vehicle penetration

**Figure 3.11** Fuelling stations spatial coverage for 8% vehicle penetration, HyWays Project.

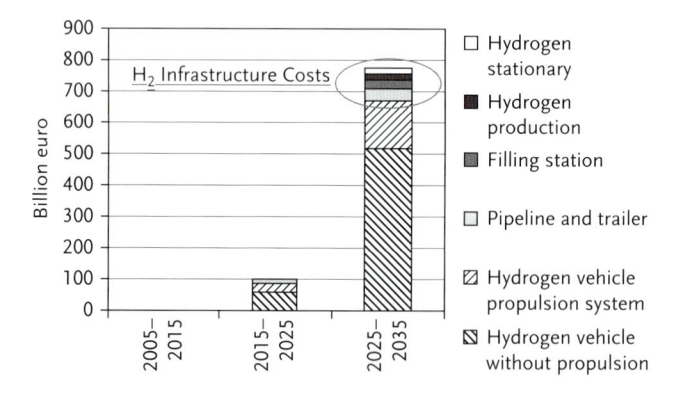

(cumulative investments for a ten-year period, hydrogen high penetration scenario, based on 6 HyWays Phase I member countries: D, F, I, GR, N, NL)

**Figure 3.12** Structure of the investments in a hydrogen economy, as an example taken for the six HyWays countries Deutschland (Germany), France, Italy, Greece, the Netherlands, and Norway (with kind permission from Springer Science and +Business Media).

3.6 Infrastructure Requirements | 95

are concerned, tracking of hydrogen in a gaseous state is a very less efficient method because of the limited amount of hydrogen which can be carried from A to B and the high requirements on the high-pressure tank vessel structures of large trucks (Figure 3.13).

**Figure 3.13** Typical hydrogen refueling station (with permission from Linde AG).

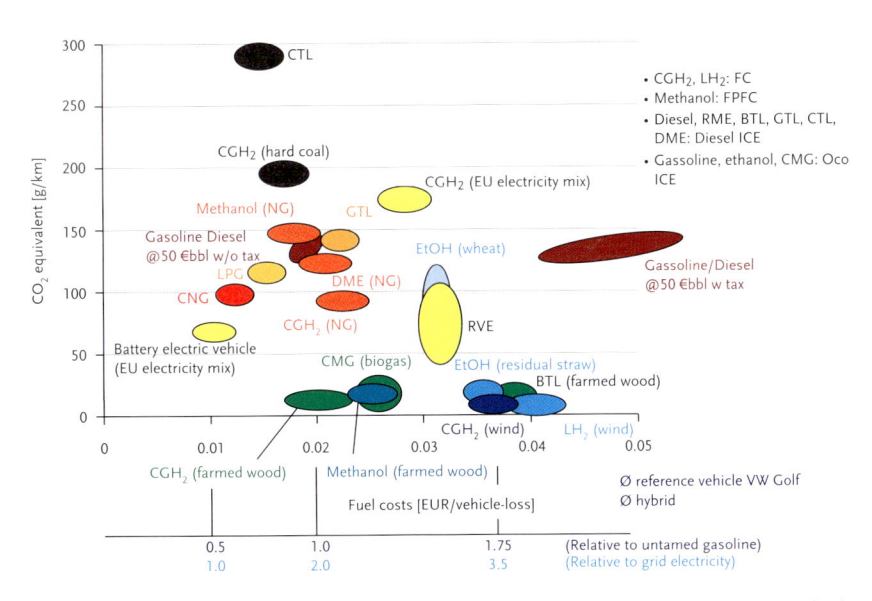

**Figure 3.14** Well-to-wheel hybridized power trains according to Ludwig Bölkow Systemtechnik.

In future energy landscape, we expect a variety of decentralized sources in combination with electric power supply. Practically, the combination of hydrogen-powered vehicles and electric or hybrid cars will be the most realistic transportation mix for the mid-term future. The costs for the realization of cryogenic storage for passenger cars and the related boundary conditions due to boil will lead to the extended use of gaseous hydrogen-powered vehicles. The storage of hydrogen locally with the help of liquefaction and liquid storage tanks will allow a much higher density of hydrogen for filling purposes. There are several options for a combination of a storage system, using both liquified and gaseous hydrogen on one side. In terms of "well-to-wheel" analysis for hybridized power trains, the $CO_2$ equivalent value and range will be point to the extraordinary quality of hydrogen, especially by taken into account the source of production. In Figure 3.14, a typical diagram of fuel cost versus GHG emissions is plotted.

## References

1 von Lueneberg, H., Welz, R. (2003) Geschichte der Luftfahrt: Luftschiffe (2). Vermittlerverlag Mannheim.

2 Hung, S., Subic, A., Wellnitz, J. (2011) Sustainable Automotive Technologies 2010: Proceedings of the 3rd International Conference, Springer Verlag.

3 Wellnitz, J., Subic, A., Leary, M. (2010) Sustainable Automotive Technologies 2010: Proceedings of the 2nd International Conference, Springer Verlag.

## Further Reading

Romm, J. J. (2006) Der Wasserstoff-Boom, Wiley-VCH, ISBN: 9783527315703.

Green, K., Miozzo, M., Dewick, P. (2005) Technology, Knowledge And The Firm: Implications For Strategy And Industrial Change, Elgar Publishing Ltd.

Wokaun, A., Wilhelm, E. (2011) Transition to Hydrogen: Pathways Toward Clean Transportation, Cambridge University Press.

# 4
# Storage of Pure Hydrogen in Different States

## 4.1
## Purification of Hydrogen

For the following described storage methods of hydrogen there are different requirements regarding the purity. In the case of liquefaction, the requirement is obvious. At 20 K, except helium and hydrogen, all other components are solid. Therefore, to avoid clogging, the acceptable level of present impurities is at about 1 ppm. The acceptable level of impurities before compression is around 5 ppm. The acceptable level before storing hydrogen as metal hydride is almost 10 ppm oxygen in the hydrogen stream, with carbon monoxide, hydrocarbons, and water at very low levels.

For hydrogen purification the following processes are applicable:

- PSA – pressure swing, adsorption
- gas permeation
- adsorption
- absorption
- distillation
- partial condensation
- catalytic oxidation (for oxygen removal).

The main implemented process in industry is the PSA, sometimes in a combination with other processes. Pressure swing adsorption (PSA) processes rely on the fact that under pressure gases tend to be attracted to solid surfaces; this is the adsorbing process. Higher the pressure, more the gas is adsorbed; when the pressure is reduced, the gas is released, or desorbed. PSA processes can be used to separate gases in a mixture because different gases tend to be attracted to different solid surfaces more or less strongly. If a gas mixture such as hydrogen with several impurities is passed under pressure through a vessel containing an adsorbent bed that attracts CO; $N_2$, $O_2$, and so on more strongly than it does hydrogen, nearly pure hydrogen will leave the adsorber bed. When the bed reaches the end of its capacity, it can be regenerated by reducing the pressure, thereby releasing the adsorbed molecules. It is then ready for another cycle of producing pure hydrogen. Typical adsorbents are activated carbon, silica gel, alumina, and zeolites.

*Hydrogen Storage Technologies: New Materials, Transport, and Infrastructure*, First Edition.
Agata Godula-Jopek, Walter Jehle, and Jörg Wellnitz.
© 2012 Wiley-VCH Verlag GmbH & Co. KGaA.
Published 2012 by Wiley-VCH Verlag GmbH & Co. KGaA

## 4.2
## Compressed Hydrogen

Storing hydrogen as compressed gas is the most often used method.

### 4.2.1
### Properties

In Table 4.1, the physical properties of hydrogen are summarized in comparison with the numbers for natural gas and gasoline.

### 4.2.2
### Compression

For the understanding of the compression of hydrogen it is important to have a look on the gas law for real gases and on the pressure–density diagram for hydrogen.

For six temperatures the curves are shown in Figure 4.1. At lower pressures, the relation is more or less proportional. At higher pressures, the density does not increase further in the same ratio. Doubling the pressure at 298 K (from 100 to 200 MPa) results in the increase of density only by a factor of 1.4. The higher the temperature the lower the density.

The corresponding equations for hydrogen are given in the following equation:

$$\left(p + \frac{a\,n^2}{V^2}\right)(V - n \times b) = nRT \tag{3.1}$$

**Table 4.1** Properties of compressed hydrogen and gasoline.

| Properties @ 293.15 K | Compressed hydrogen | Compressed natural gas | Gasoline |
|---|---|---|---|
| Upper heating value (MJ kg$^{-1}$) | 143 | 55 | 44 |
| Lower heating value (MJ kg$^{-1}$) | 120 | 50 | 41 |
| Density (kg m$^{-3}$) @ 100 kPa | 49.939 | | 720–770 @ 0.1 MPa |
| Density (kg m$^{-3}$) @ 70 MPa | 39.693 | | |
| Density (kg m$^{-3}$) @ 35 MPa | 23.651 | | |
| Density (kg m$^{-3}$) @ 20 MPa | 14.707 | 162 | |
| Volumetric energy density (MJ L$^{-1}$) @ 100 kPa | 7.14 | | 30.5 @ 0.1 MPa |
| Volumetric energy density (MJ L$^{-1}$) @ 70 MPa | 5.68 | | |
| Volumetric energy density (MJ L$^{-1}$) @ 35 MPa | 3.38 | | |
| Volumetric energy density (MJ L$^{-1}$) @ 20 MPa | 2.1 | 8.8 | |

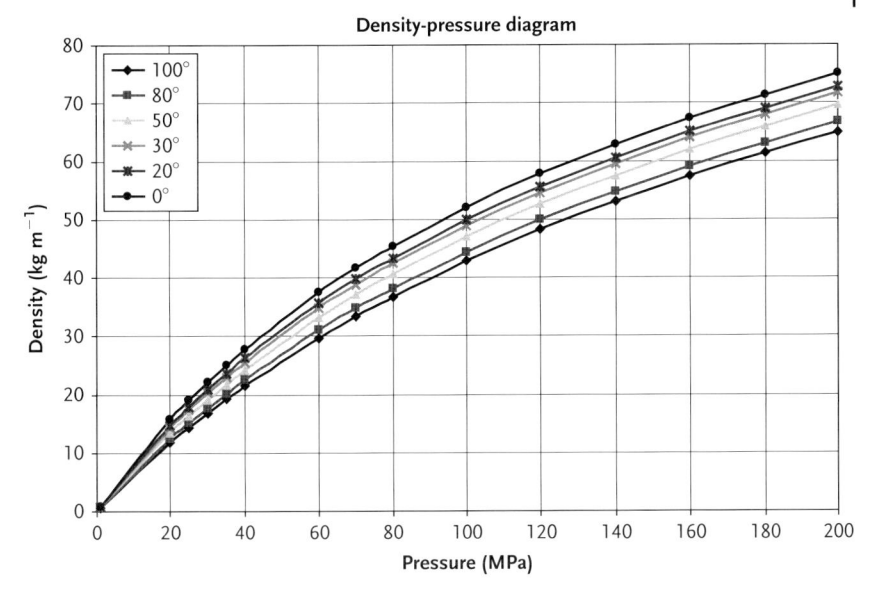

**Figure 4.1** Density of compressed hydrogen as function of the pressure for different temperatures.

where
  Measure of the attraction between the hydrogen molecules:
  $a = 24.7170 \; L^2 \; kPa \; mol^{-2}$
  Volume excluded by a mole of hydrogen: $b = 0.0270 \; L \; mol^{-1}$
  Gas pressure: $p(Pa)_s$
  Volume: $V \; (m^3)$
  Individual gas constant: $R = 4124 \; J \; K^{-1} \; kg^{-1}$
  Temperature: $T \; (K)$
  Number of moles: $n$

For the calculation of the power needed for the compression of hydrogen two different cases have to be considered. (i) The adiabatic compression that means compression without any heat exchange with the environment. (ii) The isothermal compression describes the compression at a defined temperature; therefore, a cooling of the gas has to foreseen.

The amount of work in the case of adiabatic compression is described by the following equation:

$$W = \frac{\gamma}{\gamma - 1} RT_1 \left( \left( \frac{p_2}{p_1} \right)^{\gamma - 1/\gamma} - 1 \right) \quad (3.2)$$

where
  Ideal gas constant: $R = 8.3145 \; J \; K^{-1} \; mol^{-1}$
  Temperature: $T \; (K)$

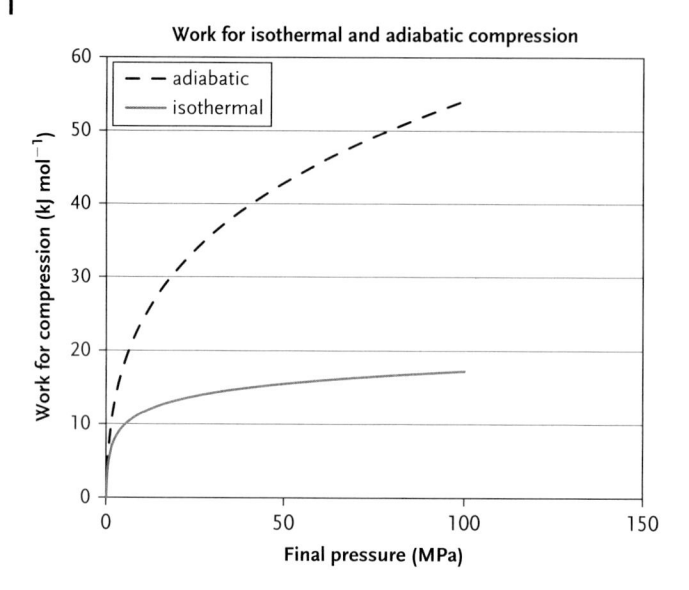

**Figure 4.2** Required work for the compression of hydrogen.

Gas pressure: $p$ (Pa)
Volume: $V$ (m³)
Specific heat ratio: $\gamma = \dfrac{C_p}{C_v}$

The amount of work in the case of isothermal compression is described in the following equation:

$$W = nRT \ln \frac{p_1}{p_2} \tag{3.3}$$

For the compression of 1 kg hydrogen and a inlet pressure of 2 MPa and a outlet pressure of 35 bar for the adiabatic compression the amount of work can be calculated as 5.45 MJ kg$^{-1}$. In the case of an isothermal compression, the resulting amount of work is 3.03 MJ kg$^{-1}$. The realistic power demand is between these two lines and could be realized with a multistage compression. The corresponding curves are shown in Figure 4.2. The compressed gas has to be cooled down to make the process more isothermal and less adiabatic. All these calculations do not include parasitic compression losses, general energy consumption of the compressor station, and so on.

Actually 35 MPa are standard for automotive applications, the first 70 MPa tanks are coming on the road for fuel-cell powered vehicles.

#### 4.2.2.1 Mechanical Compressors

The mechanical compressors are increasing the pressure by reducing the volume. The high required pressures need the use of volumetric compressors (like piston ore diaphragm compressors) instead of rotating compressors to achieve efficiency and the high pressure in one or only few steps.

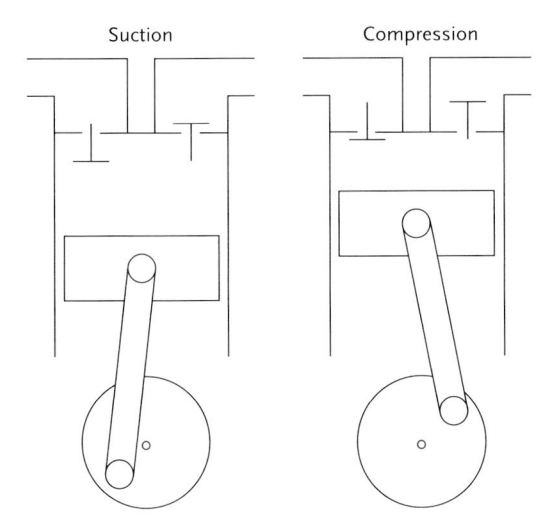

**Figure 4.3** Principle of a piston compressor.

**Piston Compressors**  There are two different possibilities to move the piston. One is the direct mechanical driving and the other the driving by a hydraulic system. The principle of such a piston compressor is shown in Figure 4.3.

Piston compressors for hydrogen are available for different pressures up to 100 MPa and different capacities up to 400 $m^3_{stand.}\ h^{-1}$.

**Diaphragm Compressor**  The membrane–diaphragm compressor is free of oil or other lubricant. This is an advantage by the compression of high chemically pure gases or explosive gases. The working principle is shown in Figure 4.4. For the compression of hydrogen a careful selection of the used materials is self-evident. The membranes are made from metals. The German company Hofer Hochdrucktechnik is building piston as well as diaphragm compressors. They have equipped several hydrogen filling stations with their compressors. There are also several companies in the United States, such as PDC machines Corp or Pressure Products Industries, who produce such compressors.

### 4.2.2.2 Nonmechanical Compressor

There are several nonmechanical compressors on a small scale already available or under intensive investigation. The advantage of such compressors is that there is a possibility to build a compressor without moving parts and to use other energy sources like thermal energy to power the compression. Mechanical compressors are heavy and huge units in which electric power has to be converted into mechanical work. From the energetic point of view and for dedicated small application other compressor seems to be more advantages. Three different types of nonmechanical compressors are therefore described in more detail.

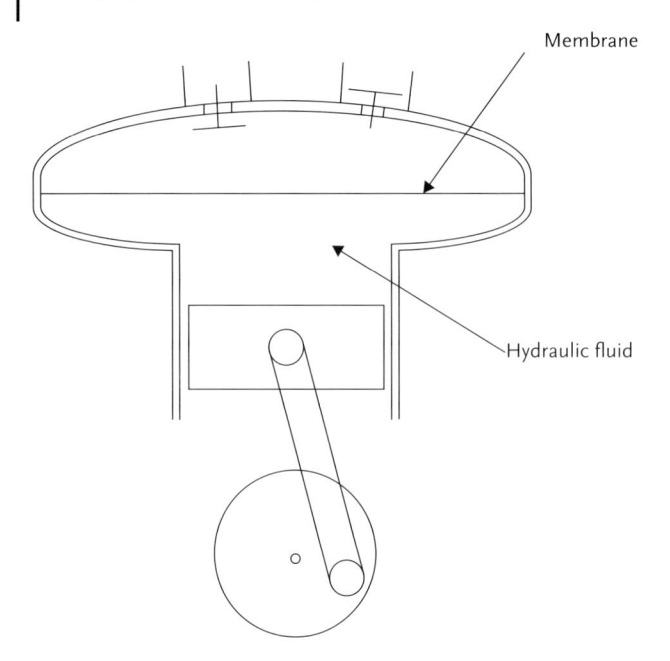

**Figure 4.4** Principle of a diaphragm compressor.

**Metal Hydride Compressor** By using metal a thermally driven absorption and desorption process a gas compression could be achieved. The process itself is a batch process and not a continuous one. In the case of compression of hydrogen, this process uses metal hydrides. The operation of a one-stage compressor consists of four process steps: (i) hydrogen adsorption at low pressure and low temperature, (ii) heating from low temperature to high temperature and producing high pressure, (iii) desorption of gas at high temperature and high pressure, and (iv) cooling from high temperature to low temperature. A semicontinuous operation can be achieved for example by using a second bed with metal hydride that is used alternately. The pressures and temperatures are according to the equilibrium of the selected metal hydride. The corresponding equation is given in the following equation:

$$\ln P_{eq} = \frac{\Delta H}{RT} - \frac{\Delta S}{R} \tag{3.4}$$

According to this equation, the pressure increases exponentially with the increasing temperature, and therefore large pressure values can be achieved by moderate temperature changes. To achieve the desired pressure a multistep process is used. For each step, the adequate metal hydride has be used. The advantage of the hydride compressor is that even with lower compression efficiency, the possible achievable cost is much lower compared with mechanical compressors. This is because for the hydride compressor relative cheap thermal energy by, for example, natural gas could be used instead of electricity that is costlier. A three-stage prototype has been presented by Popeneciu et al. [1]. Starting at a pressure of 200 kPa, a pressure of 5600 kPa could be achieved with cooling at 20 °C and heating at 80 °C. A productivity of 300 L h$^{-1}$ with a cycle time of 2 min and an amount of 360 g of metal hydrides was shown. With an accurate selection of the metal hydride and the process conditions high pressures as well as high productivities are possible. With the metal hydride thermal compressors cost-effective compression of hydrogen is possible.

**Electrochemical Compressor** The electrochemical compressor is based on a proton exchange membrane (PEM). The process is quite simple and works without any moving parts. The hydrogen is oxidized on the anode side, then due to the applied potential transported to the cathode side where the proton is reduced and hydrogen is build. On the cathode side, hydrogen is then available at a higher pressure. The principle is shown in Figure 4.5. For high pressures a cascade has to be installed. The electrochemical compression is very efficient. The power demand follows the isothermal compression. For 70 MPa about 10 MW kg$^{-1}$ are expected. Calculations have shown for a conventional PEM cell with a membrane with a thickness of 180 μm a maximum output pressure of 40 MPa (at 90 °C and 0.5 A cm$^{-2}$) can be achieved [2]. An achieved compression ratio of 40:1, from 0.36 to 14.3 MPa in a single stage was demonstrated [3].

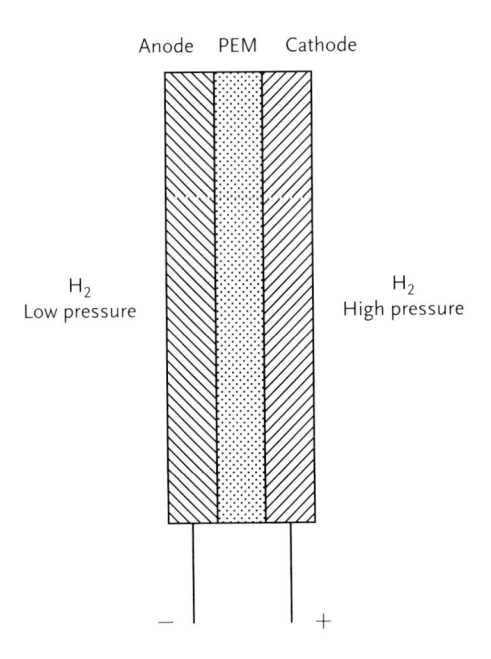

**Figure 4.5** Principle of the electrochemical hydrogen compressor.

This method for hydrogen compression for high pressures (35–70 MPa) is still on the early developmental stage. The development goals for the near future are

- increasing pressure capability
- increasing current density
- decreasing power consumption
- improving lifetime
- improving cycle stability.

**High Pressure Electrolyzer** In case hydrogen is produced by water electrolysis, there is the possibility to generate the high pressure already in the electrolysis stack. That has a big influence on the stack design. There are two methods to realize high-pressure electrolysis. One possibility is to design the stack in that way that the stack is under the high pressure and the endplates, screws, and the gaskets have to withstand the high forces. On the other side, there is a possibility to put the electrolysis stack inside a pressure vessel. Usually water is in the vessel. In this case, the requirements to the stack design are much easier to handle. From the energetic point of view, this process is very attractive. For the pressurization there is only the power for the isothermal compression needed. The potential can be calculated according to the following equation:

$$E = E_0 + \frac{3nRT}{zF}\ln\frac{p_1}{p_0} \text{ with } E_0 = \frac{\Delta G}{zF} \tag{3.5}$$

where
  Ideal gas constant: $R = 8.3145$ J K$^{-1}$ mol$^{-1}$
  Temperature: $T$ (K)
  Gas pressure: $p$ (Pa)
  Faraday constant: $F$(A s mol$^{-1}$)
  Valence: $z$ (−)
  Amount of substance: $n$ (−)

The theoretical increasing of the cell voltage as a function of pressure is given in Figure 4.6. The values are given for 298 K with the reversible cell voltage of 1.23 V.

The theoretical energy demand is the lowest achievable, but the technical realization requires a high effort to build such an electrolyzer for high pressures. The mechanical load and the functional stress are very hard. It is very hard to design such an electrolyzer and to guaranty low impurities of the product gases since the driving force for the hydrogen and oxygen permeation increases. Therefore, this technology seems only to be applicable for small capacities. An example is shown in Figure 4.7. This electrolyzer based on the alkaline technology was built for a pressure of 12 MPa and with a capacity of 100 g h$^{-1}$ by the Research Center Jülich in Germany [4]. There are also high-pressure electrolyzers based on the PEM technology built [5]. For example an PEM-based electrolyzer is described by Giner Inc. with a pressure of 8.5 MPa and a capacity of 0.333 kg h$^{-1}$ with a cell voltage of 2 V. Another PEM-electrolyzer is reported with a high efficiency for up to 13 MPa for a hydrogen

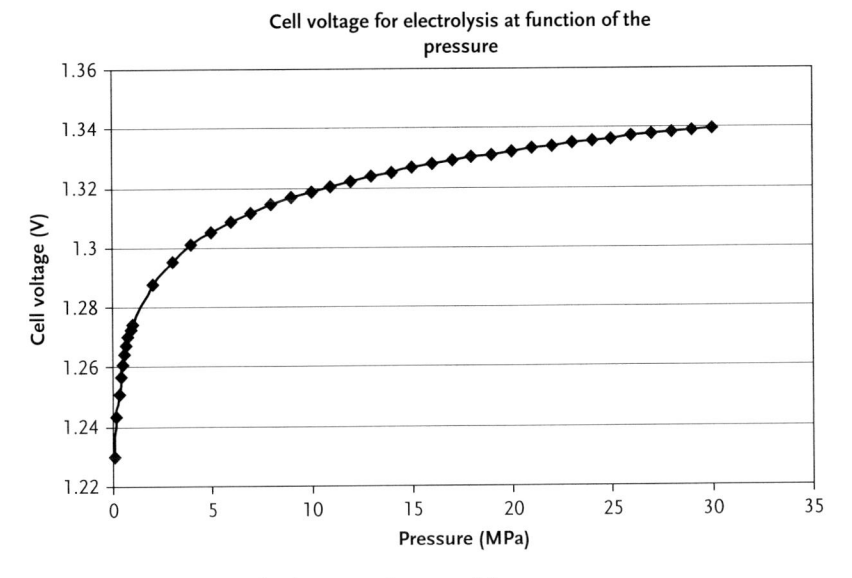

**Figure 4.6** Electrolyzer cell voltage as a function of the pressure.

**Figure 4.7** High-pressure electrolyzer from Research Center Jülich, Germany.

production rate of 0.36 Nm$^3$ h$^{-1}$ with very low cell voltages between 1.71 and 1.74 V [6]. The tests were performed with an eight cell stack. See also Chapter 2.

### 4.2.3
### Materials

For the successful use of materials for the storage of compressed hydrogen, some specialties of the behavior by use of compressed hydrogen have to be considered. Three different phenomena have to be thought of: the hydrogen embrittlement, hydrogen attack, and hydrogen permeation.

Knowing also the requirements from the application as storage pressure, cycles, environment, volume and with the well selection of materials a tank can be designed.

#### 4.2.3.1 Hydrogen Embrittlement

Hydrogen embrittlement occurs when hydrogen is in contact with a metal. This can happen already during the manufacturing process or during operation as material for hydrogen storage tanks. During production, for example, by arc welding could be a source for hydrogen atoms. Diatomic hydrogen gas can adsorb and dissociate on metal surfaces and built atomic hydrogen. Subsequently the hydrogen dissolute and hydrogen atoms can diffuse very fast into the material. Also other processes like electroplating or bating brings hydrogen in contact with metal.

The hydrogen permeation due to a hydrogen solubility of the metal leads to a certain hydrogen concentration in the metal. This hydrogen causes a reaction within the structure that will lead to the hydrogen embrittlement. The formation of a metal hydride is possible as well as the inclusion at a defect or at the grain boundary. Embrittlement leads to the hardening of the material and a significant reduction of the ductility followed by cracking well below the expected normal yield stress.

Hydrogen embrittlement increases with the partial pressure. But at a certain pressure level the effect is more or less constant. The critical value can vary between 2 and 10 MPa [7]. For steel even at a low pressure (few 100 kPa) the hydrogen embrittlement could be very pronounced. Hydrogen embrittlement effect is arrived at ambient temperatures and can often be neglected at temperatures above 100 °C. For austenitic steels the maximum effect arrives at about −100 °C, but can be neglected below 150 °C.

Materials that can be used without any specific precautions are

- brass and most of copper alloys
- aluminum and aluminum alloys
- beryllium–copper alloys.

Materials that are very sensitive to hydrogen embrittlement are

- Ni and Ni alloys
- Ti and Ti alloys.

## 4.2 Compressed Hydrogen

For steel, the hydrogen embrittlement depends on the chemical composition, heat and mechanical treatment, microstructure, impurities, and strength.

#### 4.2.3.2 Hydrogen Attack

In carbon steel, the hydrogen in the steel can react with the carbon and forms methane at grain boundaries and voids. The methane does not diffuse through the steel and is therefore collected in the voids at high pressures and causes cracks. This leads to a decarburization with a significant loss of strength. The risk of hydrogen attack increases with increasing temperature and pressure. Increasing fraction of chromium, molybdenum, titanium, and wolfram reduces risk of hydrogen attack. Other components such as aluminum, manganese, and nickel have a negative influence. Nelson curves show safe operating limits of temperature and hydrogen partial pressure for steels. Nelson curves are empirically generated curves for different steels.

#### 4.2.3.3 Hydrogen Permeation

There are two different types of hydrogen permeation to consider. First the permeation as hydrogen atoms through metals and second the permeation through nonmetals.

**Hydrogen Permeation Through Metals** For this process, first the diatomic hydrogen gas adsorbs at the metal surface and dissociates to produce atomic hydrogen. In the case of driving force, the hydrogen atoms will diffuse through the metal and then combine to diatomic hydrogen. This diatomic hydrogen then will leave the metal wall.

The hydrogen flux can be described according to the following equation:

$$J = \frac{P_H \left(p_h^{0.5} - p_l^{0.5}\right)}{d} \tag{3.6}$$

where

Flux: $J$ (m$^3$/(m$^2$ s))
Permeability: $P$ (m$^3$/(m$^2$ s Pa$^{0.5}$))
Thickness: $d$ (m)
High pressure: $p_h$ (Pa)
Low pressure: $p_l$ (Pa)

This hydrogen transport is very fast for metals such as palladium or palladium alloys at extended temperatures (>450 K). This transport mechanism could be used for the purification of hydrogen from reformats by using such membranes. In this case, composite membranes with a thin active layer deposited on a porous substrate are preferred.

**Hydrogen Permeation Through Nonmetals** Hydrogen permeation in nonmetallic materials follows the solution–diffusion process and follows the law of Fick, described by the following equation:

$$J = \frac{P_H (p_h - p_l)}{d} \tag{3.7}$$

where

Flux: $J$ (m$^3$/(m$^2$ s))
Permeability: $P$ (m$^3$/(m$^2$ s Pa))
Thickness: $d$ (m)
High pressure: $p_h$ (Pa)
Low pressure: $p_l$ (Pa)

The permeation is increasing with the increase of the temperature. It means for the storage of liquid hydrogen the hydrogen permeation is a minor issue. For structural materials for use in hydrogen storage materials with low hydrogen permeability are required. There are special alloys available with a low permeability. Also with a defined thermal treatment, the permeability can be decreased. A layer from a low permeable material inside a tank is also an option to reach an overall low permeability.

#### 4.2.3.4 Used Structural Materials

For large volume application as pipelines a carbon steel with low manganese content is commonly used. This kind of steel is cheap, which is the main reason for a mass application.

For pressure vessels mainly low-alloy steel is used. Low-alloy steel is ferritic steel that is typically used for the 10–50 L gas bottles. Usually the gas bottles are manufactured from seamless tubes and the filling pressure is typically 15 or 20 MPa. The low content of chromium, nickel, or molybdenum, which is in sum maximum 5%, makes this alloy cheap. The ferritic steels can be easily formed and welded, this is another advantage of these materials.

Austenitic stainless steel is also used for pressure vessels. The austenitic steels belong to the group of stainless steel. In addition to the common A304 and A316 the high temperature alloy A286 is used. A304 and A316 have a high content of chromium and nickel and a lower content of molybdenum or manganese. The content of the noniron components is over 25%. This makes the alloys more expensive. Type A286 alloy is an iron-based superalloy useful for applications requiring high strength and corrosion resistance up to 704 °C and for lower stress applications at higher temperatures. Type A286 alloy is a heat and corrosion resistant austenitic iron-based material that can be age hardened to a high-strength level.

The tensile strength of this alloy can be increased by heat treatment, so using this material higher pressures can be achieved.

Nickel-based alloy like Platinit, Chronin, Monel, Inconel®, and Incoloy® are not suitable for structural material for hydrogen storage. There are two reasons, one is relatively high price and second these alloys are prone to the hydrogen embrittlement.

Also aluminum alloys are suitable as a structure material. The big advantage for transportation application is the low density and the relatively high strength. But the aluminum alloys are more expensive compared to the steels. Anyway, the thickness of the wall must increase due to the lower strength of the material. The aluminum alloys have an advantage at lower pressures and therefore they are an excellent candidate for liquid hydrogen application. The aluminum plates have to be defect free; such defects could be detrimental for the storage system.

Also titanium is under investigation as possible for the use as a structural material. The β-titanium is less susceptible compared to the α-titanium. The titanium alloys have an excellent ratio of weight to strength. The machining is relatively costly as well as material is expensive. But for applications where the mass advantage is very important like the aerospace industry titanium alloys are used.

For the lightweight vessels for transportation fiber-reinforced plastic (FRP) is used. These vessels need a liner as permeation barrier. The achievable storage pressures are determined by the strength of the used fibers. In the same way, the costs are increasing.

Oxygen-free copper is usually used for piping at lower pressures for hydrogen distribution but not for tanks.

For the equipment for hydrogen storage systems copper alloys and nickel alloys could be applied as braze material or for valves, for example.

For all applications where a lightweight solution is preferred, composite material are preferably used. Composites are fiber-reinforced polymers. For the fibers, there are several options from glass fibers up to carbon fibers. For the polymer matrix, it is important to have low hydrogen permeability. But usually a liner is used as diffusion barrier

A very good overview about the compatibility of materials with hydrogen is given in the Internet-based source: "Technical Reference on Hydrogen Compatibility of Materials" [8].

### 4.2.3.5 Used Materials for Sealing and Liners

Here typically nonmetallic materials are used. These materials are often in direct contact with the hydrogen and have to withstand the thermal and mechanical loads. The temperature range is between −40 °C, during depletion, up to 90 °C, during filling. These are hard requirements to the sealing material. Especially the quick release may cause an irreversible damage of the sealing material. When the hydrogen cannot fast enough permeate to the surface, the sealing will be damaged by a explosive decomposition. This damage could not only occur at sealing, but also at other nonmetallic materials like liners.

Different kinds of elastomers (EPDM (ethylene propylene diene monomer), for example, are under investigation for gaskets for pressurized systems. The assortment of materials for liquid hydrogen is much lower because of the embrittlement of the elastomers at low temperatures. In this case, it seems that only polytertraflourethylene and polychlortriflourethylene can be used.

### 4.2.3.6 High Pressure Metal Hydride Storage Tank

A hybrid tank system, a combination of a high-pressure tank and a metal hydride tank, brings some advantages regarding the volumetric density [9]. The described tank with a total mass of 420 kg and a total volume of 180 m$^3$ reaches a hydrogen capacity of 7.3 kg. It has to be mentioned that the big advantage of this hybrid technology is that filling times like for high-pressure tanks can be achieved.

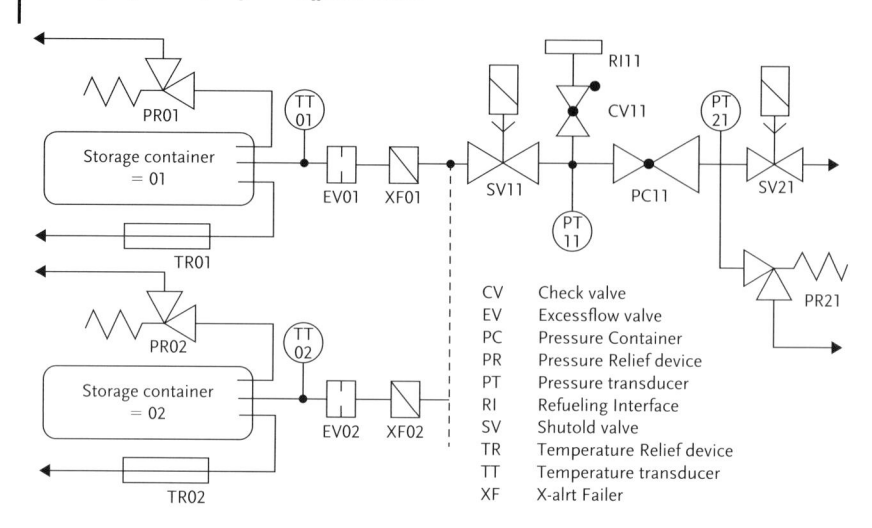

**Figure 4.8** Flow scheme of a compressed hydrogen fuel storage system.

| | |
|---|---|
| CV | Check valve |
| EV | Excessflow valve |
| PC | Pressure Container |
| PR | Pressure Relief device |
| PT | Pressure transducer |
| RI | Refueling Interface |
| SV | Shutold valve |
| TR | Temperature Relief device |
| TT | Temperature transducer |
| XF | X-alrt Failer |

### 4.2.4
**Sensors, Instrumentation**

In the flow scheme, the pressure relief valve PROn and the temperature relief device TROn can be seen in Figure 4.8. The excess valve EVOn is foreseen to minimize the hydrogen loss in the case of a break of the pipe. The filter XFOn prohibits particles to enter the vessel. For filling the valves, SV11 is opened and SV21 is closed. The compressed gas enters through the interface RI11, the storage tank. Due to the compression, the temperature inside the tank will increase. In case the temperature increases to defined maximum, tolerable temperature the filling process will be interrupted. During the release of hydrogen for feeding a consumer, the valve SV11 is open. The pressure transducer PC11 then reduces the pressure to a defined value. Out of operation all valves are closed. The shut-off valve SV21 is only used for system isolation, while the check valve CV11 inhibits the hydrogen to flow from the tank to the filling station. The pressure relief valve PR21 is able to release hydrogen safely in the case of a defect of the pressure.

### 4.2.5
**Tank Filling**

For the filling station, there are two principle options for the fuel supply: either the hydrogen is stored as liquid (or compressed hydrogen) or the hydrogen is produced on the premises. In this case, electrolysis by renewable energies such as solar power or wind energy is used, or hydrogen is reformed from biogas or natural gas. These filling stations require buffer tanks as well as high-pressure compressors to provide the hydrogen at 40 or 80 MPa.

During the filling of a high-pressure tank the internal gas temperature will increase. The temperature increase is more important for a fast filling of Type II and Type IV tanks, because there is not to much possibility for heat exchange. The compression is tendency of an adiabatic compression. The increasing of the temperature is depending on the following parameters:

- gas pressure before filling,
- final gas pressure,
- gas flow,
- the filling gas temperature,
- tank geometry,
- materials,
- and so on.

Depending on all these parameters the resulting gas temperature can easily reach the specified limit (85 °C).

The high temperature causes the following effects:

- decreasing mechanical strength;
- thermal tension due to temperature difference across the wall;
- increasing of hydrogen permeability, loss of hydrogen; and
- reduced capacity.

The decreasing of mechanical strength at higher temperature usually occurs at the same time when the pressure is the highest; this is important to know for safety aspects. The decrease of capacity can be expressed by the relation of the densities at different temperatures. For example the capacity for a 700 bar tank, which is filled at 80 °C instead of 20 °C, is decreased by (density at 20 °C − density at 80 °C/density at 20°C) or 12.5%.

The filling procedure has to be controlled to prohibit the tank from overheating and overpressure.

## 4.2.6
## Applications

### 4.2.6.1 Storage in Underground

For natural gas as well as for liquid fuels, the storage in big caverns is state-of-the-art in industrial countries. There are a lot of caverns with different volumes (up to several millions m$^3$) in use to even the gas storage and consumption. In Germany about 20% of the yearly consumption is stored in such subsurface caverns [10]. In the United States of America, hydrogen storage in caverns is realized. In Germany, it is under discussion to use hydrogen storage to make the energy peaks from wind energy useable. Gas storage in the underground is very attractive since high-energy densities could be achieved by high pressures (up to 200 bar for natural gas). For hydrogen salt caverns are very attractive; the hydrogen permeability is very low. It is expected that even for hydrogen the losses would be very low (<0.015% per year).

#### 4.2.6.2 Road and Rail Transportation
For road and rail transportation, different tanks with different sizes are available. The capacity of a trailer with the bog-standard bottles is about 300 kg hydrogen with a new developed tank the capacity can be increased up to 500 kg. The effective capacity is lower because the tanks cannot be fully be emptied.

#### 4.2.6.3 Vehicles
For automotive application from the US Department of Energy, the goals for hydrogen storage are given. The values are given in Table 4.2.

Typically for automotive application, pressure vessels with a volume of about 40 L were built. This means in the case of 70 MPa a maximum mass of 1.6 kg. Therefore, typically a bundle of pressure vessels is foreseen. The pressure vessels can be split into four types, as described in Table 4.3.

Type IV cylinders fulfill EIHP [11] requirements.

The safe and reliable operation of a compressed hydrogen storage system requires:

- system specification according to automotive requirements,
- accurate selection of parts and material,
- well-defined operational functions,
- functional safety,
- electrical safety.

For the automotive application, minimum 5 kg but preferably 10 kg of hydrogen has to be stored. Here, it is important to understand the physical limitations. Even at 70 MPa, the required volume just for hydrogen itself is about 125 m$^3$, respectively 250 L based on the density of 40 g L$^{-1}$ at 70 MPa.

The increase in the pressure and reduction in the mass require materials (fibers) with high strength. These fibers are very expensive and the tank is still produced with a low number of items. This leads to actual high costs for the pressure vessels. Costs of a Type III vessels of 5.6 € MJ$^{-1}$ assuming 1000 units with 150 liters are reported [12]. The main challenge is to combine safety criteria with technical performance and acceptable costs. The technical performance is determined by lower mass and lower volume. During STORHY-project, the vessels have been tested according to EIHP II draft regulations. That means a burst ratio of 2.35 and 15 000 cycles and a permeation rate of 1 Ncm$^3$ h$^{-1}$ L$^{-1}$. The result of the investigations showed that Type III vessels fulfill the burst pressure requirement but not the required cycles. While Type IV vessels achieve the burst pressure, the cycling requirements and they show a rather low permeation rate (0.02 Ncm$^3$ h$^{-1}$ L$^{-1}$)

Amongst others, the following manufacturers are working on storage tanks for automotive applications: Dynetek (Canada/Germany), COMAT (Germany),

**Table 4.2** DOE technical targets: on-board hydrogen storage.

| Storage parameter | Units | 2007* | 2010 | 2015 |
|---|---|---|---|---|
| Usable, specific energy from $H_2$ (net useful energy/maximum system mass) ("gravimetric capacity") | kWh kg$^{-1}$ (wt% hydrogen) | 1.5; (4.5%) | 2; (6%) | 3; (9%) |
| Usable energy density from $H_2$ (net useful energy/maximum system volume) ("volumetric capacity") | kWh L$^{-1}$; (kgH$_2$ L$^{-1}$) | 1.2; (0.036) | 1.5; (0.045) | 2.7; (0.081) |
| Storage system cost, | $/kWh net | 6; (200) | 4; (133) | 2 (67) |
| Fuel cost | $ per gallon gasoline equivalent at pump | 3 | 1.5 | 1.5 |
| Operating ambient temperature | °C | −20/50 (sun) | −30/50 (sun) | −40/60 (sun) |
| Cycle life (1/4 tank to full) | Cycles | 500 | 1000 | 1500 |
| Cycle life variation | % of mean (min) @ % confidence | N/A | 90/90 | 99/90 |
| Minimum and maximum delivery temperature of $H_2$ from tank | °C | −0.243 902 439 | −0.352 941 176 | −0.470 588 235 |
| Minimum full-flow rate | (g/s) kW$^{-1}$ | 0.02 | 0.02 | 0.02 |
| Minimum delivery pressure of $H_2$ from tank; FC=fuel cell, ICE=Internal combustion engine | Atm (abs) | 8 FC, 10 ICE | 4 FC, 35 ICE | 3 FC, 35 ICE |
| Maximum delivery pressure of $H_2$ from tank | Atm (abs) | 100 | 100 | 100 |
| Transient response 10%–90% and 0%-0% | S | Jan 75 | 0.75 | 0.5 |
| Start time to full-flow at 20 °C | S | 4 | 4 | 0.5 |
| Start time to full-flow at minimum ambient | S | 8 | 8 | 2 |
| System fill time for 5-kg hydrogen | Min | 10 | 3 | 2.5 |
| Loss of useable hydrogen | (g/h) kg$^{-1}$ $H_2$ stored | 1 | 0.1 | 0.05 |
| Permeation and leakage | Scc h$^{-1}$ | Federal enclosed area safety standard | | |
| Toxicity | | Meets or exceeds applicable standards | | |
| Safety | | Meets or exceeds applicable standards | | |
| Purity ($H_2$ from storage system) | | 98% (dry basis) | | |

Useful constants: 0.2778 kWh MJ$^{-1}$, ∼33.3 kWh gal$^{-1}$ gasoline equivalent.

*Note that some 2007 targets have been met with compressed hydrogen and liquid tanks.

**Table 4.3** Types of pressure vessels for hydrogen storage.

| Type | Characterization |
|---|---|
| I | Steel; maximum pressure: 20 MPa<br>Aluminum; maximum pressure: 17.5 MPa<br>Gravimetric hydrogen density: 1 wt% |
| II | Metal tank (aluminums/steel) with fiber composite (glass, aramid, and carbon) around the metallic cylinder; maximum pressure: 30 MPa |
| III | Composite material (glass, aramid, and carbon fibers) tank with a metal liner (aluminum, steel) maximum pressure: 44 MPa |
| IV | Composite material (carbon fiber) with a polymer liner (high density polyethylene (HDPE) maximum pressure: 70 MPa and even more<br>Gravimetric hydrogen density: 11.3 wt% |

Composites Aquitaine (France), Ullit (France), Faber(Italy), Quatum (USA), and Lincoln Composites (USA).

The filling of the tanks could be done in 3 min; the temperature of the gas will thereby increase. High-pressure storage tanks are available from a few to several hundred liters.

## 4.3
## Liquid/Slush Hydrogen

As in any gas, storing it as liquid takes less space than storing it as a gas at normal temperature and pressure, however the liquid density is very low compared to other fuels.

### 4.3.1
### Properties

At a pressure of 100 kPa, hydrogen has a boiling point of 20 K. In Table 4.4, some properties of liquid hydrogen are given in comparison to the properties of gasoline. It is important to mention that the storage densities in the table are just for the material. The storage densities for the systems are typically lower.

Looking at the values in the table, it can be seen that liquid hydrogen has worse volumetric energy than gasoline by a factor of about 4. This demonstrates the density problem for pure hydrogen. In 1 L liquid hydrogen there is 70 g hydrogen and in 1 L gasoline is even more, about 105 g $H_2$ $L^{-1}$.

### 4.3.2
### Ortho Para Conversion

Hydrogen molecules exist in two forms as the ortho and the para form. The difference is on the electron configuration. In the ortho form, the electrons of both

## 4.3 Liquid/Slush Hydrogen

**Table 4.4** Properties of liquid hydrogen, slush hydrogen, and gasoline.

| Property | Liquid hydrogen | Slush hydrogen | Gasoline |
| --- | --- | --- | --- |
| Molecular weight | 2.02 | 2 | 100–105 |
| Density (kg m$^{-3}$) | 70.79 | 85 | 720–770 |
| Boiling point at 100 kPa | 20 K | 14 K | 293–483 K |
| Heat of ortho/para conversion @boiling point (kJ kg$^{-1}$) | 703 | | |
| Heat of evaporation (kJ kg$^{-1}$) | 446 | | |
| Mass fraction hydrogen | 100% | 100% | 12–15% |
| Hydrogen density (g L$^{-1}$) | 70.79 | 85 | 100 |
| Upper heating value (MJ kg$^{-1}$) | 143 | | 44 |
| Lower heating value (MJ kg$^{-1}$) | 120 | | 41 |
| Volumetric energy density (MJ L$^{-1}$) | 8.495 | 34.9 | 30.5 |
| Octane number | | | 88–98 |

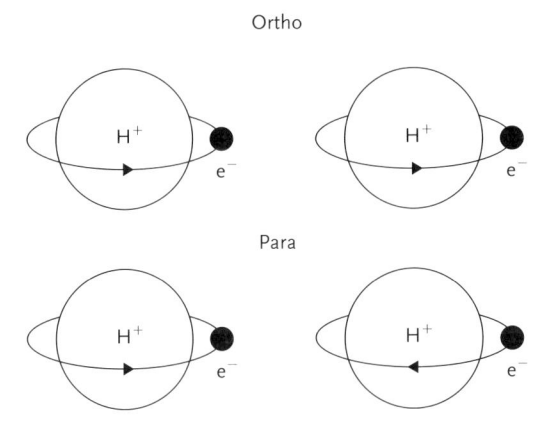

**Figure 4.9** Ortho-hydrogen and para-hydrogen.

hydrogen atoms turn in the same direction, and in the para form the two electrons turn in different directions. The ortho and para hydrogen arrangements are shown in Figure 4.9. At room temperature, 75% of the hydrogen are ortho-hydrogen and 25% are para-hydrogen. At 20 K, at hydrogen's boiling point, almost all hydrogen is para-hydrogen. The ortho-to-para-conversion is exothermic and releases a lot of energy. The heat of reaction is 527 kJ kg$^{-1}$. The noncatalyzed reaction is very slow. In short-term storage, the boil-off caused by the ortho-to-para-conversion is negligible but for long-term storage the conversion has been done during the liquefaction process. The additional heat has to be removed. In this case, a catalyst typically activated charcoal is commonly used. But there are several other catalyst materials applicable, such as copper oxide. The reaction is typically first cooled by liquid nitrogen and at deeper temperatures with liquid hydrogen.

### 4.3.3
### Liquefaction

It was in 1898, when Dewar in Scotland was the first man who successfully produced liquid hydrogen. Liquid hydrogen played an important role at the beginning of the 1950s as fuel for rockets and is still used for this application. There is still only a small need for liquid hydrogen and therefore worldwide only a small capacity of liquefiers is installed. Thereby the industrial needs are higher as the needs from aerospace applications. The worldwide capacity is 267.9 kg d$^{-1}$. Installations with a capacity of 243 kg d$^{-1}$ (96 tons year$^{-1}$) are located in America (USA; Canada, Franz. Guyana). The remaining capacity of 19.4 kg d$^{-1}$ is located in Europe (France, Germany, The Netherlands) and additional 5.5 kg d$^{-1}$ in Japan [13].

The energy demand for liquid nitrogen is less than for liquid hydrogen. The ideal work of liquefaction of nitrogen is only 745 kJ kg$^{-1}$; the ideal work of liquefaction of hydrogen is 11.62 MJ kg$^{-1}$.

The typical energy demand of the installed units for hydrogen liquefaction is between 36 and 54 MJ kg$^{-1}$. It should be mentioned that the ideal work of liquefaction of hydrogen is 11.62 MJ kg$^{-1}$, while the ideal work of liquefaction of nitrogen is only 0.338 MJ kg$^{-1}$. The goal for future efficiency-optimized installations should be 25 MJ kg$^{-1}$ for hydrogen. A power-optimized process was proposed by Quack with a calculated energy need of 23.4 MJ kg$^{-1}$ [14]. This low value could only be reached by using a well-selected flow scheme as well as the well-selected working material. The selected working material for the Brayton process is a mixture of neon and helium. For a highly efficient, large-scale liquefier an energy need of 18 MJ kg$^{-1}$ was predicted [15]. The process is based on four Joule–Brayton cycles, for a need of 10 kg s$^{-1}$. A cascade of four helium reversed, closed, and recuperative Joule–Brayton cycles were selected. The work requirement of 18 MJ kg$^{-1}$ corresponds to second law efficiency of 48%.

#### 4.3.3.1 Linde Process
The simplest liquefaction process is by the Linde or Joule–Thomson expansion cycle. First the gas is compressed at ambient temperature and after compression the compressed gas is cooled in a heat exchanger. Then the gas is passed through a throttle valve, where an isenthalpic Joule–Thomson expansion takes place. Some liquid is produced and is separated from the cold gas that goes back cooling down the compressed gas. The Lind cycle is applicable for gases like nitrogen that cool upon expansion at room temperature. For hydrogen, this process is not suitable, because hydrogen warms upon expansion at room temperature. In order for hydrogen to cool down upon expansion, the gas temperature must be below the inversion temperature (51 K at 100 kPa). Helium and hydrogen are two gases whose Joule–Thomson inversion temperatures at a pressure of one atmosphere are very low (e.g., helium and hydrogen warm up when expanded at constant enthalpy at typical room temperatures). This inversion temperature is pressure dependent (inversion temperature). A precooling of the hydrogen has to be done to reach this temperature. One option to go below the inversion temperature is to

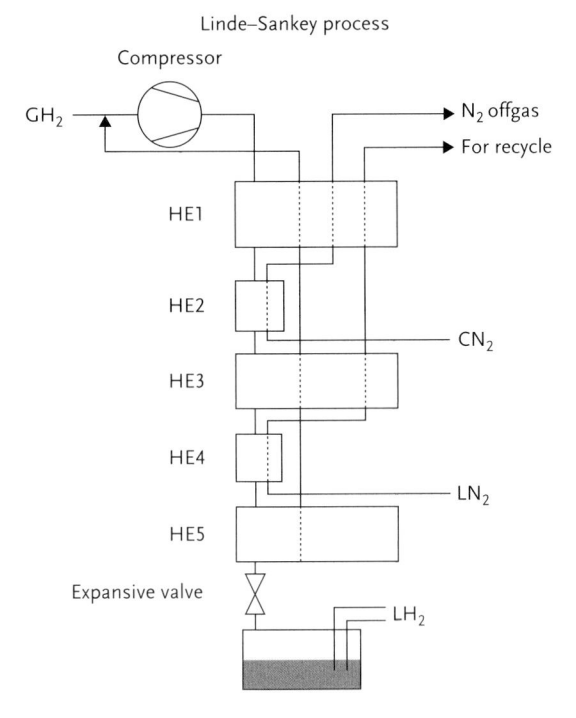

**Figure 4.10** Schematic of the Linde–Sankey Process.

use liquid nitrogen (boiling point 78 K). The Linde–Sankey process with the precooling liquid nitrogen is shown in Figure 4.10.

#### 4.3.3.2 Claude Process
In 1902, Georges Claude, a French engineer, successfully used a reciprocating expansion machine to liquefy air. The Claude process is a combination of an expansion engine and a Joule–Thomson expansion stage. Some of high-pressure gas is passed through the expansion machine. This gas passes an heat exchanger to cool down the remaining gas. Ideally the compressor is isothermal and the expansion is isenthalpic. The isenthalpic expansion is more effective than the Joule–Thomson expansion. In practice, the expansion engine cannot be used to condense because exceeding liquid formation will damage the expansion engine.

The process principle is shown in Figure 4.11.

#### 4.3.3.3 Collins Process
The Collins liquefier was originally developed for the liquefaction of helium. The compressed helium is precooled after flowing through several countercurrent heat exchangers, and then is expanded in a Joule–Thomson valve. Hereby the helium pressure drops down and the helium is partly liquefied. The still gaseous part of helium is flowing countercurrent. To achieve enough cooling the

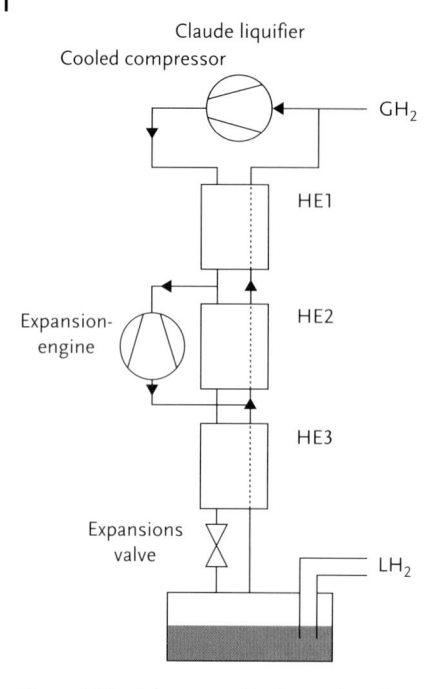

**Figure 4.11** Schematic of hydrogen liquefaction based on the Claude process.

precooling has to undershoot the inversion temperature of 30 K. For this process step, a adiabatic expansion machine is used. Where a part of the high-pressure hydrogen is cooled down and a part of the cold low-pressure hydrogen gas flows countercurrent in the heat exchanger 2. At the Collins process, two expansion machines are working at two different temperatures. The schematic is shown in Figure 4.12.

#### 4.3.3.4 Joule–Brayton Cycle
The Joule–Brayton cycle is a clockwise cycle consisting of two isotropic and two isobar changes of state. This is typically the process in gas turbine or jet engine. The counterclockwise process describes a refrigerating machine.

#### 4.3.3.5 Magnetic Liquefaction
The magnetic refrigeration is a cooling technology based on the magnetocaloric effect. The magnetocaloric effect is a phenomenon where the reversible change in temperature of a material is caused by exposing the material to a changing magnetic field. The refrigeration cycle is analogous to the Carnot cycle, thereby at the beginning the working material is introduced into a magnetic field. The working fluid starts in thermal equilibrium with the refrigerated environment. In addition, a magnetic material (magnetic dipoles) is added to the working fluid. The refrigeration cycle is consisting of the following four steps.

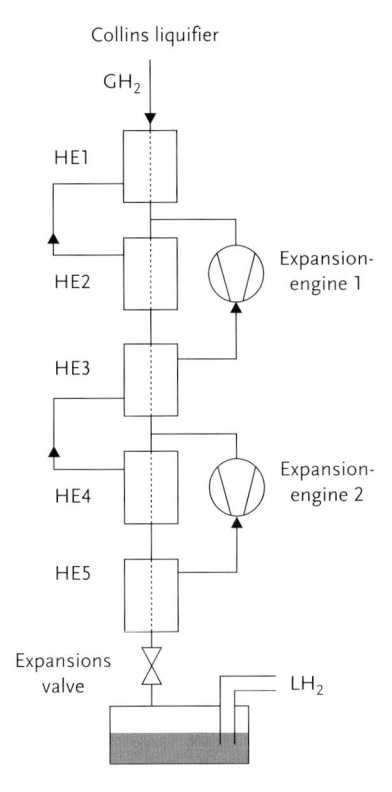

**Figure 4.12** Schematic of hydrogen liquefaction based on the Collins liquifier.

1. **Adiabatic magnetization:** The working material is placed in a thermally insulated environment. The increasing external magnetic field causes magnetic dipoles of the atoms to align, thereby the materials entropy and heat capacity decreases.
2. **Isomagnetic enthalpic transfer:** The magnetic field is kept constant to prevent the dipoles absorbing the heat. The added heat is then removed by another fluid.
3. **Adiabatic demagnetization:** The working material is put in the adiabatic condition. This means the total enthalpy remains constant. The magnetic field is decreased; therefore, the thermal energy causes the magnetic moments to overcome the field, and thus the working material cools. Energy transfers from the thermal entropy to magnetic entropy.
4. **Isomagnetic entropic transfer:** The magnetic field is held constant in order to prevent the material from heating back. The material is placed in thermal contact with the environment being refrigerated. The temperature of the environment is higher than the temperature of the working material; therefore, the energy migrates into the working material.

The first magnetic refrigerators were built in 1930s. This technology was the first that allowed temperatures below 0.3 K. In 1997, the first near room temperature magnetic refrigerator was built. Only for special application, as cooling below 4 K, magnetic refrigerators are available. As working material alloys of gadolinium [16, 17] were investigated, there are also other promising materials under investigation.

The magnet refrigeration process promises higher efficiencies comparative with other cooling processes. For the liquefaction of hydrogen, high efficiencies are expected [18].

#### 4.3.3.6 Thermoacoustic Liquefaction

A liquefier for natural gas with the thermoacoustic principle was build and tested [19]. This principle allows liquefaction without any moving parts. The prototype had the capacity of 1.9 m$^3$ day$^{-1}$. With a thermo acoustically driven orifice pulse tube reactor, a cooling power of 2.070 kW at 133.1 K could be achieved. An efficiency of 23% revered to Carnot performance was shown. This process could be applied for liquefaction of natural gas and air, there is no application for hydrogen, known so far.

### 4.3.4
### Hydrogen Slush

A liquid–solid mixture of hydrogen (hydrogen slush) is characterized by a higher density and heat capacity than the liquid hydrogen at the boiling point, as presented in Table 4.1. These characteristics are advantageous for application in space, for aircrafts and for cars, but require an effective process for the production of the slush. Typically the hydrogen slush has 50% mass fraction of solid. This increases the density by 15.5% and the heat capacity by 18.3% compared to the properties of hydrogen at the boiling point. On the other hand the improvement of the properties costs only an increasing of the energy demand of 12%.

Three different productions methods are known.

1. **Freeze-thaw method:** This method uses the latent heat of vaporization to freeze the hydrogen. By removing hydrogen gas at the triple point above liquid hydrogen, the formation of a porous solid layer at the liquid surface can be observed. After mixing of solid with liquid, the process can be repeated. To obtain a high quantity of solid this has to be repeated several times. One disadvantage of this process is that it is a batch process. The other disadvantage is that the triple-point pressure is at 6.95 kPa. This means in the case of leakage, air, oxygen gets into the reactor. This is not desired from the safety point of view. The recovery of the triple point hydrogen gas is also costly.
2. **Heat exchanger with an auger:** Another option to produce slush hydrogen is a continuous process using an auger inside a heat exchanger. The idea is to use a liquid helium cooled tube; there at the wall of the tube the solid hydrogen is produced. This solid hydrogen is continuously removed by the auger. This continuous process also has an advantage in terms of energy compared with the freeze thaw method. For the production of 50% mass fraction slush

hydrogen, the increase in the energy demand for the auger method was about 12% and for the freeze-thaw method 17–22.5% [20].
3. **Slush gun:** Another continuous method for the production of slush hydrogen is by using the slush gun [21]. The liquid hydrogen is precooled by cold gaseous helium and fed through a nozzle into the tank. In the tank is a cold gaseous helium atmosphere that produces solid hydrogen particles that settle on the triple-point liquid hydrogen surface resulting in slush hydrogen. These slush guns allow a continuous production of high-quality hydrogen slush.

### 4.3.5
### Boil-Off

The main issue in liquid hydrogen storage is hydrogen boil-off, which can lead to a hydrogen loss without benefit. Boil-off is a function of thermal insulation, tank size, tank shape, and ortho-to-para state ratio of hydrogen. Boil-off rates have to be concerned, especially for the storage of liquid hydrogen over longer periods. The boil-off rates vary between 0.03% for big storage tanks to 1–2% for small ones (5–10 kg hydrogen). It is very important to understand the boil-off phenomena to apply the correct countermeasures. Vehicle tanks must include efficient boil-off minimizing systems; otherwise evaporation losses will be unacceptable even when the car is parked for some days. In other cases like sea transportation, it might be possible to use the boil-off gas for the power drive. Zero boil-off possibilities are also described in this chapter.

There are several mechanisms known:

- **Ortho-to-para conversion:** The conversion of ortho-hydrogen to para-hydrogen is an exothermic reaction. In Section 3.3.2, this phenomena is well described. If unconverted liquefied hydrogen is placed in a tank, the heat of conversion will subsequently start the evaporation of the hydrogen. Therefore, the ortho-to-para conversion is a must for a longer storage. The desired storage time usually determines the optimum amount of conversion.
- **Heat leak:** Shape and size effects, conduction, convection, radiation, and cool down. A part of the liquid hydrogen boils due to heat transfer through the tank wall. The gaseous hydrogen then leaks, because there is an opening (pressure release valve) in the tank, otherwise the pressure would increase. The boil-off is dependent on the heat transfer through the tank wall, tank size, and shape. There are high requirements regarding the thermal insulation of a $LH_2$-tank. Bigger the tank size, better the ratio of surface to volume (S/V). For very small tanks with 0.1 $m^3$ the boil-off is about 2%, for large tanks of 100 $m^3$ it could be 0.06.%. But it is not only the boil-off that is high for small tanks, it is also the cost. It is the surface that determines the costs. The best ratio can be achieved with spheres. This means with big spheres the boil-off effect can be minimized. The heat transfer from the environment to the liquid hydrogen is via convection, conduction, and radiation. The conduction effect can be limited by using for the tank a material with low heat conductivity. The convection effect can be

limited by evaporating the space between the double wall. A multilayer insulation facing the inner wall of the tank limits the heat transfer by radiation. The main design parameters for the storage of LH$_2$ are: hydrogen temperature, operating pressure, and insulation quality. There are two different concepts for thermal isolation. One option is to use closed cell foam between the two walls; this foam could be improved by different metallic layers. The other design involves a multilayered system with low emissive, high reflective layers separated by the fiber glass. This design not only allows low thermal conductivity due to the partial vacuum between the layers and low heat transfer by radiation due to the behavior of the layers, but also the connections to the tank as valves and connectors have to be designed for low heat transfer. Even an advanced thermal insulation cannot prohibit boil-off. Therefore, venting of the gaseous hydrogen is necessary to control the tank pressure. The venting line has to be protected for air coming into the line, because the air will suddenly freeze and block the vent line.

- **Thermal stratification, thermal overfill:** In a tank when heat is moving to the liquid hydrogen the warmer hydrogen with the lower density will flow on top of the vessel to the liquid–vapor interface. This liquid layer with the higher temperature compared with the bulk temperature determines the vapor pressure. This vapor pressure is higher compared to the vapor pressure of the bulk liquid. This reduces the time the hydrogen can be stored without venting the vapor. The stratification is stable to the fact that liquid hydrogen has poor thermal conductivity. The thermal overfill describes the situation in which a tank has been filled with a liquid whose saturation pressure exceeds the maximum operating pressure of the tank. When a liquid hydrogen tank is thermally overfilled with a warm liquid, the surface layer phenomena establishes a liquid surface temperature corresponding to the operational pressure of the tank rather than the higher vapor pressure of the bulk liquid. If the surface layer is disturbed, the underlying bulk liquid may be brought into rapid equilibrium, which causes a rapid boil-off.
- **Sloshing:** It is the motion of liquid hydrogen in the tank due to acceleration or deceleration. Some of the thermal energy then is transformed in thermal energy, which causes evaporation of the liquid.
- **Flashing:** Whenever liquid hydrogen is transferred from a tank with a higher pressure to a tank with lower pressure liquid, hydrogen will evaporate. For the liquid hydrogen tanks, for automotive application, one option is to burn the gaseous hydrogen coming from the tank after pressure raised in the tank to the limited pressure due to boil-off. The hydrogen can be catalytically burned by the oxygen from the air, the only product is then water vapor.

#### 4.3.5.1 Zero Boil-Off Solutions

A system for the recovery of the boil-off gas has been described by Fuura *et al.* [22]. It was shown that a metal hydride storage system can be used to keep the hydrogen gas without any loss of hydrogen or energy. The used metal hydride vessel had a mass of 10 kg and a hydrogen capacity of 0.13 kg.

Another system with zero boil-off based on cryocoolers and passive insulation was described [23]. The goal of the project was to demonstrate zero boil-off for hydrogen storage for in-space storage of propellants as hydrogen and oxygen. A cryocooler was foreseen to intercept and reject the cryogenic storage system heat leak such that boil-off and venting is excluded. The selected cryocooler was a commercial unit based on a two-step Gifford–McMahon cycle refrigerator. The cryocooler extracts 30 W at 20 K. The cooler has a Carnot efficiency of 4%. Another system using a heat pipe system was investigated [24]. The analyzed system for zero boil-off consists of a liquid pump feeding a nozzle. The hydrogen then flows around the heat pipe which then removes the heat from the storage tank. The heat sink from the heat pipe is realized by a cryocooler with a radiator.

For reducing boil-off of liquid hydrogen tanks, a method based on liquid nitrogen was proposed by Linde the CooLH2® process [25]. The idea for this process is derived from the road transportation of liquid helium. Helium is a relatively scarce element, and therefore a precious good, so the boil-off losses have to be minimized.

The idea is to use a liquid nitrogen shield around the well-insulated wall of the liquid hydrogen tank. The liquid nitrogen then is able to cool the wall and this reduces drastically the heat transfer into the liquid hydrogen. Linde claims to allow a storage time without any boil-off losses for 12 days. There is a difference in the helium system. The cooling media has not to be refueled. In the cooling media, liquid air will be produced directly from dried air in a heat exchanger cooled by liquid hydrogen. Carbon dioxide as well as moisture has to removed because otherwise the heat exchanger will be blocked.

### 4.3.6
**Materials**

The main requirements regarding the structural material for the liquid storage of hydrogen is the applicability at low temperatures (20–30 K). At these temperatures hydrogen embrittlement is a minor issue, because hydrogen solubility increases with the temperature. The low temperatures set limits to the selection of materials. Some storage concepts for liquid hydrogen take higher pressures in consideration, this is a way to reduce the boil-off problems. In this case, the requirements same as for the storage of compressed hydrogen are valid.

#### 4.3.6.1 Tank Material
Standard materials for liquid hydrogen storage are metals as stainless steel and aluminum. For space and automotive application new lightweight and freeform tanks are under investigation. For the lightweight-version fiber-reinforced material with a metallic inner coating as permeation barrier was chosen.

#### 4.3.6.2 Insulation
The superinsulation for the insulation of cryogenic liquids has been developed for space application. This kind of insulation requires a low pressure. The pressure

must be low to leave the region of pressure independent gas heat transfer and to reach the region where the heat transport goes via the molecular flow. About 20–30 layers of alumina (silver or gold)-coated polyester films helps to prohibit the heat transfer by radiation. A spacer between the foils improves the impact. Liquid hydrogen tank systems are mainly double-walled austenitic stainless steel vessels that are evacuated between the walls. For proper insulation of the liquid hydrogen tank, the quality of the vacuum is essential. A very good insulation can be achieved at minimum $10^{-4}$ hPa. Attention to the suspensions as well as to the connectors is requested. These are sources for heat transfer by heat conduction. So the design and material selection has to be done carefully. With a vacuum multilayer insulation heat conductivity of $10^{-5}$ W m$^{-1}$ K$^{-1}$ could be achieved. For heat conductivity lower as $10^{-2}$ W m$^{-1}$ K$^{-1}$ vacuum is required. Nonevacuated insulations, that is typically foam, and fibrous insulation have heat conductivities above $10^{-2}$ W m$^{-1}$ K$^{-1}$. The thermal conductivity of polyurethane foam is about $2 \times 10^{-2}$ W m$^{-1}$ K$^{-1}$.

#### 4.3.6.3 Braze Materials
For the liquid hydrogen based on stainless steel (1.4401/1.4404) different braze materials, such as Ni- and Cu-based materials, have been evaluated [26]. Showing that neither of braze joints has any phenomenon of hydrogen embrittlement.

### 4.3.7
#### Sensors, Instrumentation

For a safe operation for monitoring and controlling several sensors for measuring temperature, pressure, fuel level are required. Electrical power in the environment with air with possible hydrogen obligates appropriate precautions. All sensors have to be designed according to the guidelines for intrinsic safety.

In addition, several valve heat exchangers, pumps have to be implemented for a proper function of a liquid hydrogen storage system. The flow scheme of the liquid hydrogen fuel storage system is shown in Figure 4.13 [27].

During filling, valves SV01 and SV02 ore open, so that liquid hydrogen can flow from the filling station into the storage tank. Due to a possible high-surface temperature in the tank, hydrogen vaporizes immediately. This vaporized hydrogen then will leave the tank through valve SV02 back to the filling station. The boil-off valve opens at a determined pressure. In this case, it is foreseen to convert the hydrogen with the oxygen from the air in the catalytic reactor BS01 to water vapor. For operation, the hydrogen can be released as liquid or gas. At a tank pressure close to the boil-off pressure, the gaseous hydrogen flows through the opened SV02 and check valve CV02 through the heat exchanger HE01 where the hydrogen is heated up by waste heat from the hydrogen consumer. Finally through the open valve SV03 the hydrogen flows to the consumer.

At lower tank pressures, the hydrogen will be released and flows directly through the open SV01 to the consumer the same way as described. At the time when the tank pressure decreases to a defined margin, heated hydrogen will be pumped by IPG1 into the tank to increase the pressure. The tank is equipped with

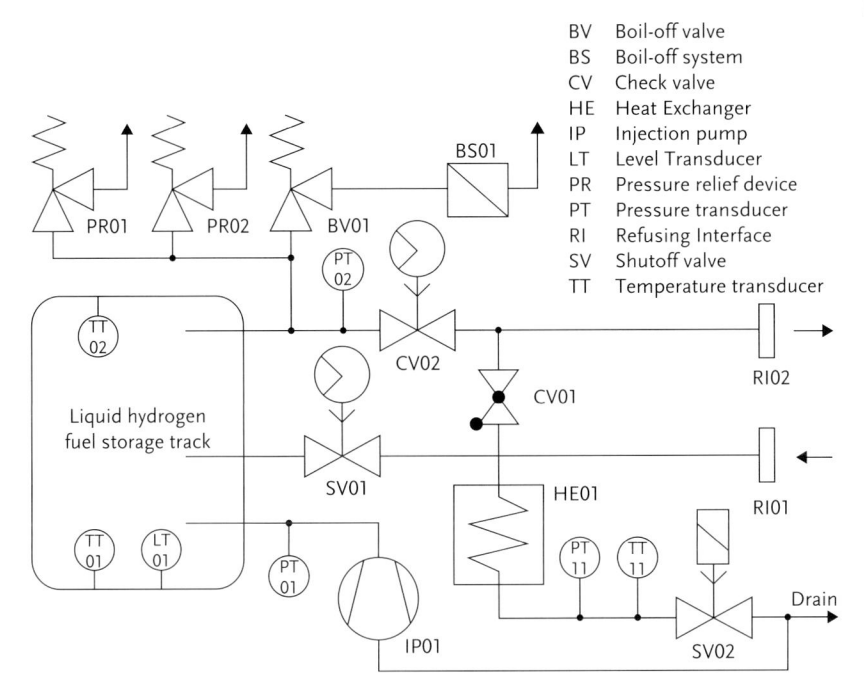

**Figure 4.13** Flow scheme of a liquid hydrogen storage system for the automotive application.

a redundant pressure relieve valve. In the case of an accident, all valves are closed immediately.

For the functionally safe operation an electronic control unit is implemented. This unit receives data from the sensors and guaranties the interactions with the actuators as valves and pumps. The operations have to be well defined. A failure mode and effect analysis and a failure mode, effects and diagnosis analyses have to be performed. And the hardware, beginning from the sensors to valves, has to be developed and selected. All these have to be carried out very carefully to achieve a safe and proper operation of a liquid hydrogen storage system

### 4.3.8
### Applications

Besides the application of liquid hydrogen as rocket fuels, liquid hydrogen is attractive for storage and transportation purposes.

#### 4.3.8.1 **Storage**
The sphere with the highest content of liquid hydrogen has a volume of about 3200 m$^3$ and keeps about 230 tons liquid hydrogen and is situated at the Kennedy Space Center Cape Canaveral, Florida, USA. The liquid hydrogen is foreseen as fuel for the center engine for the launchers for the space shuttles. The required amount for one shuttle launch is about 1500 m$^3$ equivalents to 106 tons.

#### 4.3.8.2 Sea Transportation

Hydrogen is anticipated to be one of the promising energy carriers used in the future. One practical way for the worldwide distribution of hydrogen is by tanker as liquid hydrogen. Therefore, the study of the large-scale sea transportation was performed [28]. Described project had three phases:

- basic study of hydrogen technology and the overseas researches on hydrogen tankers;
- conceptual design of large-scale hydrogen tanker;
- preliminary study of the basic elements of using a liquid hydrogen tank such as tank insulation and tank support system.

For the transportation of liquid hydrogen from Canada to Germany in the Euro-Quebec Hydro-Hydrogen Pilot Project (EQHHPP) ships with numbers of liquid hydrogen tanks have been foreseen. The total volume of the LH2 tanks was 15 000 $m^3$.

Liquid hydrogen sea transportation can use the experience from the liquid natural gas (LNG) transportation. The LNG sea transportation started already in 1959 by the "Methane Pioneer" (LNG cargo capacity 5000 $m^3$) with self-supporting prismatic aluminum tanks. The LNG capacity increased in the last 50 years up to 210 000 $m^3$. For LNG, aluminum as well as stainless is suitable as tank material. The LNG technology is different to the liquid hydrogen, because LNG has different properties. The boiling point is at 110 K for LPG and 20 K for hydrogen. The density of the liquid at the boiling point is 424 kg $m^{-3}$ for LPG and 71 kg $m^{-3}$ for hydrogen. Therefore, for the hydrogen the requirements for the thermal insulation are major. Two different designs have been elaborated for a total capacity of 200 000 $m^3$. One design is based on four spherical tanks and the second on a prismatic design. As a result from the study, it could be shown that the independent tank systems from the LNG ship (spherical and prismatic) are basically applicable to the liquid hydrogen tank. The insulation system is the most essential issue for the hydrogen tanker. It was suggested that in any section the insulation system shall be provided with the redundancy to prevent the bursting evaporation by accidental temperature increase.

#### 4.3.8.3 Road and Rail Transportation

For road and rail transportation different tanks with different sizes are available. The size of the tank for road transportation is in the range of 30–60 $m^3$ (equivalent to 2100–4200 kg) and for rail transportation about 115 $m^3$ (equivalent to 8000 kg) to. It is not the mass that limits the capacity, it is the maximum possible volume that limits the transportation capacity. The values are given in Table 4.5.

Taking these numbers in account, it is clear that a road delivery for mass application of compressed hydrogen is not suitable. For transportation of hydrogen over longer distances, the transportation in the liquid state is self-evident. Actually for the transportation for a distance of 100 km energy consumption of 1.8 GJ has to be taken into account.

**Table 4.5** Comparison of deliverable amount of energy with a standard 40 to truck (26 to/maximum 45 m³).

|  | Compressed 200 bar | Liquid | Liquid/slush | Gasoline |
| --- | --- | --- | --- | --- |
| Density (kg m$^{-3}$) | 20 | 70.79 | 85 | 720–770 |
| Stored mass (kg) | 350 | 2100–4000 | 2450 | 26 000 |
| Stored energy | 14.5 GJ | 300–572 GJ | 420 GJ | 1144 GJ |

#### 4.3.8.4 Vehicles

For a typical vehicle with a cruising range of about 500 km for hydrogen-fuelled fuel-cell vehicles is requested, this corresponds to a minimum amount of hydrogen of 5 kg. For the fuel itself the gravimetric as well as the volumetric numbers are looking good, but for the system including the tank the specific values are different. This results in a volume of the liquid hydrogen of 60 L. For the storage of liquid hydrogen different materials have been used. Tanks can be made from steel and aluminum ore fiber-reinforced polymers (FRP). Because of the low pressure inside the tank the form is open to be adapted to the vehicle. With different materials different specific values could be achieved. For example, following gravimetric densities were achieved for:

- steel 6%,
- aluminum 15%,
- FRP 17.5%.

The state-of-the-art liquid hydrogen storage consists of a double-wall cylindrical tank with a storage mass of about 9 kg. Material used for this tank is stainless steel, since it has a low hydrogen permeability and is resistant to hydrogen embrittlement. There is some experience on hydrogen tanks made from aluminum, and first tanks with FRP have been already built. With aluminum the weight of the tank can be reduced to the half. Within the STORHY project [29] a free-form liquid hydrogen fuel storage system, made of coated composite material, has been developed and tested. The free-form design allows integrating the system better into the existing car structure. The heat entry by radiation, convection, and conduction was minimized by using about 40 layers of highly reflective aluminized polyester foils. These foils are separated by glass fiber spacers. To avoid the heat transfer by convection, vacuum pressure below $10^{-2}$ Pa was foreseen. To keep low pressure as a measure to avoid the outgassing of the materials, a novel coating of the wall was implemented. The support structure to keep the inner tank in position to the outer tank was made from the glass-fiber-reinforced material.

The mass comparison for different hydrogen tanks with a gasoline tank is given in Table 4.6.

Another tank was built by Airliquide for the GM Zafira fuel-cell car [30]. The tank was tailored as per the available space under the floor of the vehicle. The diameter was therefore 400 mm and the length 1000 mm. It can store 5.4 kg of hydrogen at a pressure of 0.5 MPa. The weight of the complete module is 85 kg without the valve box 50 kg. According to the EIHP draft standard, a new storage system was

**Table 4.6** Mass comparison for different liquid hydrogen storage tanks and gasoline tanks

|  | Gasoline 38 L | L-H$_2$ BMW hydrogen 7 | L-H$_2$ STORHY cylinder | L-H$_2$ STORHY Free-form |
|---|---|---|---|---|
| Fuel | 30 kg | 10 kg | 10 kg | 10 kg |
| Tank | 10 kg | 100 kg | 40 kg | 38 kg |
| Auxiliary | 3 kg | 60 kg | 25 kg | 8 kg |
| Total | 43 kg | 170 kg | 75 kg | 56 kg |
| Stored energy (MJ) | 1320 | 1430 | 1430 | 1430 |
| Volumetric density (MJ L$^{-1}$) | 47.3 | 4.7 | 4.7 | 5.1 |
| Energy density (MJ kg$^{-1}$) | 30.7 | 8.4 | 19.0 | 25.5 |

developed. It stores 12 kg of hydrogen; it supports a supply flow of 20 kg h$^{-1}$. The achieved evaporation rate is less than 3% per day and autonomy of 3 days without any evaporation losses. With building a storage vessel based on the aluminum design reached a weight reduction of 50% compared with the steel design.

Researchers at the Lawrence Livermore National Laboratory have reported a pressurized cryo store for 10.7 kg hydrogen with the gravimetric energy density of 6.1 wt% or 7.32 MJ kg$^{-1}$ and 3.96 MJ L$^{-1}$ with a closed tank. This means that with increasing temperature the pressure in the tank increases. After 6 days of storing liquid hydrogen, the temperature increases up to 80 K and the corresponding pressure to 35 MPa [31].

The tank built by the LLNL is shown in Figure 4.14.

**Figure 4.14** Tank built by LLNL.

Several promising prototypes have been build; however there is still the need to optimize the storage tanks from the technical point of view and what is very important from the cost side.

For a city bus Linde AG built a liquid hydrogen storage system [32]. The tank system consists of three oval tanks with a capacity of 190 L. The system is characterized by the arrangement of three communicating tanks in line avoiding cold shut-off valves; depletion of liquid transfer by means of bubble lifting and fueling with a cold disconnectable filling and venting coupling and coaxial filling and venting line.

The data of different storage systems for liquid hydrogen are summarized in Table 4.7.

In Figure 4.15, a schematic of liquid hydrogen storage system from Linde is shown.

**Table 4.7** Liquid hydrogen storage systems from Linde.

|  | "70L" | "120L" | "115LF" | "600LF" | "500L" |
| --- | --- | --- | --- | --- | --- |
| $H_2$-amount (kg) | 4.0 | 6.8 | 6.5 | 34.0 | 32.0 |
| Water volume (L) | 70 | 120 | 115 | 600 | 500 |
| Operating pressure (bar) | 3.0 | 3.0 | 3.0 | 3.0 | 3.0 |
| Filling pressure (bar) | 3.5 | 3.5 | 3.5 | 3.5 | 3.5 |
| Length (mm) | 1000 | 915 | 1080 | 5500 | 2125 |
| Diameter (mm) | 400 | 540 | 400 × 560 | 500 | 710 |
| System weight (kg) | 90 | 120 | 115 | 480 | 450 |
| Weight % (basis: system) | 4.4 | 5.7 | 5.7 | 7.1 | 7.1 |
| Refueling time | <2 min | <2 min | <2 min | <12 min | ~10 min |
| Maximum extraction amount (kg h$^{-1}$) | ~5.8 | ~15 | ~5.8 | ~10 | ~10 |
| Heat input (W) | <1.5 | <1.5 | <1.5 | <3.5 | <3.5 |
| Dormancy time (h) | >36 Std. | >48 Std. | >48 Std. | >100 Std. | >100 Std. |
| Boil-off rate (%/day) | <4% | <3% | <3% | <2% | <2% |

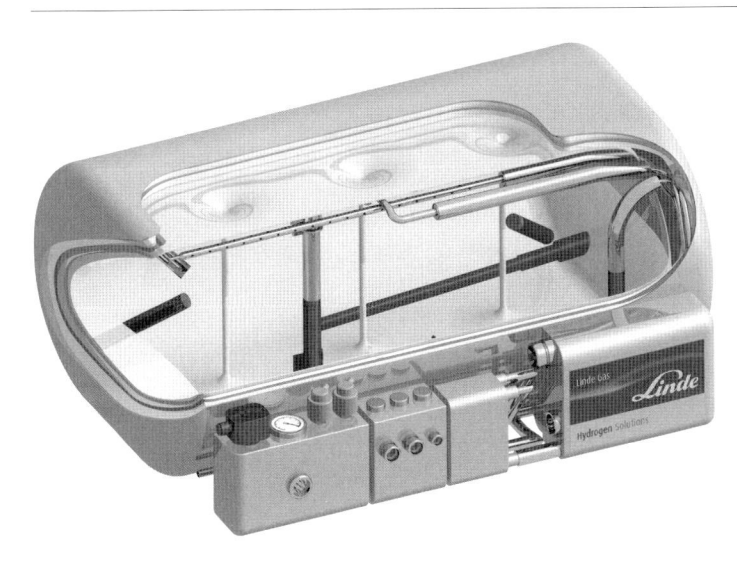

**Figure 4.15** Linde hydrogen tank.

#### 4.3.8.5 Aircraft

Hydrogen has the potential as a future fuel for aircrafts. It is expected that in the future the jet fuel would be replaced by another fuel, since a more carbon free fuel is requested or simple the availability is not any longer guaranteed. Actually there are three fuel options under discussion, a synthetic carbon-based fuel, liquid natural gas (LNG), and hydrogen.

The change from jet fuel to hydrogen requires some changes of engines. The combustion chamber, fuel pump, as well as the fuel heat exchanger have to be adapted and respectively developed. Hydrogen has a big advantage as fuel for aircrafts because of the high mass-related energy density (144 mJ $kg^{-1}$); this is an important parameter. For aircraft application, the required mass of the fuel storage system is more important as the volume. Therefore, for aircrafts, only liquid hydrogen storage can be considered. Assuming the same stored energy the hydrogen storage requires about three times the volume of a jet fuel storage system, but the weight of the hydrogen system will be about a quarter.

The replacement of jet fuel by hydrogen requires a production of hydrogen and a corresponding infrastructure, as pipelines, storage at the airport.

The environmental impacts have to be considered very carefully. On burning jet fuel, the products are mainly carbon dioxide and water (plus soot, $NO_x$, CO, $SO_2$, and $HC_s$) and in case of hydrogen about the factor of 2.5 of water will be produced and $NO_x$. The higher water content in the exhaust gas contributes to the formation of contrails. These contrails are considered to contribute to the anthropogenic greenhouse effect. But these contrails can be avoided by avoiding the critical air masses, where atmosphere is supersaturated over ice.

The first turbojet engine on hydrogen started in 1937 [33]. In 1988, the first flight of a Tupolev TU 155 laboratory aircraft proved the principle feasibility of transport aircraft flying on LNG and hydrogen [34]. On the TU-155 experimental aircraft, the right-hand power plant, including the engine and its fuel and control system was modified for operation on liquid hydrogen or liquid natural gas. A cryogenic fuel tank was installed in the rear section of the cabin. The other two power plants remained unchanged and were operated with jet fuel. The aircraft was successfully tested, operating on both hydrogen and liquid natural gases. The cryogenic engine was tested under all aircraft maneuvers. The maximum engine operation time per flight was limited to 1 h.

In 1990, the German–Russian Cooperation was initiated. In 1990/1993, the feasibility study CRYOPLANE based on Airbus A310 was defined, followed by tank tests (1994/1999) demonstrator studies based on Do 328. In 2000, the system studies in the frame of a European project started. In a European project for different categories of aircraft the practical configuration was shown.

It could be shown that the take-off weight of long-range aircrafts is some 15% smaller as for conventional aircrafts. The operational weight increases some 20–25%. Finally the energy consumption increases by 8–15% because of the bigger wetted area and the higher mean flight weight.

The total required amount of hydrogen for the full intra-European air traffic based on 1999 was expected to 30 000 tons day$^{-1}$. This number has to be compared with the actual production capacity of liquid hydrogen that is in Europe 19 tons day$^{-1}$. It will take a lot of time and investments to build up the liquid hydrogen infrastructure for aircrafts.

#### 4.3.8.6 Rockets

The first rocket with hydrogen-based propulsion was started in 1963. But many years before the investigations started and it was clear that liquid hydrogen is an ideal fuel for rocket engines. The RL10 rocket engine developed by Pratt and Whitney Rocketdyne was used in the Saturn I/S-IV second stage as well as in the Centaur upper stage [35]. Modified R10 rocket engines are still in use.

#### 4.3.8.7 Solar Power Plants

Liquid hydrogen storage for small-size solar power plants was investigated [36]. The assumed power of the solar plant was 4.5 MW. Hydrogen can act as an energy carrier that allows local energy storage. Especially when the energy source is renewable, the solar plant is investigated and the power production does not naturally fit with the power demand of the consumer. The production of hydrogen and oxygen is proposed with the subsequent storage of the two products. A liquid storage was proposed, because of the acceptable energy density. The proposed storage system must provide the right coupling between the hydrogen availability that depends on the sun and the hydrogen consumption that depends on the required load. The thermal management and the combination of regasification with the liquefaction leads to a strong decrease of power demand for the liquefaction with respect to traditional liquefaction plants (2.8 MJ kg$^{-1}$ instead of 21 MJ kg$^{-1}$). Since hydrogen and oxygen for liquefaction are not identical with the flows for regasification, a number of regenerative heat exchangers are needed to store in cold.

## 4.4 Metal Hydrides

Energy storage is an essential part in all scenarios of a future hydrogen economy, in particular those connected with sustainable energy sources. One attractive and important option is energy storage via the reversible storage of hydrogen in metal hydrides that can then be employed by fuel cells. For mobile applications, metal hydrides have to fulfill certain requirements related to material properties. Certainly an advantage in using metal hydrides is the safety aspect. Even during severe accidents there are no potential fire hazards from hydrogen release, for example, because hydrogen will remain within the metal structure.

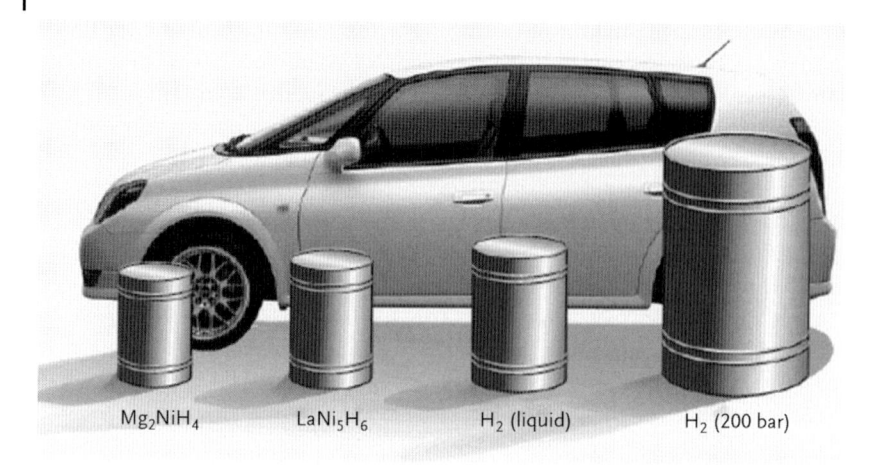

**Figure 4.16** Volume of 4 kg of hydrogen compared in a different ways with size relative to the car (here Toyota) [37], reprinted by permission.

The biggest advantage of metal hydrides is their high volumetric storage density, far above compressed, gaseous, or liquid hydrogen storage density.

Figure 4.16 [37] shows the comparison of volume of 4 kg of hydrogen stored in different ways, as liquid, in compressed form, and in metal hydrides.

Liquid and compressed hydrogen is presently an option for prototype cars or even larger commercial vehicles, but the potential of solid hydrogen storage of hydrogen might be a viable option for the future with respect to hydrogen storage capacity, both gravimetric and volumetric.

Figure 4.17 shows the volumetric and gravimetric densities of selected metal hydrides. According to the results, it can be seen [38] that the highest volumetric density of 150 kg m$^{-3}$ is shown in the case of complex metal hydrides, $Mg_2FeH_6$ and $Al(BH_4)_3$. This is double the volumetric density of liquid hydrogen. Metallic hydrides are in the range of 115–120 kg m$^{-3}$. The best achievable gravimetric density for a simple metallic hydride is around 7.6 wt% of $H_2$ for $MgH_2$. By contrast, among complex hydrides, $LiBH_4$ shows the highest gravimetric density of 18 wt%. The volumetric and gravimetric density values of metal hydrides are compared with those of carbon nanotubes, petrol, and other hydrocarbons in Figure 4.17. The lower part of the diagram presents the pressurized hydrogen storage for steel (left part of the figure) and in hypothetical composite containers (right part of the figure).

Solid-metal hydrogen compounds can be formed as a result of the reaction between hydrogen and metals, intermetallic compounds, and alloys. Metal hydrides are formed by metals and alloys that can reversibly store (absorb) large amounts of hydrogen simultaneously releasing heat in the process. Conversely,

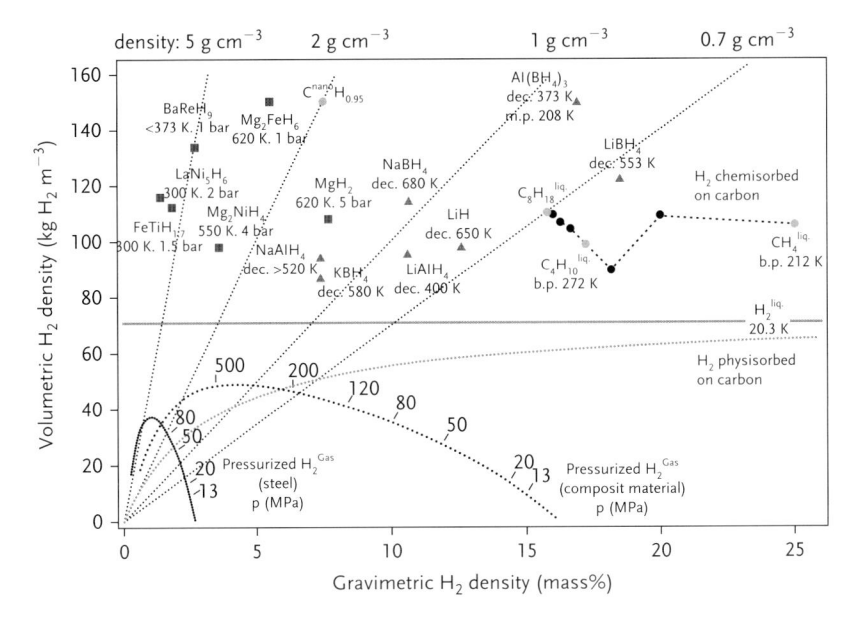

**Figure 4.17** Volumetric and gravimetric hydrogen density of some selected hydrides [38], reprinted by permission.

hydrogen is released (desorbed) after heat is delivered to the compound. This is shown by the following equation:

$$\text{Me} + \frac{x}{2} H_2 \leftrightarrow \text{MeH}_x \qquad (3.8)$$

Recently a lot of interest has been focused on complex metal hydrides. An alternative to the reversible metal hydrides are kinetically stabilized hydrides, which do not decompose (or the process is very slow) due to kinetic limitations. Such hydrides include for example $AlH_3$, $LiAlH_4$, $Li_3AlH_6$, $Mg(AlH4)_2$, $Ca(AlH_4)_2$. An illustration of the metal hydride family tree is shown in Figure 4.18 [39], showing that plenty of hydrides and solid solutions can be formed with hydrogen.

Light metals such as Li, Na, Mg, Al, B, Be form a wide variety of metal hydrogen compounds. They are especially interesting due to their light weight and the ability to accommodate one or more hydrogens per metal atom. Detailed information on intermetallic compounds and their hydrogen storage properties including specific elements, alloys, and engineering properties can be found in the databases created by the International Energy Agency (IEA) with support from the US DOE and the Sandia National Laboratories (SNL) [40, 41].

The thermodynamics of the formation of hydrides from gaseous hydrogen, taking $LaNi_5$ as an example (Figure 4.19), has been described by Schlapbach and Züttel [37] using pressure-composition isotherms in the following way.

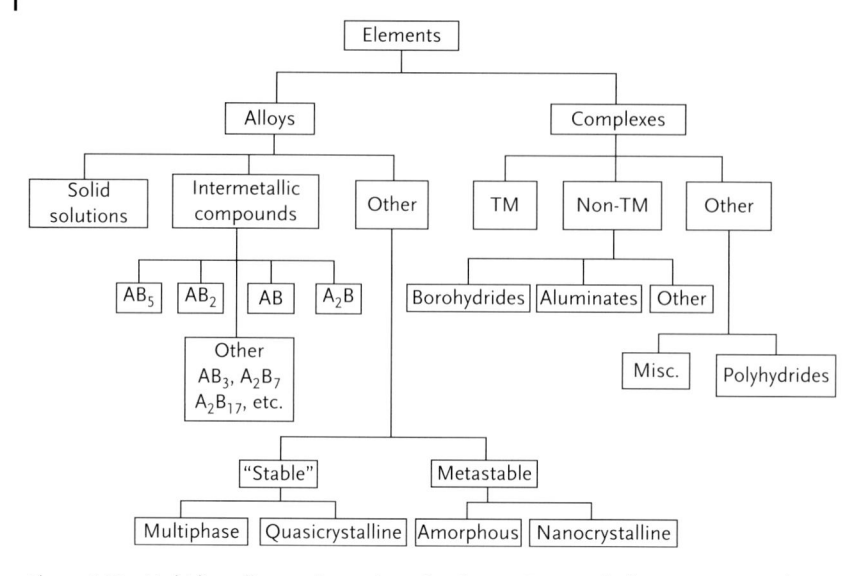

**Figure 4.18** Hydriding alloys and complexes family tree, TM stands for transition metals, non-TM stands for nontransition metals [39], reprinted by permission.

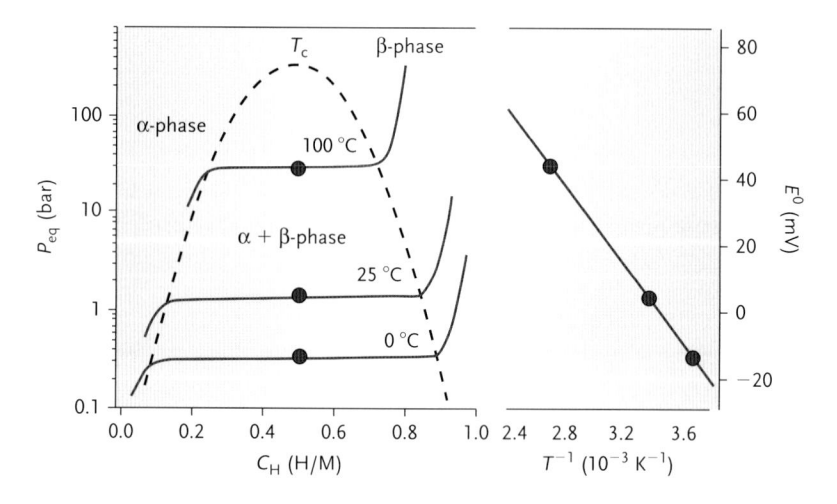

**Figure 4.19** Pressure versus composition versus temperature curves for hydrogen absorption in intermetallic compound (LaNi$_5$) showing solid solution ($\alpha$-phase), hydride solution ($\beta$-phase), and an area where two phases coexist in equilibrium ($\alpha+\beta$-phase). The slope of the van't Hoff plot is corresponding to the enthalpy of hydride formation [37] reprinted by permission.

The nonequilibrium $\alpha$-phase region is an area where the host metal (LaNi$_5$) dissolves hydrogen as a solid solution. With an increase of hydrogen pressure and hydrogen concentration in the host metal, the nucleation and the growth of a hydride occurs, being represented by the β-phase. The coexistence of both phases, solid solution and hydride (α+β) at the pressure flat plateau, indicates (from the length of the plateau) the amount of hydrogen that can be reversibly stored. Here the reactants remain in thermodynamic equilibrium during the absorption of more hydrogen. At the critical point, $T_c$, the transition from phase α to β is a continuous process. The equilibrium pressure is described by the van't Hoff equation and is related to the changes of enthalpy and entropy of hydride formation.

$$\text{van't Hoff equation}: \ln p = \frac{\Delta H}{RT} - \frac{\Delta S_0}{R} \qquad (3.9)$$

where $\Delta H$ is the enthalpy of hydride formation in kJ mol$^{-1}$, $\Delta S_0$ is the entropy of hydride formation in J K$^{-1}$ mol$^{-1}$ H$_2$ at the pressure $p = 10^5$ Pa, $R$ is the gas constant, and $T$ is the absolute temperature.

The primary requirement is that metal hydrides must be useful under practical conditions, meaning operation temperatures in the range of 243 and 393 K (−30 °C to 120 °C) and pressures around 105 Pa.

### 4.4.1
### Classical Metal Hydrides

Many interesting and attractive materials for hydrogen storage are hydrides based on rare earth–transition metal intermetallics as well as light metal compounds related to Mg–Ni intermetallics. Many of them have been developed recently.

#### 4.4.1.1 Intermetallic Hydrides (Heavy Metal Hydrides)

Intermetallic compounds consist of two or more metallic elements in an alloy in a certain defined proportion. The most important set of hydride-forming intermetallic compounds including the prototype and the structure is presented in Table 4.8 [38].

Element A is usually a rare earth or alkaline earth metal with a high affinity to hydrogen, typically forming a stable binary hydride. Element B is often a

**Table 4.8** Metallic hydrides of intermetallic compounds including the prototype and the structure [2] reprinted by permission.

| Intermetallic compound | Prototype | Hydrides | Structure |
|---|---|---|---|
| AB$_5$ | LaNi$_5$ | LaNiH$_6$ | Haucke phases, hexagonal |
| AB$_2$ | ZrV$_2$, ZrMn$_2$, TiMn$_2$ | ZrV$_2$H$_{5.5}$ | Laves phases, hexagonal or cubic |
| AB | TiFe, ZrNi | Ho$_6$Fe$_{23}$H$_{12}$ | Cubic, CsCl- or CrB-type |
| A$_2$B | Mg$_2$Ni, Ti$_2$Ni | Mg$_2$NiH$_4$ | Cubic, MoSi$_2$- or Ti$_2$Ni-type |
| AB$_3$ | CeNi$_3$, YFe$_3$ | CeNi$_3$H$_4$ | Hexagonal, PuNi$_3$-type |
| A$_2$B$_7$ | Y$_2$Ni$_7$, Th$_2$Fe$_7$ | Y$_2$Ni$_7$H$_4$ | Hexagonal, Ce$_2$Ni$_7$-type |
| A$_6$B$_{23}$ | Y$_6$Fe$_{23}$ | Y$_2$Ni$_7$H$_3$ | Cubic, Th$_6$Mn$_{23}$-type |

transition metal with a low affinity to hydrogen and forms only unstable hydrides. A careful choice of A and B elements can provide a source of hydrogen across a wide range of pressure. Main intermetallic compounds are described below.

The $AB_5$ family of hydrides, composed of A – rare metal and B – Co or Ni has been studied by Zijlstra and Westendorp [42] who reported that the hexagonal magnetic alloy $SmCo_5$ can absorb up to 2.5 atoms of hydrogen at a pressure around 2 MPa at room temperature. After reducing the pressure, the absorbed hydrogen was quickly desorbed by the material. van Fucht et al. [43] performed a study of the hydrogen uptake by this hydride as well as $LaNi_5$ and $La_{1-x}Ce_xNi_5$ confirming that these materials are able to absorb large quantities of hydrogen at moderate pressures and at room temperature, and that after reducing the pressure hydrogen is desorbed. The reported hydrogen capacity for $LaNi_5$ was $7.6 \times 10^{22}$ atoms $cm^{-3}$ (1.40 wt%), nearly twice that of liquid hydrogen. It was observed that after the absorption–desorption of hydrogen the volume of the lattice expanded by 24%. The classic example of an $AB_5$ hydride is $LaNi_5$. The absorption–desorption of hydrogen on $LaNi_5$ can be described by the following equation:

$$LaNi_5 + 3H_2 \leftrightarrow LaNi_5H_6 \quad \Delta H = -31 \text{ kJ mol}^{-1} \tag{3.10}$$

In general, in addition to Ni, B can be elements such as Co, Al, Mn, Fe, Sn, Si, Ti, and many others. Modern alloys of this family are mainly based on a lanthanide mixture with different metals like Ce+La+La+Nd+Pr (element A) and Mn+Co+Al+Fe+··· (element B). In general, $AB_5$ compounds do not show high gravimetric hydrogen capacities, but they are easy to activate and are tolerant to small amounts of oxygen and water in hydrogen. $AB_5$ hydrides are used for Ni-MH (nickel-metal hydride) batteries as anodes. The stability of metal hydrides is usually presented in the form of a van't Hoff plot. The lines for $AB_5$ hydrides are shown in Figure 4.20.

It can be seen that isothermal pressure-composition hysteresis is located in the lower part of the van't Hoff plot with the exception of $MnNi_5$. Mechanically grinding a small amount of palladium or platinum with $AB_5$ type intermetallic hydrogen storage alloys has been shown to greatly improve the hydrogen absorption and desorption performances of the alloys. After grinding palladium with $LaNi_{4.7}Al_{0.3}$, $CaNi_5$, and $Mg_2Ni$ in air, Shan et al. [44] found that the absorption rate of the treated alloys was hundred times faster than the untreated ones, and the treated alloys retained both low activation pressures and fast absorption and desorption rates even after more than 2 years of exposure to air.

The $AB_5$ family represents with a group of 330 compounds, $A_2B$ 103 compounds, and AB 156 compounds [40].

The $AB_2$ group of intermetallics represents the widest range of possible materials with nearly 509 compounds. These $AB_2$ alloys are more difficult to activate than other intermetallic hydrides, but once activated they show good hydrogenation–dehydrogenation kinetics. Element A can be from the IVA group

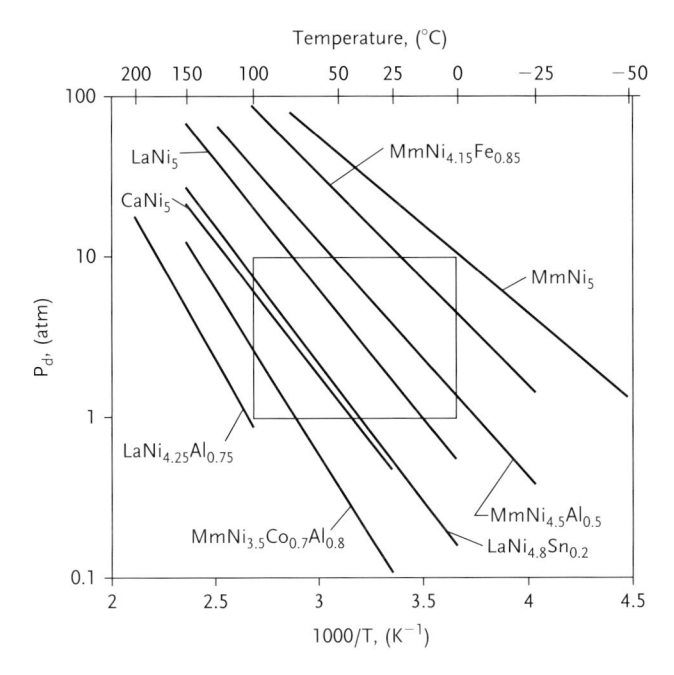

**Figure 4.20** van't Hoff plots of AB$_5$ metal hydrides [39], reprinted by permission.

(such as Ti, Hf, Zr) and/or can include rare earth metals with atomic number 57–71. B-elements are represented by a large variety of transition and non-transition metals. According to Sandrock [39], who successfully melted quantities of TiFe, Ti (Fe, Mn), and ZrMn$_2$, these alloys seem to be sensitive toward impurities in hydrogen and show passivation. Gravimetric hydrogen capacities are typically slightly higher than AB$_5$. Some examples are in the range of 1.5 wt% (ZrFe$_{1.5}$Cr$_{0.5}$) to 2.43 wt% (TiCr$_{1.8}$).

### 4.4.1.2 Magnesium-Based Hydrides

Magnesium hydride, MgH2, has received a lot of attention due to its high storage capacity of 7.7 wt%, light weight and low costs. Magnesium hydride shows the highest energy density, 9 MJ kg$^{-1}$, when compared to other reversible metallic hydrides suitable for hydrogen storage. This hydride is thermodynamically stable and therefore has a high desorption enthalpy ($\Delta H = -74.5$ kJ mol$^{-1}$) [41, 45]. Stored hydrogen is released at a desorption temperature of 330 °C at 0.1 MPa [45, 46]. This temperature is rather too high for most applications because filling the tank would take several hours. Using high-energy ball milling for the preparation of the hydride and by addition of catalysts based on metal oxides, it is possible to obtain fast desorption kinetics over times of several minutes. The tetragonal unit cell of magnesium hydride is shown in Figure 4.21 [47].

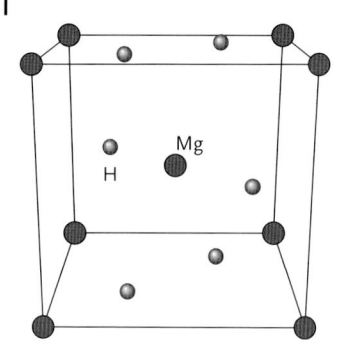

**Figure 4.21** Unit cell of magnesium hydride [47].

Depending on the synthesis method, the hydrogen storage capacities of magnesium-based hydrides varies between 7.7 wt% ($MgH_2$) and 3.5 wt% (Mg+20 wt% Mn) and desorption temperatures between 180 °C and 450 °C [48].

Various binary and ternary magnesium alloys have been studied in order to reduce the desorption temperature and enhance the hydrogenation–dehydrogenation kinetics. If the temperature is below 200 °C, hydrogen storage capacity is significantly reduced [49]. Shang *et al.* [47] have modified the hydrogen storage properties of $MgH_2$ by its mechanical alloying with 8 mol% of various metals (Al, Ti, Fe, Ni, Cu, and Nb). Dehydrogenation at 300 °C under vacuum showed that the ($MgH_2$+Ni) mixture gives the highest level of hydrogen desorption coupled with the most rapid kinetics, followed by $MgH_2$ with Al, Fe, Nb, Ti, and Cu. $MgH_2$ as received and the milled $MgH_2$ showed relatively slow kinetics compared with the alloyed mixtures. Theoretical predictions showed that the ($MgH_2$+Cu) system is the most unstable, followed by ($MgH_2$+Ni), ($MgH_2$+Fe), ($MgH_2$+Al), ($MgH_2$+Nb), and ($MgH_2$+Ti) [47].

The highest amount of $H_2$ desorption for this doped alloys was found in $MgH_2$+Ni (7.7 wt%), followed by $MgH_2$+Al (5.7 wt%), $MgH_2$+Fe (5.3 wt%), $MgH_2$+Ti (2.6 wt%), $MgH_2$+Nb (2.6 wt%), and $MgH_2$+Cu (1.7 wt%). This is illustrated in Figure 4.22. The alloying elements evidently improved the level of hydrogen desorption at 300 °C.

Liang *et al.* [50] have mechanically milled Mg and magnesium hydride in order to prepare a ternary alloy Mg–Ni–La for hydrogen storage. By milling the powder size reduces resulting in faster absorption and slower desorption kinetics. It was found that the ternary Mg–La–Ni composite ($MgH_2$+$LaH_3$+$Mg_2NiH_4$) has better hydrogen absorption–desorption kinetics than the Mg–La and Mg–Ni binary composites. The $LaH_3$ had positive effects on absorption and less effect on desorption. At high temperature, $Mg_2Ni$ has better catalytic effects than $LaH_3$ on both absorption and desorption (Figure 4.23).

It was observed [51, 52] that sorption kinetics of magnesium-based hydrides are significantly enhanced not only by fine microstructure, but also by catalyst additions. Significant reaction rates can be obtained at reduced temperatures. Using $Cr_2O_3$, the complete desorption was reached after less than 5 min, whereas without catalyst the desorption of hydrogen needs about 30 min. The desorption

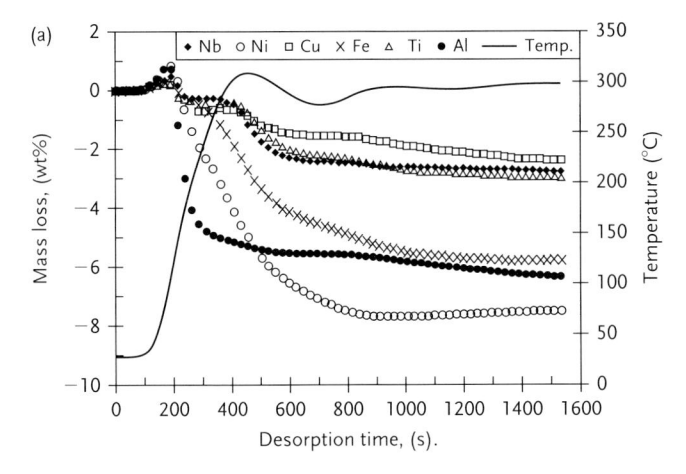

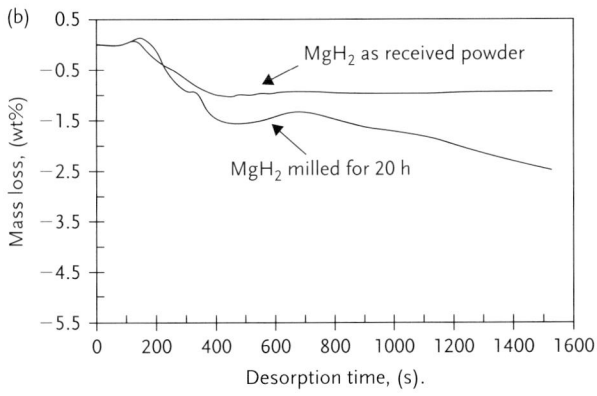

**Figure 4.22** (a) Weight loss of dehydrogenation versus time at 300 °C under vacuum for mechanically alloyed magnesium hydride mixtures (b) and as-received and mechanically alloyed MgH$_2$[47].

rate was determined to be equivalent to 29 kW per kg of powder. Long-term cycling stability tests (1000 cycles) showed that the capacity decreased only from 7 to 6.6 wt% of hydrogen, being five times better than commercial rechargeable battery cells. MgH$_2$ doped with 5 wt% V did not indicate disintegration of the material and change in absorption and desorption isotherms even after 2000 cycles [53].

### 4.4.2
### Light Metal Complex Hydrides

#### 4.4.2.1 Alanates
Complex hydrides often contain light atoms such as Na, Li, K, Be, B, Mg, Al, Si, and Ca. A novel group of reversible hydrogen storage compounds is alanates (alkali metal aluminum hydrides). The alanates are ternary (or higher) hydride

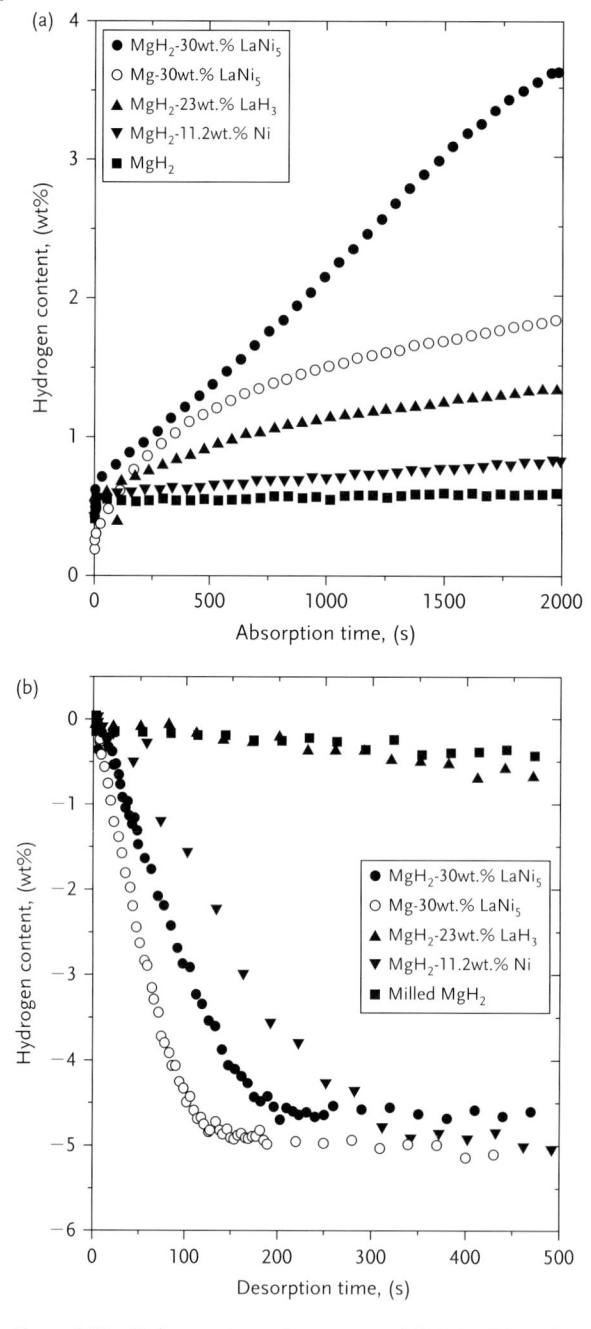

**Figure 4.23** Hydrogen absorption curves at 302 K (29 °C) and pressure 1.0 MPa (a) and hydrogen desorption curves at 573 K (300 °C) and pressure 0.015 MPa (b)[50], reprinted by permission.

compounds, containing aluminum, stored hydrogen, and one (or more) other metal(s).

Plenty of alanates have been investigated with respect to hydrogen storage. Complex "chemical" hydrides, such as $NaAlH_4$ and $LiAlH_4$, are promising materials in this regard because of their high concentration of useful hydrogen (reversible hydrogen capacities 5.6 and 7.9 wt%, respectively), which is typically much higher than most metal hydrides (0.5–2.0 wt%) and carbon materials at room temperature (0.5–1.0%), see also in Section 7.2. Important limitation in the application of light metal hydrides is mainly slow kinetics. Presently $NaAlH_4$ is among the most advanced and intensively investigated hydrogen storage materials. This is because of the reversible hydrogen content (5.6 wt%), commercial availability and the ability to release hydrogen at temperatures close to 353 K (80 °C), which is in the operational range of a PEM fuel cell, capable of releasing energy from hydrogen in automobile applications. The thermodynamics and kinetics of the reversible dissociation of metal-doped $NaAlH_4$ as a hydrogen (or heat) storage system have been investigated in some detail [54]. The experimentally determined enthalpies for the first (3.7 wt% of H) and the second dissociation step of Ti-doped NaAlH (3.0 wt% H) of 37 and 47 kJ $mol^{-1}$ are in accordance with low- and medium-temperature reversible metal hydride systems, respectively.

The decomposition of sodium alanate is performed at temperatures above 200 °C according to the following equations:

$$3NaAlH_4 \leftrightarrow Na_3AlH_6 + 2Al + 3H_2 \quad \Delta H = 37 \text{ kJ mol}^{-1} \quad 3.7 \text{ wt\%} \quad (3.11)$$

$$Na_3AlH_6 + 2Al + 3H_2 \leftrightarrow 3NaH + 3Al + 4.5H_2$$
$$\Delta H = 47 \text{ kJ mol}^{-1} \quad 1.8 \text{ wt\%} \quad (3.12)$$

In the first step, the pressure is about 0.1 MPa and the temperature 30 °C. The second step occurs at 0.1 MPa and at the temperature of about 100 °C [54]. Under normal operational conditions of a PEM fuel cell (80 °C) only the first step can be used to release hydrogen at a sufficient pressure. The undoped sodium alanate has a very slow hydrogen release, rate and rehydrogenation to a reasonable extent is not possible at moderate temperatures and pressures. This can only be changed if a catalyst is used to increase the reaction rates for both reactions. Transition metal catalysts such as titanium, iron, vanadium, or zirconium are known to enhance the rates, but not to a sufficient level [55–57]. It was shown that samples doped with 4 mol% of Ti exhibited the best behavior with respect to hydrogen desorption kinetics, followed by 4 mol% of zirconium and then 4 mol% of iron, confirming that titanium itself is the best catalyst to improve the kinetics [56, 58]. An interesting synergistic behavior was revealed when doping $NaAlH_4$ with 1 mol% Fe and 3 mol% Ti.

Scandium and cerium have been found to be highly active catalysts that can achieve the same rates of hydrogenation as titanium colloids. Substantial rehydrogenation of the order of 3.5 wt% was possible at pressures of 5 MPa within 1 h, which is far better than the performance of titanium-doped systems [59]. Wet chemical synthesis and ball milling of the titanium precursor ($TiCl_3$) into sodium alanate

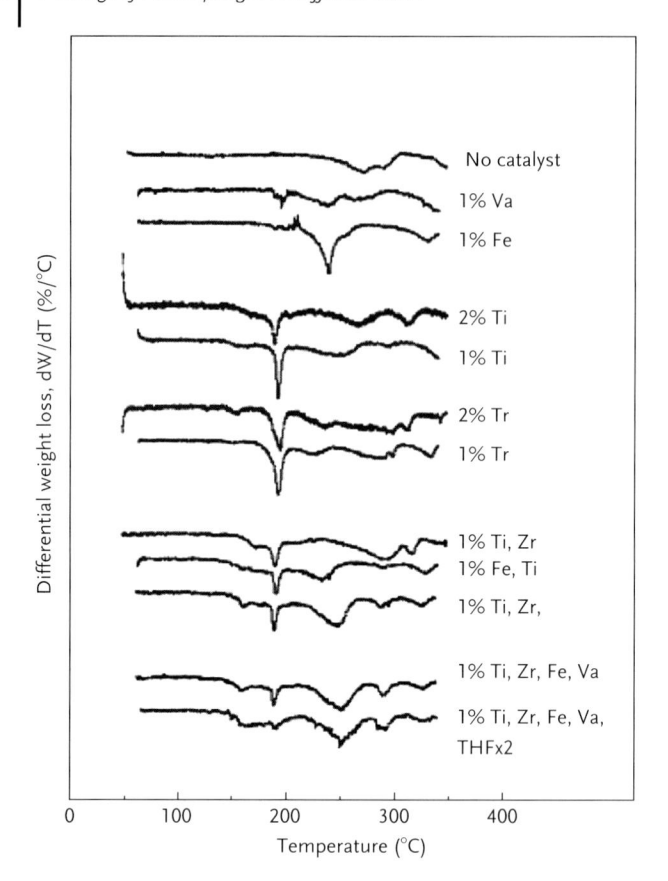

**Figure 4.24** Differential TGA analyses of NaAlH$_4$ doped with varying amounts of single and multiple metal catalysts [57].

can accelerate the sorption rates of hydrogen [60]. The test results of sodium alanate doped by different transition metal catalysts are shown in Figure 4.24.

Without catalyst, the initial dehydrogenation temperature of NaAlH$_4$ was 240 °C. The TGA (thermogravimetric analyses) results revealed that Ti- and Zr-doped systems had their effective hydrogen desorption temperature lowered by approximately 85 °C (to around 150 °C), while systems with only Fe or V had their effective desorption temperature lowered by about 55 °C (to 180 and 189 °C, respectively). These results have proved that doping NaAlH$_4$ with transition metals improves the dehydrogenation kinetics.

According to Bogdanovic et al. [61], it seems that sodium alanate system does not have a sufficient gravimetric storage capacity for transport applications. 5.6 wt% may be enough for the whole system, but insufficient for the storage material itself because it has to be kept in a container that is moderately pressure resistant and a powerful heat exchanger is needed to remove the heat released during refueling of the system for automotive applications. Therefore, higher capacities are needed for that purpose. Outside transport applications sodium alanate could be used in low-power units where high capacities are less important.

Other similar hydrides of interest are lithium alanate LiAlH$_4$, calcium alanate Ca(AlH$_4$)$_2$, and magnesium alanate Mg(AlH$_4$)$_2$. The decomposition products of these alanates are presented by the respective reactions:

$$LiAlH_4 \rightarrow \frac{1}{3}Li_3AlH_6 + Al + H_2 \qquad (3.13)$$

$$Li_3AlH_6 \rightarrow 3LiH + Al + \frac{3}{2}H_2 \qquad (3.14)$$

$$Ca(AlH_4)_2 \rightarrow CaAlH_5 + Al + \frac{3}{2}H_2 \qquad (3.15)$$

$$Mg(AlH_4)_2 \rightarrow MgH_2 + 2Al + 3H_2 \qquad (3.16)$$

The reaction enthalpies and H$_2$ evolution rates for a number of kinetically stabilized metal hydrides AlH$_3$, LiAlH$_4$, Li$_3$AlH$_6$, Mg(AlH$_4$)$_2$, and Ca(AlH$_4$)$_2$ and undoped NaAlH$_4$ have been evaluated [62]. It was found that despite fast kinetics at low temperature and high hydrogen capacities, these hydrides are not reversible at reasonable temperatures and pressures and are not practical storage media for most applications. Synthesis of these hydrides is complex and expensive and new methods have to be found. According to the authors [62] "the critical technical challenge lies in overcoming the thermodynamic barrier to reforming the hydride from the used fuel and will require a new cost-effective efficient method to regenerate or recycle the hydride from used fuel and reaction products."

Figure 4.25 presents differential scanning calorimerty (DSC) results for three metastable hydrides: LiAlH$_4$, Ca(AlH$_4$)$_2$, and Mg(AlH$_4$)$_2$. The reactants form a more stable compound by undergoing a structural transition and releasing hydrogen.

Vajo and Olson [63] modified the thermodynamics and kinetics of hydrogen sorption reactions in light-metal hydrides by employing additives to lower the enthalpy and increase the equilibrium pressure. The authors have investigated destabilization of LiBH$_4$ using MgH$_2$ and various other Mg compounds, MgX$_n$ (where X=F, Cl, OH, O, S, Se, Si, Ge, or Ni). The equilibrium pressure of the destabilized LiBH$_4$/MgH$_2$ was about 10 times greater than the equilibrium pressure calculated for pure LiBH$_4$. Vajo and Olson [63] stated that although destabilization by alloy or compound formation are productive ways to change the thermodynamics of the reactions, it is still far away from the system targets set by the US DOE. This is shown in Figure 4.26.

#### 4.4.2.2 Amides-Imides (Li$_3$N−Li$_2$NH−LiNH$_2$)

Lithium nitride shows a significant potential for on board hydrogen storage. Chen et al. [64] demonstrated that in fresh lithium nitride, Li$_3$N, the absorption of hydrogen began at a temperature of around 100 °C and if the temperature was maintained below 200 °C, the sample could absorb a significant amount of

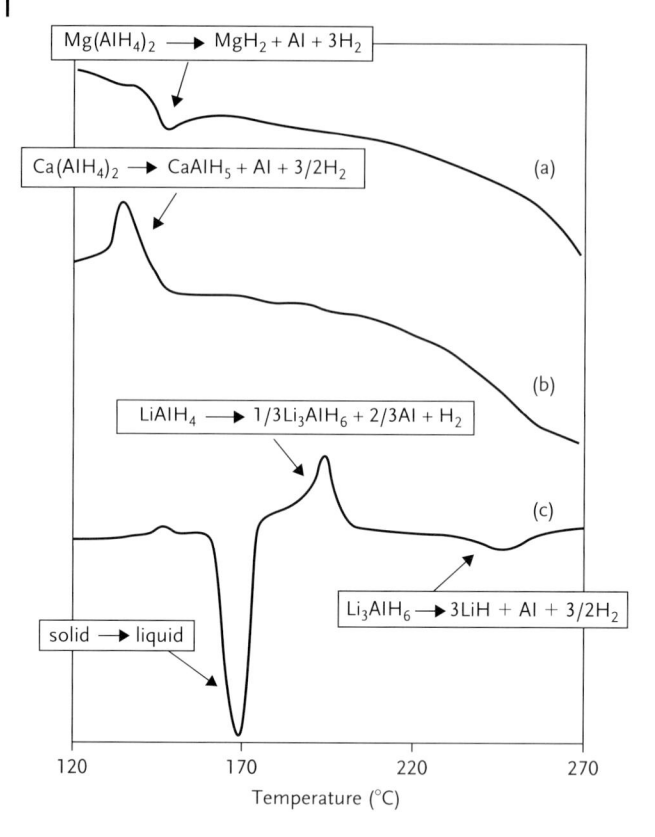

**Figure 4.25** DSC traces from (a) Mg(AlH$_4$)$_2$, (b) Ca(AlH$_4$)$_2$, (c) LiAlH$_4$ [62], reprinted by permission.

hydrogen. It is assumed that the reversible hydrogen storage process in Li$_3$N occurs according to the two step reaction:

$$Li_3N + 2H_2 \leftrightarrow Li_2NH + LiH + H_2 \leftrightarrow LiNH_2 + 2LiH \quad \Delta H = -161 \text{ kJ mol}^{-1} \quad (3.17)$$

The theoretical gravimetric storage capacity is 10.4 wt%, corresponding to the molar ratio of hydrogen atoms to the nitride molecule equal to 4. Practically the hydrogen stored was about 9.3 wt% at 255 °C [64]. As only the second reaction step between imide Li$_2$NH and amide LiNH$_2$ is reversible, theoretically 7.0 wt% of hydrogen can be reversibly stored as a result. This proceeds according to the following equation:

$$Li_2NH + H_2 \leftrightarrow LiNH_2 + LiH; \quad \Delta H = -45 \text{ kJ mol}^{-1} \quad (3.18)$$

However, based on pressure-composition-temperature measurements it was found that in fact it was possible to reversibly store 6.5 wt% of hydrogen at 255 °C [64, 65]. Similar to lithium nitride, phenomenon for hydrogen storage was also

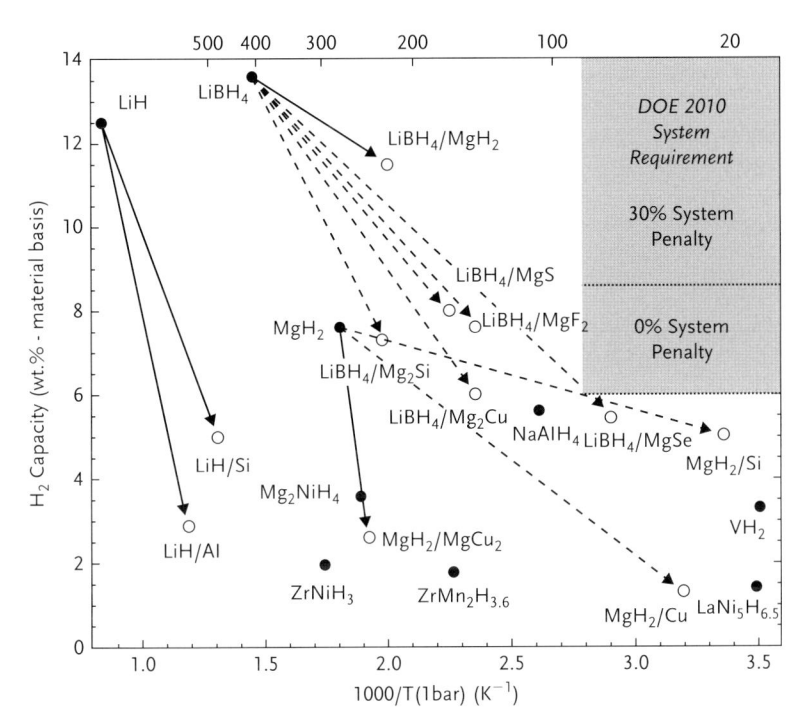

**Figure 4.26** Gravimetric hydrogen capacity versus $1/T$ (1 bar) for several transition metal, low, complex and destabilized hydrides. [63], open circles represent destabilized systems, filled circles represent individual hydrides, solid arrows represent experimental results, and dashed represent calculated ones. Reprinted by permission.

found in the Ca−N−H system, by reaction of calcium nitride hydride, $Ca_2NH$ to calcium imide, CaNH. Theoretically 2.1 wt% of hydrogen can be stored reversibly in $Ca_2NH$. In practice, the maximum capacity, experimentally demonstrated, was 1.9 wt% at 550 °C [64]. Ichigawa et al. [66] have investigated the hydrogen storage properties of $Li_2NH$ as one of the suitable hydrogen storage materials for onboard use. To improve the reaction kinetics, 1 mol.% of $TiCl_3$ has been added as catalyst and thermogravimetry measurements have shown ∼6 wt% of hydrogen desorbed in the range of 150–250 °C, indicating that the reaction kinetics were significantly improved; however, the desorption temperature was too high for onboard storage. For that purpose, a mixture composed of LiH and $Mg(NH_2)_2$ has been developed, desorbing at around 7 wt% of hydrogen in the range from 120 °C to 200 °C. The hydrogen desorption equilibrium pressure was higher than 5 MPa at 200 °C indicating that such a system has a high potential for onboard application. This results confirmed previous investigations by Fuji and Ichikawa [67]. Nine possible metal (Li, Mg, Ca)−H−N systems for hydrogen storage has been designed as

**Table 4.9** New possible metal–N–H systems for hydrogen storage [66], reprinted by permission.

| Possible reversible reaction for hydrogen storage | Theoretical hydrogen capacity (mass%) |
| --- | --- |
| $2LiH + Mg(NH_2)_2 \leftrightarrow Li_2NH + MgNH + 2H_2$ | 5.58 |
| $8LiH + 3Mg(NH_2)_2 \leftrightarrow 4Li_2NH + Mg_3N_2 + 8H_2$ | 6.93 |
| $12LiH + 3Mg(NH_2)_2 \leftrightarrow 4Li_3N + Mg_3N_2 + 12H_2$ | 9.15 |
| $MgH_2 + Mg(NH_2)_2 \leftrightarrow 2MgNH + 2H_2$ | 4.88 |
| $2MgH_2 + Mg(NH_2)_2 \leftrightarrow Mg_3N_2 + 4H_2$ | 7.40 |
| $3MgH_2 + 4LiNH_2 \leftrightarrow Mg_3N_2 + 2Li_2NH + 6H_2$ | 7.08 |
| $2LiH + Ca(NH_2)_2 \leftrightarrow Li_2NH + CaNH + 2H_2$ | 4.58 |
| $MgH_2 + Ca(NH_2)_2 \leftrightarrow MgNH + CaNH + 2H_2$ | 4.10 |
| $4CaH_2 + 3Mg(NH_2)_2 \leftrightarrow Mg_3N_2 + 4CaNH + 8H_2$ | 4.78 |

promising hydrogen storage material [66]. They are presented in Table 4.9, including possible reverse reactions for hydrogen storage.

Hydrogen absorption and desorption characteristics in the Mg–Na–N–H and Mg–Li–N–H systems were investigated among others by Xiong et al. [68, 69]. In the case of interaction between $Mg(NH_2)_2$ and NaH, hydrogen was released at temperature of 120 °C. The mixture was then converted to a new Mg–Na–N–H mixture able to reabsorb hydrogen at 60 °C. After 10 cycles of testing no obvious deterioration in hydrogen absorption–desorption performance was noticed. A ternary imide, $Li_2MgN_2H_2$ was reported to be able to reversibly take up more than 5.5 wt% of hydrogen at 180 °C with desorption equilibrium pressure higher than 2 MPa [68].

Dam et al. [70] described the high potential of a novel optical method, hydrogenography, for screening hydrogen storage properties of metal alloys. Using this, both the kinetics and the thermodynamics of complex metal hydrides can be measured by using the change in optical properties created by the fast thin film solid-state reaction between the metallic thin film components and a reactive gas. Hydrogenography can be useful in the search for cheap catalysts for hydrogen absorption. In this case, the kinetics of absorption is measured by applying a constant hydrogen pressure above the equilibrium value.

### 4.4.2.3 Borohydrides

A large number of borohydrides (aka hydroborates) exist depending on the stoichiometry of metal, boron, and hydrogen such as $M[BH_4]_n$, $M[B_3H_8]_n$, $M[B_6H_6]_n$, $M[B_9H_9]_n$, and $M[B_{12}H_{12}]_n$ [71]. In the case of alkali metal borohydrides $M[BH_4]_n$ the metal can be Li, Na, K, Rb, or Cs and in alkaline earth borohydrides M=Be, Mg, Ca. Alkali metal and alkali earth metal borohydrides belong to a class of materials having the highest gravimetric hydrogen densities (from 7.42 wt%, 87.1 kg $H_2$ m$^{-3}$ for $K[BH_4]$ up to 18.36 wt%, 122.5 kg $H_2$ m$^{-3}$ for $Li[BH_4]$ and 20.67 wt%, 145.1 kg $H_2$ m$^{-3}$ for $Be[BH_4]_2$). They are ionic, white, crystalline compounds, sensitive toward moisture [72]. Decomposition of borohydrides by reaction with water occurs according to the following reaction:

$$MBH_4 + xH_2O \rightarrow (MBO_2, x - 2H_2O) + 4H_2$$

$x$ is dependent on the nature of the metal, temperature, and composition of the solution [73].

Sodium borohydride (tetrahydroborate), $NaBH_4$, has been intensively investigated as a hydrogen storage media as it is available and relatively inexpensive, and more stable material than lithium borohydride. Sodium borohydride is commonly used as a reducing agent in organic chemistry and the pharmaceutical industry, or as a bleaching agent in paper manufacturing. In addition, $NaBH_4$ has been used as fuel in a direct borohydride fuel-cell (DBFC) [74, 75]. DBFC has been investigated for example for unmanned underwater vehicles (UUV) propulsion power [76] and for portable application [77]. Medis Technologies' [78] 1 kW Power Pack can be used for a number of consumer, military, and light industrial applications such as mobile phones, portable music, portable lighting, and so on.

Sodium borohydride generates hydrogen in a hydrolysis reaction that can be described as follows:

$$NaBH_4 + (2 + x)H_2O \rightarrow 4H_2 + NaBO_2(aq) \, xH_2O + heat$$

Storage capacity of sodium borohydride reported by Wee et al. is 10.7 wt% [79]. For this stoichiometric reaction, the storage medium is ($NaBH_4 + 2\,H_2O$), and half of hydrogen is provided by water. In reality excess water is required (expressed as "$x$ – excess hydration factor" in Eq. (4.20)). This excess of water is essential to the gravimetric hydrogen storage capacity. Usually if $x$ is equal to 2 or 4, the capacity is then 7.3 or 5.5 wt%, respectively [73]. According to Marrero-Alfonso et al. [80], in an ideal case 2 moles of water is needed for liberating 4 moles of hydrogen. In the case of ideal hydrolysis ($x=0$), on the pure reactant basis the specific energy is 0.01084 kg $H_2$ per kg of reactant, which meets the FreedomCAR 2015 requirement set as 0.09 kg $H_2$ per kg of the hydrogen storage/production system [80].

Lithium borohydride $Li[BH_4]$ with high weight content (18.36 wt%) of hydrogen belongs to the highest energy content borohydrides. Typically used as a reducing agent for aldehydes, ketones, esters, and epoxides, it is also regarded as a potential hydrogen storage material. Thermal desorption of lithium borohydride into LiH and boron is described by the following equation:

$$LiBH_4 \rightarrow LiH + B + \frac{3}{2}H_2$$

The reaction liberates three of four hydrogens from the material above the melting temperature (280 °C) [71]. The thermal desorption spectrum of lithium borohydride exhibits four endothermic peaks attributed to a polymorphic transformation around 110 °C, melting at 280 °C and desorption of three of the four hydrogen at 680 °C. The calculated enthalpy $\Delta H$ is $-177.4$ kJ mol$^{-1}$ and the entropy $\Delta S$ is 238.7 J K$^{-1}$ mol$^{-1}$ [72]. Figure 4.27 shows the energy levels for the hydrogen desorption of $Li[BH_4]$.

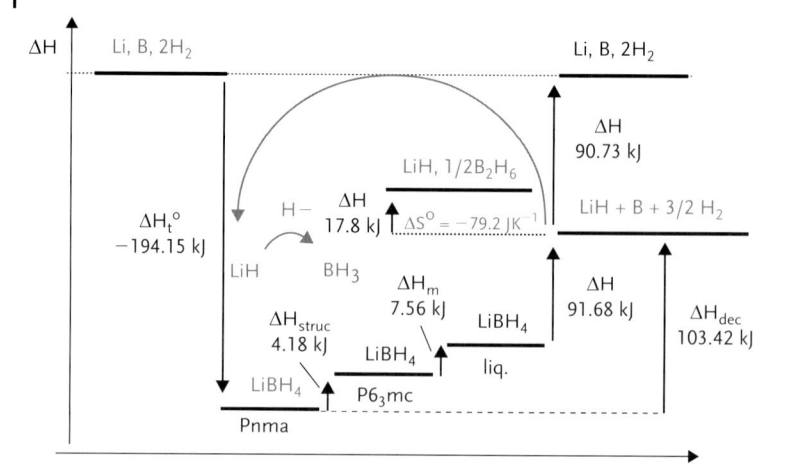

**Figure 4.27** Enthalpy diagram of the phases and intermediate products of Li[BH]$_4$ [72], reprinted by permission.

The most stable state for Li[BH$_4$] is the low temperature phase, which is transformed into the high-temperature modification at 118 °C, followed by melting at 280 °C. As LiH is very stable, the desorption occurs at above 727 °C and therefore is not useful in technical applications [72]. Magnesium tetraborohydride, Mg[BH$_4$]$_2$, is among the possible borohydride candidates for hydrogen storage material, with high gravimetric and volumetric hydrogen densities (14.9 wt% and 112 g L$^{-1}$, respectively) and a favorable enthalpy of decomposition estimated theoretically as about 40 kJ mol$^{-1}$ H$_2$ [81]. The dehydrogenation of magnesium tetraborohydride can be described by the following reaction:

$$Mg(BH_4)_2 \rightarrow MgH_2 + 2B + 3H_2$$

Pure Mg[BH$_4$]$_2$ has not yet been investigated as a material for hydrogen storage [82]. Chlopek et al. [82] have studied several possibilities to synthesize solvated and desolvated material by wet chemical and mechanochemical methods.

In general, borohydrides are not very easy materials for handling with regard to hydrogen storage. Thermal decomposition (dehydrogenation) is a highly energy consuming process, as borohydrides need high temperatures to release hydrogen. Soloveichik et al. [81] reported that some key features of the hydrogen release from Mg[BH$_4$]$_2$ remain unexplained and that due to the existence of multiple steps with different enthalpies it is very complicated to rehydrogenate Mg[BH$_4$]$_2$. The tests to completely recharge products of decomposed Mg[BH$_4$]$_2$ at 250–350 °C at pressure of 13.4 MPa were unsuccessful, and only magnesium metal could be rehydrided to MgH$_2$. According to the authors, the most important requirement for practical application of magnesium borohydride as a hydrogen storage medium is to develop effective catalysts enabling hydrogenation/dehydrogenation reactions of intermediate reaction steps (polyborane and boride phases).

## 4.4.3
**Application**

Metal hydrides are important energy storage materials. They are still extensively investigated for onboard applications. Metal hydrides can be used not only as a potential storage hydrogen materials, like for example $MgH_2$, but also can serve as the precursors for the production of future complex hydrides. For example, $LiAlH_4$ can be produced on a large scale by the reaction of LiH with $AlCl_3$ dispersed in the solvent or by milling $AlH_3$ with NaH [83, 84]. There is a wide spectrum of metal hydrides application in industry. They are used for hydrogen storage in stationary and mobile applications, in electrochemistry (battery, catalysts), in hydrogen processing for the purpose of separation from mixed gas streams [85], purification, separation of isotopes, compression, and gettering (e.g., inert gas purification or vacuum maintenance). Metal hydrides are also used for thermal applications, such as heat storage, heat pumping, refrigeration, or in materials processing like for example magnet processing. Several examples are presented below.

The core activity of Hystore Technologies Ltd. located in Cyprus is the production and characterization of metal hydride ($AB_5$ type) storage units and systems ranging from 5 to 500 000 NL of hydrogen or even more [86]. They are hydrogen compressors based on metal hydrides, metal hydride-based air conditioning systems (heat and cool), and hydrogen purification units. Hystore Technologies Ltd. constructs metal hydride tanks (MHT) of different shapes (tubular, rectangular, etc.) and different hydrogen capacities. The hydrogen charging pressure could be designed to be from 1 to 100 bar or more. Depending on the desired hydrogen charging or discharging rate, MHT can be constructed as air-cooled (heated) or water-cooled (heated) with the expected life of the MHT more than 10 000 cycles. The pure hydrogen gravimetric density of the MHT, including the metal hydride and the tank, can vary from 0.7 to 1.1 wt%, depending on the operating pressure and the overall MHT design. The hydrogen volumetric density can be as high as 530 NL of hydrogen per liter of MHT. Examples of different capacity and shape products that are produced by Hystore Technologies Ltd. are presented below (Figures 4.28 and 4.29).

Hydrogen Components Inc., Colorado, [87] produce hydrogen-storage containers SOLID-H™ that are filled with metal powders that absorb and release hydrogen (type $AB_2$ and $AB_5$ based on alloys A, L, M, H). Alloy A composition is nearly 1 : 1 TiFe. It has some minor additives to adjust the pressure–temperature characteristics and to help the hydride to "activate" (charge for the first time with hydrogen). Alloy A falls short of $AB_2$ in terms of rechargeable hydrogen content (1.6 wt% of hydrogen). Alloys L, M, and H are rare earth pentanickel alloys having the formula $AB_5$. These type of containers have been used in the satellite navigation system called GALILEO. A typical container is shown in Figure 4.30.

In the frame of the GALILEO European GPS (global positioning system) constellation, SPECTRATIME has developed an atomic clock space based on the Passive Hydrogen MASER (microwave amplification by stimulated emission of

**Figure 4.28** Hystore metal hydride tank 35 000 NL $H_2$, reprinted with permission of Hystore Technologies Ltd (www.hystoretechnologies.com).

radiation) principle using SOLID-H™ BL-30 from Hydrogen Components Inc. with a fine purified $LaNi_5$ metal hydride to store the hydrogen (consumed by the Passive Hydrogen MASER) inside a low volume and under a relatively low pressure. The Hydrogen MASER is shown in Figure 4.31. The overall physical package (PP) performs the function of a very narrow-band microwave amplifier and frequency discriminator, featuring extremely stable central frequency. The main component of the PP is the microwave cavity and shield assembly (MCSA), which includes the storage bulb, the microwave cavity, its magnetic shielding, and the thermal control devices. A low vacuum enclosure is built over the MCSA to simulate the final flight thermal exchange conditions. Other fundamental items are the hydrogen beam assembly, which includes the hydrogen purifier, the dissociator bulb, beam collimator, and state selector. The hydrogen supply is provided by the hydrogen storage assembly (HSA), which is constituted by a solid-state hydrogen storage vessel, with relevant pressure and temperature devices. The high vacuum is assumed by the ion pump and the getters [88]. The Hydrogen MASER is used in several applications with the same objective: frequency stability (as reference) in research institutes, astronomy (telescope), telecommunication (synchronization), space (GALILEO, ACES), and geodesy (displacement, positioning).

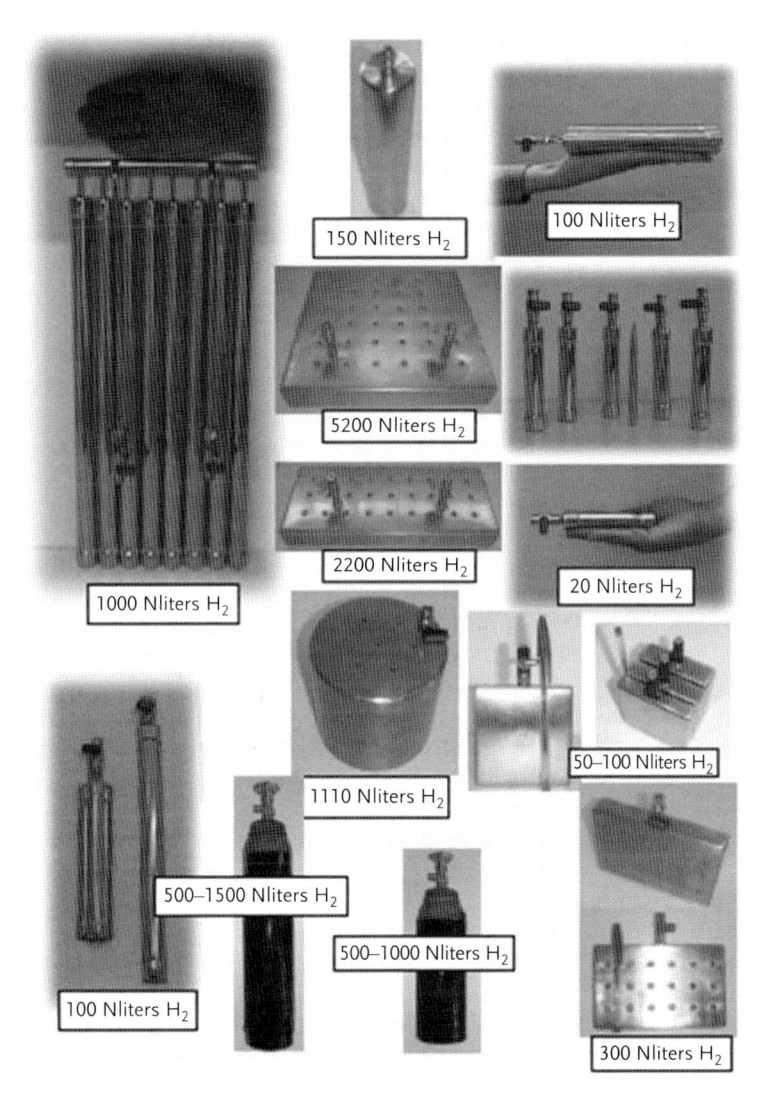

**Figure 4.29** Different capacity and shape of example products based on metal hydrides, reprinted with permission of Hystore Technologies Ltd (www.hystoretechnologies.com).

A hydrogen storage container is filled with a solid-state metal hybrid alloy. The amount of hydrogen is sufficient for continuous operation during the whole duration of the mission. Solid-state storage of hydrogen is superior to conventional storage in a high-pressure vessel both from the point of view of safety and mass economy. Upon heating the hydrogen storage assembly (HSA) by controlling a resistor heater, a few bars of molecular hydrogen are released in the pipe leading to the hydrogen flow control device (Figure 4.32).

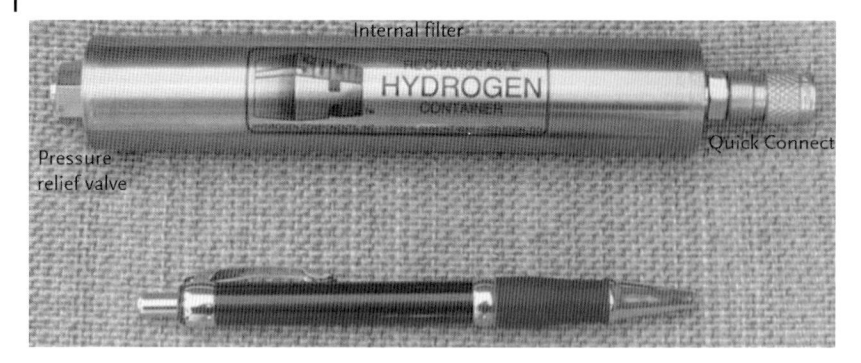

**Figure 4.30** SOLID-H model BL-H metal hydride container holds 30 SL of $H_2$ gas [87], reprinted with permission of Hydrogen Components Inc., Colorado, USA.

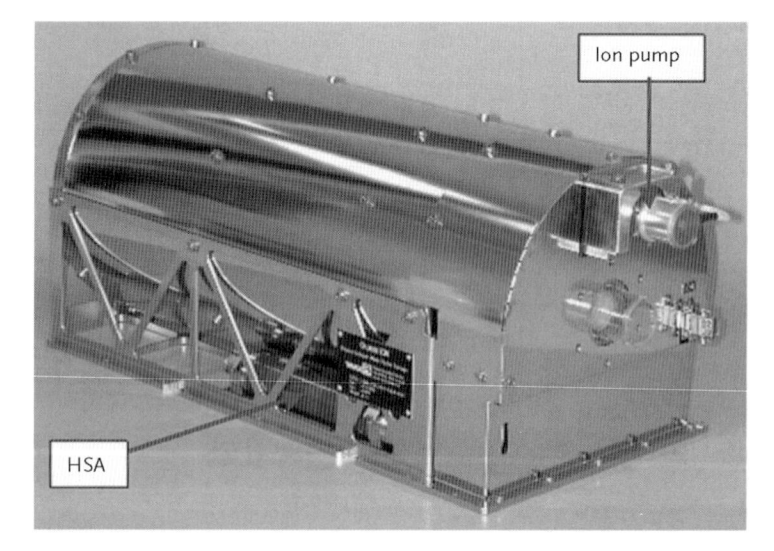

**Figure 4.31** An overview of the hydrogen MASER produced by the SPECTRATIME, reprinted with permission of SPECTRATIME.

At the Hannover Fair 1998 a Siemens Nixdorf laptop computer was demonstrated, which was powered by a laboratory PEM fuel cell (FhG ISE Freiburg, Germany) and a commercial metal hydride tank SL002 (GfE Metalle und Materialien GmbH, Germany), which has been developed for demonstration models. The total energy density of the system was 222 Wh $L^{-1}$ and 73 Wh $kg^{-1}$, respectively [49] (Figure 4.33).

In recent years, the US Office of Naval Research and the Naval Sea Systems Command have developed new technologies for diving and life-support systems for underwater swimming operations and diving in contaminated water. Nuckols et al. [89] have presented an overview of recent advances in diver thermal

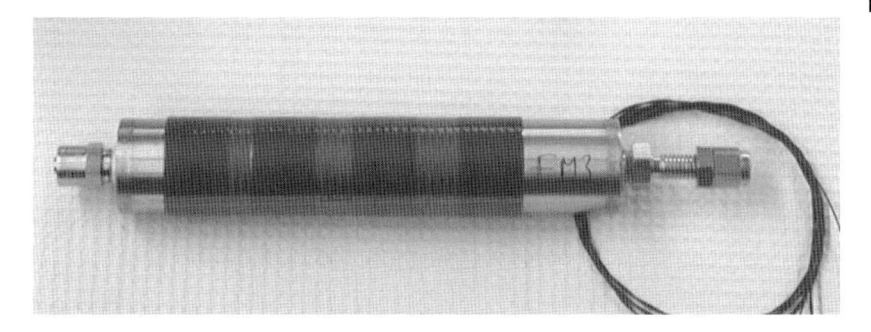

**Figure 4.32** Hydrogen container of MASER by SPECTRATIME, reprinted with permission of SPECTRATIME.

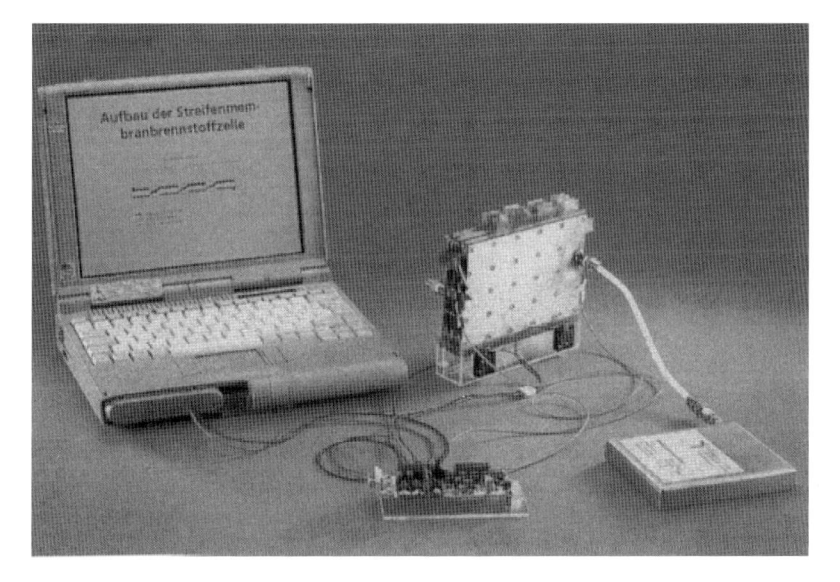

**Figure 4.33** Siemens Nixdorf Notebook powered by a PEM fuel cell/metal hydride tank [49], reprinted by permission.

protection systems. The efforts made to transfer these new technologies into the civilian sector have been discussed. Key technology areas included catalytic heating using hydrogen as an energy source and portable heating and/or cooling sources using metal hydrides. Lindler and Nuckols [90] have reported that less than 0.4 kg of hydrogen could supply 2 kW of heat needed to sustain up to six divers in 0 °C water for up to 6 h. Recent research sponsored by the US Office of Naval Research with diver heating systems using hydrogen catalytic reactions has proved that certain metal alloys can be used as an attractive alternative to high-pressure flasks or cryogenic systems for hydrogen storage. When utilized in pairs they are able to produce a compact, fully regenerative cooling source without a need of external power supply during operation. By exposing high-pressure

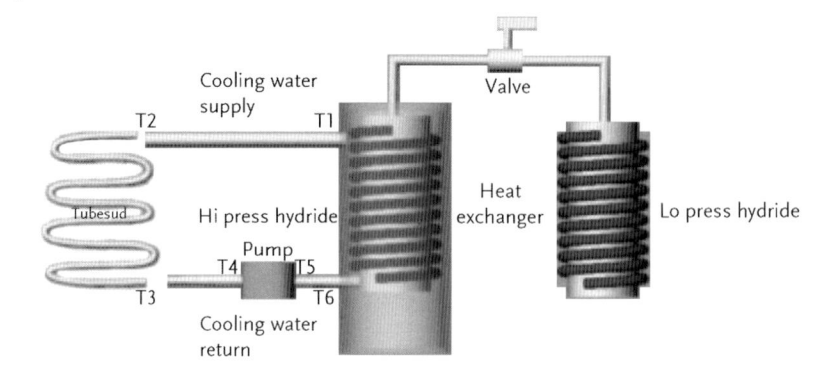

**Figure 4.34** Schematic of diver heater–cooler concept (reprinted by permission).

hydrogen gas to these metal alloys, metal hydride heats up during hydrogen absorption, generating heat. A heat sink is created by reducing the storage pressures, resulting in hydrogen desorption from the metal hydride. A concept of a cooling system for warm water diving by coupling two different metal hydrides having different absorption/desorption isotherms has been developed (Figure 4.34) and patented by Nuckols [91]. A container of metal hydride with low-temperature storage isotherm is coupled to another hydride with high-temperature isotherm by a single gas line. Both hydride containers are at the same temperature. The container with the low-temperature hydride has low pressure relative to the high-temperature hydride. When the valve is opened, hydrogen from the high-pressure metal hydride is released into the low-pressure metal alloy. As a result of releasing hydrogen from the high-pressure container the hydride is cooled down, creating a heat that can be used to cool the protective garment. Absorption of hydrogen into the low-pressure container results in heating up the hydride, generating a heat source that could be used to dump heat to the surrounding environment. The heating and cooling canisters of metal hydrides are presented in Figure 4.35 [89]. The canister on the right contains a metal alloy whose equilibrium pressure is higher than the ambient pressure (the presence of frost on the canister as hydrogen is released out of the solution from the hydride). The canister on the left produces heat as the alloy absorbs hydrogen.

In 2005, the Office of Naval Research has verified the feasibility of using hidden heats of absorption and desorption of hydrogen from various metal alloys pairs as a compact cooling source. The possibility of this cooling method for other submerged and/or surface applications has been suggested. Metal hydrides as a cooling source have been found suitable for fully regenerative cooling capability. Presently a prototype (Figure 4.36) of heating–cooling circuit with a tube and shell heat exchanger to contain two metal hydride alloys is built as a cooling source for contaminated water diving applications [89].

Supplement heat provided by the hydrogen catalytic reactions can significantly increase a time of diving (up to 145% longer dive duration is then reported without

**Figure 4.35** The heating and cooling potentials for canisters of metal hydrides (reprinted by permission).

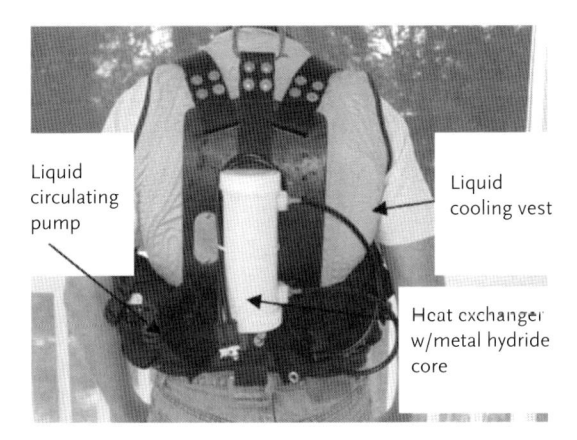

**Figure 4.36** Prototype metal hydride cooling–heating system (reprinted by permission).

heating) [89]. The hydrogen catalytic heater system can either be fueled by injecting a small stream of hydrogen into the breathing gas supply upstream of the diver's suit or by providing a premix of noncombustible hydrogen and air through the gas supply unit. Nuckols *et al.* [89] reports that the possibility of introducing commercially available metal hydride storage system (as described above) as alternative to high-pressure gas cylinders for storing the hydrogen in a catalytic heater has been evaluated (Figure 4.37). At present, there is an ongoing work on developing a storage medium based on gadolinium–nickel–tin alloy for this application in order to

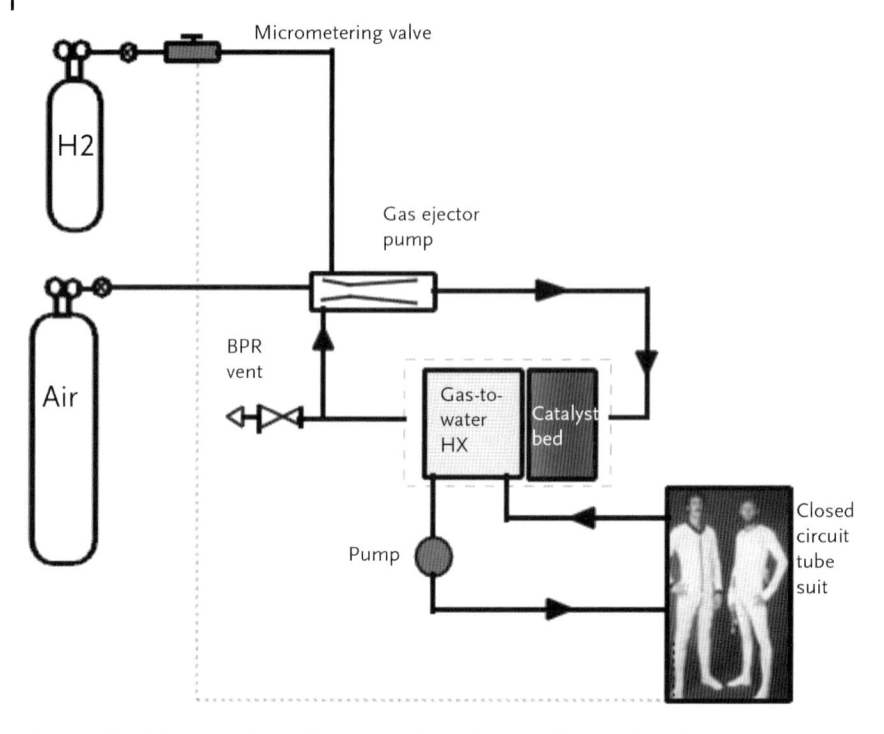

**Figure 4.37** Schematic of multidiver heater circuit (reprinted by permission).

provide a storage pressure in excess of 0.7 MPa at a temperature of 0 °C. The hydrogen catalytic water heater integrates this technology into a heater design to provide whole body thermal protection for surface-supplied diving applications, multidiver applications, and free swimming diving [89].

Pfeifer et al. [92] have evaluated the possibilities of a thermal coupling of alanate hydride tanks with a 1 kW hydrogen-fuelled high-temperature PEM fuel-cell (HT-PEMFC) operating at 160–200 °C. Sodium alanate ($NaAlH_4$) doped with a cerium catalyst was chosen due to the fast kinetics for hydrogen absorption and desorption as well as high gravimetric storage density. The weight of the hydride tank (constructed by TU Hamburg-Harburg) was 20 kg with a heat capacity of 500 J $kg^{-1}$ $K^{-1}$. Figure 4.38a shows the sketch and Figure 4.38b the hydride tank construction.

The kinetics of the hydride decomposition was found to have the major influence on the performance of the fuel-cell system. Due to the decomposition step at above 110 °C, a low-temperature PEM fuel-cell (LT-PEMFC) with a working temperature of 85–90 °C was not suitable to access full potential of complex hydride storage. It was found out that an optimum fuel-cell system temperature of 185 °C and fuel-cell total power of 1 kW fits to a 2 kg sodium alanate hydride tank. Complete alanate discharging was obtained at the minimum operational

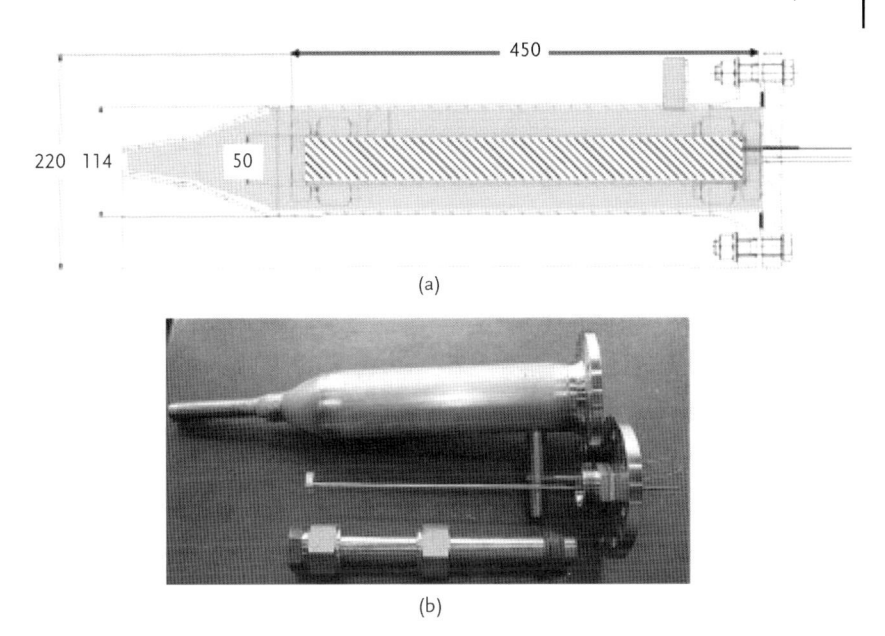

**Figure 4.38** Sketch of the hydride tank construction (a) and photo of the tank (b) [92].

temperature of 120 °C and a cumulative output of 0.8 kWh was reached. The schematic of a complete 1 kW hydrogen-fuelled high-temperature PEM fuel cell with alanate hydride tanks is presented in Figure 4.39.

Botzung et al. [93] presented a hydrogen storage system using metal hydrides (MH) for a combined heat and power (CHP). A 0.1 kg prototype has been build and tested. Hydride technology was selected for high volumetric capacity, low pressures (<0.35 MPa), and low temperatures (<75 °C). The CHP system is shown in Figure 4.40.

For CHP applications a hydride composition $La_{0.90}Ce_{0.05}Nd_{0.04}Pr_{0.01}Ni_{4.63}Sn_{0.32}$ was used with absorption and desorption pressures of 2 and 0.185 MPa, respectively, at 75 °C. Plate-fin rectangular hydrogen storage system was build with the following parameters: effective storage capacity 1.2 $Nm^3$ (106 g), metal hydride weight 10.3 kg, total system weight 54.1 kg, external size: 480×280×116 mm, volume of MH 2.4 L, volume of tank 15.6 L. The metal hydride CHP tank can be seen in Figure 4.41.

The metal hydride tank experimental results with a constant flow rate were as follows: storage of 1.12 $Nm^3$ of hydrogen in 10 kg of metal hydride, absorption rate of 1.15 $Nm^3$ (104 g) in 15 h 30 min (flow rate 75 NL $m^{-3}$) at 65 °C, desorption rate of 1.33 $Nm^3$ of hydrogen (119 g) in 10 h 48 min (flow rate 125 NL $h^{-1}$) at 85 °C. This is represented in Figure 4.42.

The metal hydride tank results with constant pressure have reached storage of 1.12 $Nm^3$ of hydrogen in 10 kg, absorption in 20 min (at pressure 0.35 MPa) at 20 °C, desorption in 1 h 50 min (at pressure 0.15 MPa) at 75 °C.

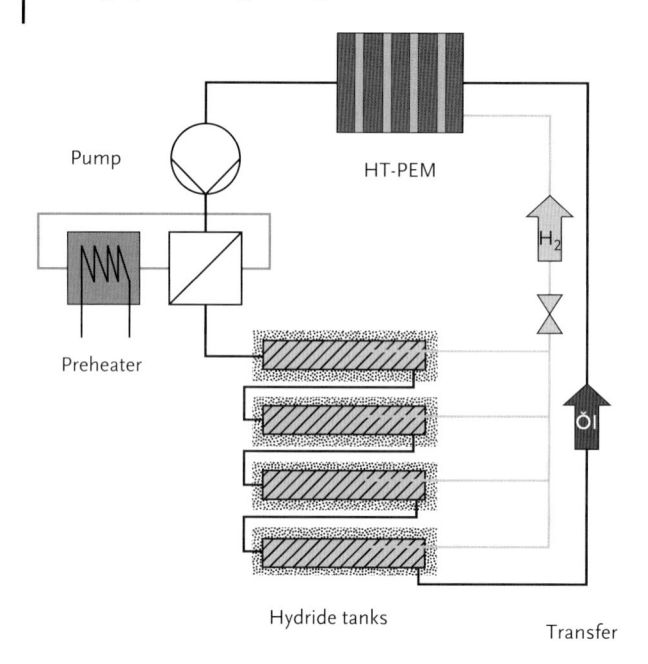

**Figure 4.39** Schematic of the entire system for the thermal coupling alanate hydride tanks and an HT-PEMFC [92].

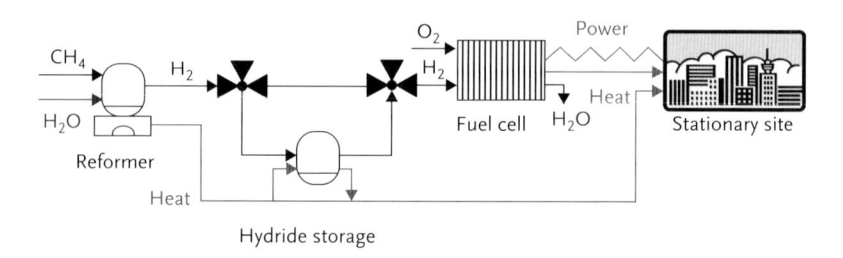

**Figure 4.40** The combined heat and power (CHP) system with a metal hydride tank [93].

For the continuous production of electricity with solar heat power plants the storage of heat is essential. Felderhoff and Bogdanovic [94] reported chemical-based methods for thermal solar energy storage based on metal hydrides. A system based on magnesium hydride/magnesium was suitable for that application because in magnesium hydride the reversible hydrogen storage capacity is 7.6 wt% and in addition a large amount of heat (75 kJ mol$^{-1}$ of H$_2$, 0.9 kW kg$^{-1}$ Mg) is liberated at temperatures between 200 °C and 500 °C upon reaction of magnesium with hydrogen. Several installations realized as laboratory prototypes have been established. First, a facility for a heat accumulator based on magnesium hydride was built and tested at the Max-Planck-Institute for Coal Research

**Figure 4.41** Experimental set up — MH tank [93].

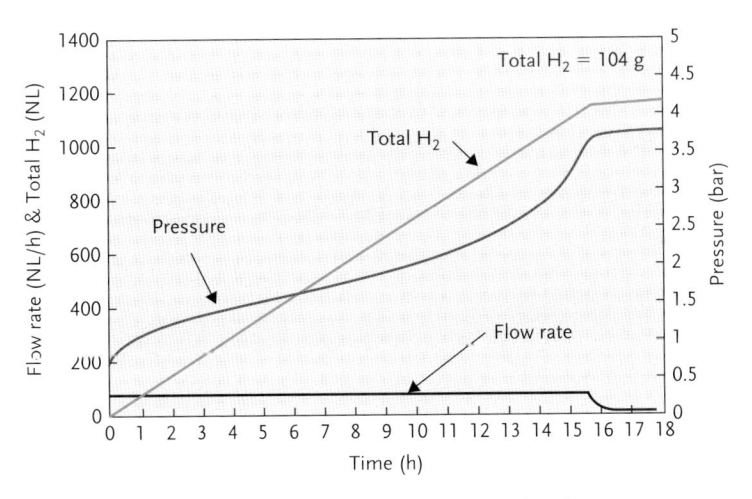

**Figure 4.42** Absorption cycle with a flow rate of 125NL h$^{-1}$ H$^{-1}$, a fluid circulation at 65 °C and a total H$_2$ uptake of 104 g. [93].

and was in operation for more than 1 year. The heat accumulator was able to produce 4 kW of heat up to 90% at the same temperature level. The first model of a small solar power station was built within the frame of a joint project between both Max-Planck-Institutes in Mülheim, the IKE Institute at Stuttgart University and the HTC Solar Company in Lörrach. The main components were a "fixed focus concentrator" for solar radiation, a cavity radiation receiver, a heat pipe system for heat transfer, a Stirling engine, an MgH$_2$ store, and a hydrogen

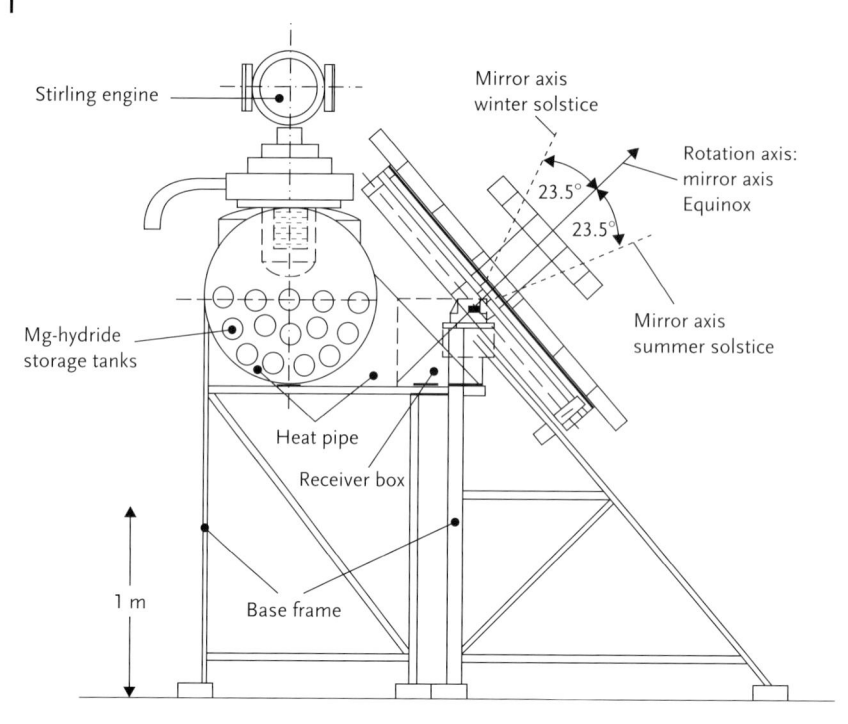

**Figure 4.43** Cross section of the model of the solar power station, with $MgH_2$ heat storage units and fix-focus solar concentrator [94], reprinted by permission (http://www.mdpi.com/1422-0067/10/1/).

pressure or low temperature (LT) hydride store. For supply of the whole installation with solar heat, a solar mirror collector with 6.5 m² area of a light weight type construction was built. The concentrated solar heat is used to drive the Stirling engine and produce power. Simultaneously, the remainder of the solar heat is absorbed by the $MgH_2$ heat store, and the released hydrogen feeds the pressure container or, alternatively, the LT hydride store. In periods of little or no solar irradiation, the $MgH_2$/Mg system cools down and the hydrogen flows back to the $MgH_2$ store and produces heat and then electric power via the Stirling engine [94] (Figure 4.43).

Lohstroh et al. [55] have performed safety test evaluations for the case of a complex hydride tank containing $NaAlH_4$ doped with titanium. Heat exchanger tubes containing 80–100 mL of the material were sealed under an argon atmosphere operating with a burst disc with an opening pressure of 1 MPa. From 130 °C, when hydrogen is first released, the behavior was monitored. The set up and reactor used for the investigation can be seen in Figure 4.44.

It was observed that without an external source of ignition the powder–hydrogen mixture did not ignite spontaneously. A pressure burst of 10 bar led to powder ejection over several meters with spread of a powder cloud (see Figure 4.45).

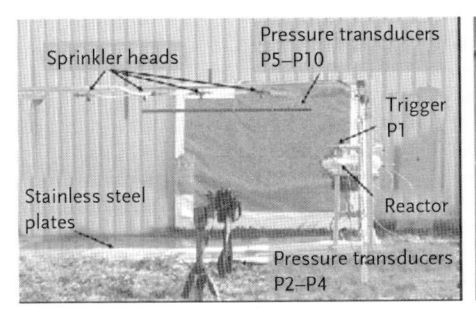

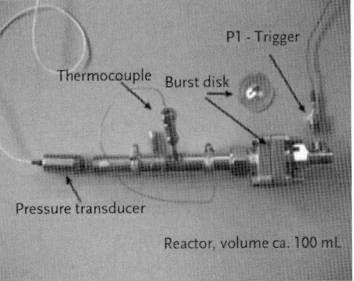

**Figure 4.44** Photograph of the used set-up and reactor [55].

**Figure 4.45** Powder expulsion cloud without external source of ignition [55].

Another experiment by the authors [55] was performed with intentional ignition of the powder, the solid-material–gas mixture started to burn and the temperature of the reaction cloud exceeded 500 °C. According to the authors, the major risk for mixtures of complex hydride and hydrogen originates from the free hydrogen; however, these tests need to be repeated in order to yield reproducible outcomes that represent real conditions.

Ovonics developed and demonstrated a low cost, platinum free 1.5 kW metal hydride fuel-cell (MHFC) stack producing 1.5 kW at 150 W kg$^{-1}$ and 150 W L$^{-1}$. MHFC stacks provide the power of stationary PEMFC stacks at a fraction of the current cost. Currently Ovonics and ETI are working on demonstrating an operating prototype 1 kW fuel-cell power generator and have started commercialization toward a military power system for applications in the 5–20 kW power range [95]. A picture and the specification of 1.5 kW metal hydride fuel cell is shown in Figure 4.46.

The performance curves for such a MHFC stack are shown in Figure 4.47.

German research and development on fuel-cell powered submarines started in 1985. Having successfully proven the tests on 205 class submarines powered by Siemens PEM fuel cells using GfE metal hydride tank for storing hydrogen in

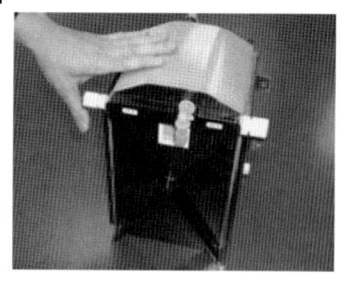

| Power | 1.5 kW |
|---|---|
| Voltage | 28 V |
| Current | 54 A |
| Dimensions (w/o fittings and tabs) | 20 cm × 18.1 cm × 27 cm (7 7/8" × 7 1/8" × 10 5/8") |
| Volume | 9.8 L |
| Weight | 9.8 kg |
| Specific power | 153 W kg$^{-1}$ |
| Power density | 153 W L$^{-1}$ |
| Platinum content | None |
| Fuel | Hydrogen |

**Figure 4.46** Specification and photo of 1.5 kW MHFC developed by Ovonics for military applications (by permission of Energy Conversion Devices Inc. Ovonics).

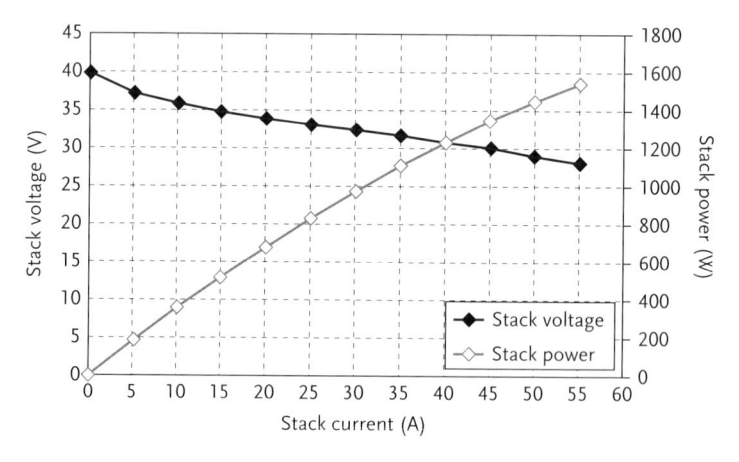

**Figure 4.47** Performance curves for 1.5 kW metal hydride fuel cell stack (by permission of Energy Conversion Devices Inc. Ovonics).

1989, Siemens and HDW (Howaldtwerke Deutsche Werft AG) started a new production line with fuel-cell modules of a higher power of 120 kW in order to optimize the capabilities of the fuel-cell technology. These modules are now implemented on the U214 class export submarines of TKMS [49, 96]. U212A, a nonnuclear submarine, has been developed for use of an extremely silent fuel-cell plant running on a set of nine 34 kW Siemens PEM hydrogen fuel cells. Oxygen is stored in the liquid form (LOX) in tanks. The chosen hydrogen storage method on board U212A is metal hydride. HDW pioneered the construction of these hydride storage cylinders and readied them for series production. During loading, the metal hydride cylinders can be cooled by means of a land-based cooling device. In order to liberate the hydrogen during a submarine's mission, the cylinders are heated by using the fuel-cell cooling water. In this way, the fuel-cells' waste heat is almost completely consumed; therefore, hardly any heat energy has to be transferred to the surrounding sea water. The use of metal hydrides in this context is

**Figure 4.48** Submarine class 212A at sea (by kind permission of Juergen Giefer, Federal Ministry of Defence, BMVg – Rü VII 3).

the safest way to store hydrogen as it almost eliminates any free-flowing gas. In addition, the system is completely maintenance free and can therefore be located in the outer hull area of the submarine. A sophisticated double piping system with respective security measures is installed to prevent leakages in the $H_2$ system onboard. A submarine class 212 A at sea and its cross-section are presented in Figures 4.48 and figure 4.49, respectively. The metal hydride storage system for the U212A submarine is presented in Figure 4.50.

The U212A was contracted for the German Navy in September 2008 and the delivery is expected in 2012/2013.

The schematic of air-independent propulsion (AIP) system used by U212A is shown in Figure 4.51. LOX tanks, metal hydride tubes, and reactant water storage interface with the control system and several auxiliary systems. The electrical energy output is supplied to the main switchboard.

### 4.4.4
**Outlook**

Eberle *et al.* [46] have previously discussed the properties of one of the most promising solid-state storage systems in sodium alanate, $NaAlH_4$, and compared it with liquid and compressed gaseous hydrogen technologies taking into account the targets of the US DOE. The biggest advantage of metal hydrides is their high volumetric storage density ($6-11 \times 10^{22}$ atoms cm$^{-3}$) when compared with liquid

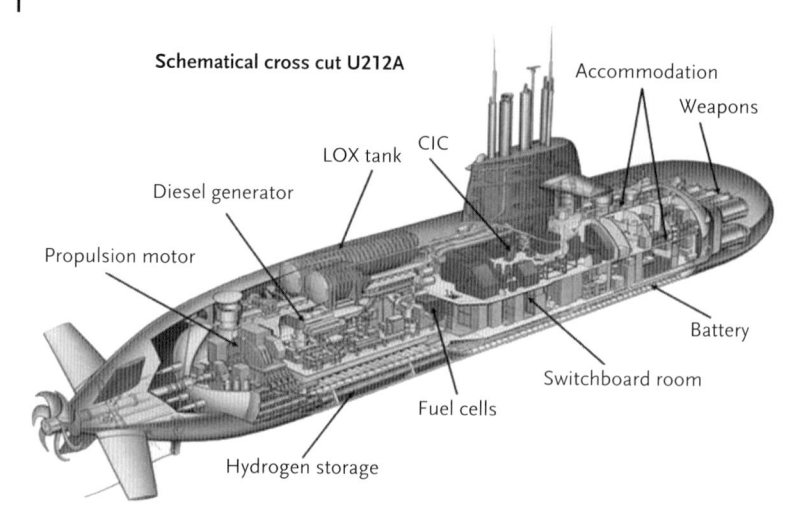

**Figure 4.49** Submarine 212A cross-section [96] (by kind permission of Juergen Giefer, Federal Ministry of Defence, BMVg – Rü VII 3).

**Figure 4.50** Hydrogen storage (based on metal hydride) for supplying hydrogen to Siemens PEM fuel cell [96] (by kind permission of Juergen Giefer, Federal Ministry of Defence, BMVg – Rü VII 3).

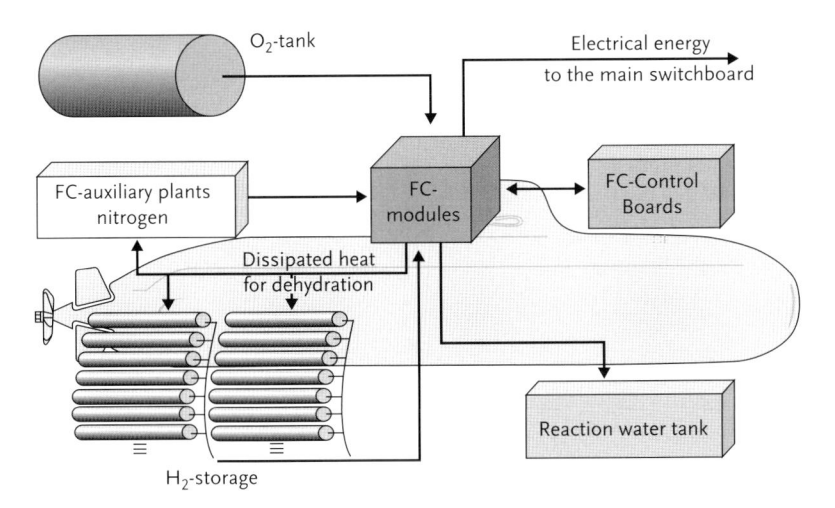

**Figure 4.51** Air independent propulsion system of submarine 212A (by kind permission of Juergen Giefer, Federal Ministry of Defence, BMVg — Rü VII 3).

(20 K) hydrogen ($4.2 \times 10^{22}$ atoms cm$^{-3}$) and compressed hydrogen (70 MPa — $2.3 \times 10^{22}$ atoms cm$^{-3}$, 35 MPa — $1.3 \times 10^{22}$ atoms cm$^{-3}$, and 25 MPa — $1.0 \times 10^{22}$ atoms cm$^{-3}$, respectively). It was stated that sodium alanate NaAlH$_4$ offers currently the best compromise between storage capacity (5.6 material-wt %) and a favorable absorption–desorption behavior.

The authors made clear that weight and volume (including a pressure vessel, thermal insulation, and heat exchanger) can reduce significantly the hydrogen storage density on a system base and stated that it was misleading to compare the system values of liquid or compressed hydrogen with the material-based value for a hydride storage system. It was stated finally that hydrides are not mature enough at the present time for automotive applications [46]. Sakintuna et al. [48] have compared and summarized recent developments of metal hydrides and carbon-based materials taking into account storage capacity, operating conditions, and thermal effects. Although magnesium hydride, MgH$_2$, has a high hydrogen storage capacity of 7.7 wt%, good functional properties and low costs combined with good recycling and high thermodynamic stability results in a relatively high desorption enthalpy corresponding to an unfavorable desorption temperature of 300 °C at 1 bar hydrogen pressure. Magnesium-based metal hydrides have storage capacities varying from 3.5 to 7.3 wt% and desorption temperatures from 180 to 450 °C. Use of complex hydrides for hydrogen storage might be challenging due to their kinetic and thermodynamic limitations. Sodium alanate (a low cost, readily available material, already with a gravimetric capacity of 5.6 wt%) has been proven to have fast kinetics for hydrogen absorption and desorption when doped with a cerium catalyst [92]. According to Sakintuna et al. [48], complex aluminum hydrides cannot be considered as rechargeable hydrogen carriers generally due to their irreversibility and poor kinetics. The gravimetric hydrogen storage capacity is in the range of 2.5–5.0 wt%. In contrast, lithium-based hydrogen storage compounds have gravimetric capacities from 3.1 to 10.0 wt%. $Li_3N$ can theoretically store 10.4 wt% of hydrogen and the adsorbed hydrogen content in the case of $LiBH_4$ can reach up to 18 wt%, but not reversibly. The most popular $LaNi_5$-based alloy of the intermetallic compounds does not exceed 1.4 wt% of hydrogen capacity at moderate temperature.

## References

1 Popeneciu, G., Almasan, V., Goldea, I., Lupu, D., Misan, I., and Arelean, A. (2009) Investigation on a three-stage hydrogen thermal compressor based on metal hydrides. *J. Phys.*, **182**, 0182–012053, Conference Series.

2 Grigoriev, S.A., Djous, K.A., Millet, P., Kalinikov, A.A., Porembskiy, V.I., Fateev, V.N. *et al.* (2008) Characterization of PEM electrochemical hydrogen compressors. Fundamentals and Developments of Fuel Cells Conference FDFC, Nancy, France, 10–12 December.

3 Ludwig, L. (2008) Development of High Efficient Solid State Electrochemical Hydrogen Compressor (EHC).

4 Janßen, H., Emonts, B., and Stolten, D. (2008) Alkalische Hochdruck-Elektrolyse, Status am Forschungszentrum Jülich. NOW Workshop, Ulm, Germany, 7 July.

5 www.ginerinc.com/ (23 January 2010).

6 Millet, P., Ngameni, R., Grigoriev, S.A., Mbemba, N., Brisset, F., and Ranjbari, A. (2009) PEM water electrolyzers: from electrocatalysis to stack development. *Int. J. Hydrogen Energy*, **35** (10), 5043–5052.

7 Barthelemy, H. (2007) Compatibility of Metallic Materials with Hydrogen. Review of the Present Knowledge Proceedings 2nd International Conference on Hydrogen Safety, San Sebastian.

8 www.ca.sandia.gov/matlsTechRef (7 February 2010).

9 Mori, D. and Hirose, K. (2009) Recent challenges of hydrogen storage technologies for fuel cell vehicles. *Int. J. Hydrogen Energy*, **34**, 4569–4574.

10 Crotogino, F. and Hamelmann, R. (2007) Wassertsoffspeicherung in Salzkavernen. Tagungsband 14. Symposium zur Nutzung regenerativer Energiequellen und Wasserstofftechnik Nov, Stralsund, Germany, 8–10 November.

11 www.eihp.org/ (7 February 2010).

12 STORHY (2008) Publishable Final Activity Report. Hydrogen Storage Systems for Automotive Application.

13 Altmann, M., Gaus, S., Landinger, H., Stiller, C., and Wurster, R. (2001) Wasserstofferzeugung in offshore Windparks, Killer Kriterien, grobe Auslegung und Kostenabschätzung www.Lbst.de

14 Quack, H. (2010) Die Schlüsselrolle der Kryotechnik in der Wasserstoff-Energiewirtschaft, www.tu-dresden.de/die_tu_dresden/fakultaeten/fakultaet_maschinenwesen/iet/kkt/forschung/0k_bis_50k/schluesselrolle_quack, (22 May).

15 Valenti, G. and Macchi, E. (2008) Proposal of an innovative, high-efficient, large-scale hydrogen liquifier. *J. Hydrogen Energy*, **33**, 3116–31211

16 Barclay, J.A. (2008) *Active Magnetic Regenerative Liquefier*, www.hydrogen.energy.gov/pdfs/progress11/iii_17_barclay_2011.pdf, (22 May).

17 Utaki, T., Kamiya, K., Nakagawa, T., Yamamoto, T.A., and Nurmazawa, T. (2006) Research on a Magnetic Refrigeration Cycle for Hydrogen Liquification. 14th International Cryocoolers Conference, Annapolis, MD, 14–16 June.

18 Waichi, I. (2003) Magnetic refrigeration technology for an international clean energy network using hydrogen energy (WE-NET). *Int. J. Hydrogen Energy*, **28**, 559–567.

19 Wollan, J.J., Swift, G.W., Backhaus, S., and Gardner, D.L. (2002) Development of a thermo acoustic natural gas liquefier. Proceedings of AIChE Meeting, New Orleans.

20 Voth, R.O. (1978) Producing Liquid–Solid Mixtures of Hydrogen Using an Auger. NBSIR 78–875.

21 Brunnhofer, K. (2003) Slush hydrogen-Kraftstoff für die Zukunft. *DKV Tagungsbericht*, **30**, 45–62.

22 Fuura, T., Tsunokaka, S., Hirotani, R., Hashimoto, T., Akai, M., Watanabe, S. et al. (2004) Development of a $LH_2$ vehicle tank boil-off gas recovery system using hydrogen storage alloys. 15th World Hydrogen Energy Conference, Yokohama, Japan, 27 June–2 July.

23 Hedayat, A., Hastings, L.J., Bryant, C., and Plachta, D.W. (2002) Large scale demonstration of liquid hydrogen storage with zero boiloff. Proceedings of the Cryogenic Engineering Conference, Vol. **47**, Grenoble, France.

24 Ho, S.H. and Rahman, M.M. (2008) Three-dimensional analysis for liquid hydrogen in a cryogenic storage tank with heat pipe–pump system. *Cryogenics*, **48**, 31–41

25 Reijerkerk C.J.J. (2004) Potential of cryogenic hydrogen storage in vehicles, National Hydrogen Association Conference, 26–30 April.

26 Michler, T., Naumann, J., and Leistner, W. (2007) Evaluation of braze materials used for automotive liquid hydrogen tanks with special respect to hydrogen embrittlement and low temperature toughness. *Materialwiss. Werkstofftech.*, **38**, 43–50.

27 Eichberger, B. (2008) The electronic requirements on automotive hydrogen storage. FISTA. World Automotive Congress., Munich, Germany, 14–19, September.

28 Abe, A., Nakamura, M., Sato, I., Uetan, H., and Fujitani, T. (1998) Studies of the large-scale transportation of liquid hydrogen. *Int. J. Hydrogen Energy*, **23** (2), 115–121.

29 www.STORHY.net (7 February 2010).

30 Friedel, M., Heinrich, F., and Laurent, A. (2006) Liquid hydrogen technologies for mobile use. Proceedings of the WHEC, Lyon, France, 16/13–16 June.

31 Salvador, M.A., Gene, B., Francisco, E.L., Elias, L.O., Tim, R., Vernon, S. et al. (2008) Automotive Cryogenic capable pressure vessels for compact, high dormancy $(L)H_2$ storage. FY Annual Report.

32 Knorr, H., Held, W., Prümm, W., and Rüdiger, H. (1998) The MAN hydrogen propulsion systems for city buses. *Int. J. Hydrogen Energy*, **23** (3), 201–208.

33 Reinhard, F. (2001) Cryoplane Flugzeuge mit Wassertsoffantrieb. DLRG-Bezirksgruppe Hamburg.

34 Pohl, H.W. and Valentin, V.M. (1997) Hydrogen in future civil aviation. *Int. J. Hydrogen Energy*, **22**(10/11), 1061–1069.

35 Sutton, G. (2005) *History of Liquid Propellant Rocket Engines*. American Institute of Aeronautics and Astronautics, Reston, VA

36 Giovanni, C., Claudio, C., and Coriolano, S.(2005) Liquid $H_2$ storage for small-size solar power plants. Proceedings of GT 2005, GT 2005–68038, Reno-Tahoe, NV, 6–9 June.

37 Schlapbach, L. and Zuettel, A. (2001) Hydrogen-storage materials for mobile application. *Nature*, **414**, 353–358.

38 Zuettel, A. (2004) Hydrogen storage methods. *Naturwissenschaften*, **91**, 157–172.

39 Sandrock, G. (1999) A panoramic overview of hydrogen storage alloys from a gas reaction point of view. *J. Alloys Compd.*, **293–295**, 877–888.

40 Sandrock, G. and Thomas, G. (2001) The IEA/DOE/SNL on-line hydride databases. *Appl. Phys. Mater. Sci. Process.*, **2** (72), 153–155.

41 Hydride Information Center for Metal-Hydrogen Systems, Properties, Applications and Activities. Sandia National Laboratories. http://hydpark.ca.sandia.gov/, September 2007.

42 Zijlstra, H. and Westendorp, F.F. (1969) Influence of hydrogen on the magnetic properties of $SmCo_5$. *Solid-State Commun.*, **7** (12), 875–859.

43 van Fucht, J.H.N., Kuijpers, F.A., and Bruning, C.A.M. (1970) Reversible room-temperature absorption of large quantities of hydrogen by intermetallic compounds. *Philips Res. Rep.*, **25**, 133–141.

44 Shan, X., Payer, J.H., and Jennings, W.D. (2009) Mechanism of increased performance and durability of Pd-treated metal hydriding alloys. *Int. J. Hydrogen Energy*, **34**, 363–369.

45 Soerensen, B. (2005) *Hydrogen and Fuel Cells: Emerging Technologies and Applications*. Elsevier Academic Press, London, UK, ISBN: 0-12-655281-9.

46 Eberle, U., Arnold, G., and von Helmolt, R. (2006) Hydrogen storage in metal–hydrogen systems and their derivatives. *J. Power Sources*, **154**, 456–460.

47 Shang, C.X., Bououdina, M., Song, Y., and Guo, Z.X. (2004) Mechanical alloying and electronic simulations of ($MgH_2$+M) systems (M=Al, Ti, Fe, Ni, Cu and Nb) for hydrogen storage. *Int. J. Hydrogen Energy*, **29**, 73–80.

48 Sakintuna, B., Weinberger, B., Lamari-Darkrim, F., Hirscher, M., and Dogan, B. (2006) Comparative study of hydrogen storage efficiency and thermal effects of metal hydrides vs. carbon materials. WHEC, Lyon, France, 16/13–16 June.

49 Guenther, V. and Otto, A. (1999) Recent developments in hydrogen storage applications based on metal hydrides. *J. Alloys Compd.*, **293–295**, 889–892.

50 Liang, G., Huot, J., Boily, S., Van Neste, A., and Schulz, R. (2000) Hydrogen storage in mechanically milled Mg–$LaNi_5$ and $MgH_2$–$LaNi_5$ composites. *J. Alloys Compd.*, **297**, 261–265.

51 Martinez-Franco, E., Oelerich, W., Klassen, T., and Bormann, R. (2002) Light metal hydrides for hydrogen storage in the zero emission vehicles. International Congress on Advanced Materials, Materials Week 2001 – Proceedings, ed. Werkstoffwoche-Partnerschaft GbR, publ. Werkstoff-Informationsgesellschaft mbH, Frankfurt.

52 Dehouche, Z., Klassen, T., Oelerich, W., Goyette, J., Bose, T.K., and Schulz, R. (2002) Cycling and thermal stability of nanostructured $MgH_2$–$Cr_2O_3$ composite for hydrogen storage. *J. Alloys Compd.*, **347**, 319–323.

53 Dehouhe, Z., Djaozandry, R., Hout, J., Boily, S., Goyette, J., Bose, T.K. et al. (2000) Influence of cycling on the thermodynamic and structure properties of nanocrystalline magnesium based hydride. *J. Alloys Compd.*, **305**, 264–271.

54 Bogdanovic, B., Brand, R.A., Marjanovic, A., Schwickardi, M., and Tolle, J. (2000) Metal-doped sodium aluminium hydrides as potential new hydrogen storage materials. *J. Alloys Compd.*, **302**, 36–58.

55 Lohstroh, W., Fichtner, M., and Breitung, W. (2009) Complex hydrides as solid storage materials: first safety tests. *Int. J. Hydrogen Energy*, **34** (14), 5981–5985.

56 Wang, J., Ebner, A.D., Edison, K.R., Ritter, J.A., and Zidan, R. (2003) Performance of metal-doped sodium aluminium hydride for reversible hydrogen storage. *Fuel Chem. Div. Prepr.*, **48** (1), 279–280.

57 Ritter, J.A., Riggleman, R.A., Ebner, A.D., and Zidan, R. (2002) Development of reversible hydrogen storage material from metal-doped sodium aluminum hydride. *Fuel Chem. Div. Prepr.*, **47** (2), 786–787.

58 Wang, J., Ebner, A.D., Edison, K.R., Ritter, J.A., and Zidan, R. (2003) Metal-doped sodium aluminium hydride as a reversible hydrogen storage material. Adsorption Science and Technology, Proceedings of the Third Pacific Basin Conference, Kyongju, Korea, pp. 301–305.

59 Bogdanovic, B., Felderhoff, M., Pommerin, A., Schuth, F., and Spielkamp, N. (2006) Advanced hydrogen storage materials based on Sc, Ce and Pr doped $NaAlH_4$. *Adv. Mater.*, **18**, 1198–1201.

60 Jensen, C.M., Zidan, R., Mariels, N., Hee, A., and Hagen, C. (1999) Advanced titanium doping of sodium aluminum hydride: segue to a practical hydrogen storage material?. *Int. J. Hydrogen Energy*, **24** (5), 461–465.

61 Bogdanovic, B., Eberle, U., Felderhoff, M., and Schuth, F. (2007) Complex aluminum hydrides. *Scr. Mater.*, **56**, 813–816.

62 Graetz, J. and Reilly, J.J. (2007) Kinetically stabilized hydrogen storage materials. *Scr. Mater.*, **56**, 835–839.

63 Vajo, J.J. and Olson, G.L. (2007) Hydrogen storage in destabilized chemical systems. *Scr. Mater.*, **56**, 829–834.

64 Chen, P., Xiong, Z., Luo, J., Lin, J., and Tan, K.L. (2002) Interaction of hydrogen with metal nitrides and imides. *Nature*, **420**, 302–304.

65 Gregory, D.H. (2008) Imides and amides as hydrogen storage materials in *Solid-State Hydrogen Storage, Materials and Chemistry* (ed. G. Walker), Woodhead Publishing Limited and CRC Press LLC, Cambridge, UK.

66 Ichigawa, T., Leng, H.Y., Isobe, S., Hanada, N., and Fujii, H. (2006) Recent development on hydrogen storage properties in metal–N–H systems. *J. Power Sources*, **159**, 126–131.

67 Fuji, H. and Ichikawa, T. (2004) *Lithium, Graphite and Magnesium-Based Hydrogen Storage Systems Prepared by Mechanical Milling*. UK–Japan International Workshop Solid-State Hydrogen Storage, Birmingham, UK.

68 Xiong, Z., Wu, G., Hu, J., Chen, P., Luo, W., and Wang, J. (2006) Investigations on hydrogen storage over Li–Mg–N–H complex – the effect of compositional changes. *J. Alloys Compd.*, **417**, 190–194.

69 Xiong, Z., Hu, J., Wu, G., and Chen, P. (2005) Hydrogen absorption and desorption in Mg–Na–N–H system. *J. Alloys Compd.*, **395**, 209–212.

70 Dam, B., Gremaud, R., Broedersz, C., and Griessen, R. (2007) Combinatorial thin film methods for the search of new light weight metal hydrides. *Scr. Mater.*, **56**, 853–858.

71 Nakamori, Y. and Orimo, S. (2008) Borohydrides as hydrogen storage materials in *Solid-State Hydrogen Storage, Materials and Chemistry* (ed. G. Walker), Woodhead Publishing Limited and CRC Press LLC, Cambridge, UK.

72 Zuettel, A., Borgschulte, A., and Orimo, S.-I. (2007) Tetrahydroborates as new hydrogen storage materials. *Scr. Mater.*, **56**, 823–828.

73 Leon, A. (2008) Hydrogen storage in *Hydrogen Technology, Mobile and Portable Applications* (ed. A. Leon), Springer-Verlag, Berlin, Heilderberg.

74 Cheng, H., Scott, K., and Lovell, K. (2006) Material aspects of the design and operation of direct borohydride. *Fuel Cells*, **6** (5), 367–375.

75 Ma, J., Choudhury, N.A., and Sahai, Y. (2010) A comprehensive review of direct borohydride fuel cells. *Renewable Sustainable Energy Rev.*, **14** (1) 183–199.

76 Lakeman, J., Rose, A., Pointon, K.D., Browning, D.J., Lovell, K.V., Waring, S.C. et al. (2006) The direct borohydride fuel cell for UUV propulsion power. *J. Power Sources*, **162** (2), 765–772.

77 Liu, B.H., Li, Z.P., and Suda, S. (2008) Development of high-performance planar borohydride fuel cell modules for portable applications. *J. Power Sources*, **175** (1), 226–231.

78 Finkelshtain, G. (2008) Commercialization of Direct Borohydride Fuel Cells. Fuel Cell Seminar and Exposition, Phoenix, AZ, USA, 27–30 October.

79 Wee, J.-H., Lee, K.-Y., and Kim, S.H. (2006) Sodium borohydride as the hydrogen supplier for proton exchange membrane fuel cell systems. Fuel Process. Tech., **87** (9) 811–819.

80 Marrero-Alfonso, E., Gray, J.R., Davis, T.A., and Matthews, M.A. (2007) Hydrolysis of sodium borohydride with steam. *Int. J. Hydrogen Energy*, **32** 4717–4722.

81 Soloveichik, G.L., Gao, Y., Rijssenbeek, J., Andrews, M., Kniajanski, S., Bowman, R.C., Jr et al. (2009) Magnesium borohydride as a hydrogen storage material: properties and dehydrogenation pathway of unsolvated $Mg(BH_4)_2$. *Int. J. Hydrogen Energy*, **34**, 916–928.

82 Chlopek, K., Frommen, C., Leon, A., Zabara, O., and Fichtner, M. (2008) Synthesis and properties of magnesium tertahydroborate, $Mg(BH_4)_2$. *J. Mater. Chem.*, **17**, 3496–3503.

83 Eigen, N., Keller, C., Dornheim, M., Klassen, T., and Borman, R. (2007) Industrial production of light hydrides for hydrogen storage. *Scr. Mater.*, **56**, 847–851.

84 Eigen, N., Kunowsky, M., Klassen, T., and Bormann, R. (2007) Synthesis of $NaAlH_4$-based hydrogen storage material using milling under low

pressure hydrogen atmosphere. *J. Alloys Compd.*, **430** (1–2) 350–355.
85 Sheridan, J.J., Eisenberg, F.G., Greskovich, E.J., Sandrock, G.D., and Huston, E.L. (1983) Hydrogen separation from mixed gas streams using reversible metal hydrides. *J. Less-Common Met.*, **89** (2) 447–455.
86 Christodoulou, C. (2009) Metal Hydrides Application at HyStore. Cyprus, July.
87 Lynch, F. (2009) *Metal Hydride Containers and Application*, Hydrogen Components, Inc., Colorado, USA.
88 Mosset, P. Spectratime, September 2009.
89 Nuckols, M.L., Wood-Putnam, J.L., and VanZandt, K.W. (2009) New developments in thermal technologies for hot and cold exposures. Special Operations Forces Industry Conference (SOFIC), Tampa, FL, 1–4 June.
90 Lindler, K.W. and Nuckols, M.L. (2002) Hydrogen fuel storage/delivery for a diver heater using hydrogen catalytic reactions. *Oceans*, **2** (29–31), 919–924.
91 Nuckols, M.L. (2007) Reconfigurable hydrogen transfer heating/cooling system. U.S. Patent 7,213,409.
92 Pfeifer, P., Wall, C., Jensen, O., Hahn, H., and Fichtner, M. (2009) Thermal coupling of a high temperature PEM fuel cell with a complex hydride tank. *Int. J. Hydrogen Energy*, **38** (4), 3457–3466.
93 Botzung, M., Chaudourne, S., Gillia, O., Perret, C., Latroche, M., Percheron-Gueganb, A. *et al.* (2008) Simulation and experimental validation of a hydrogen storage tank with metal hydrides. *Int. J. Hydrogen Energy*, **33**, 98–104.
94 Felderhoff, M. and Bogdanovic, B. (2009) High temperature metal hydrides as heat storage materials for solar and related applications. *Int. J. Mol. Sci.*, **10**, 325–344.
95 Fok, K., English, N.J., Privette, R.M., Wang, H., Wong, D.F., Lowe, T.D. *et al.* (2008) Powering up metal hydride fuel cells for military applications. 2008 Fuel Cell Seminar and Exposition, Phoenix, AZ, USA, 27–30 October.
96 Giefer, J.W. , Sr (2008) German submarine class 212A – fuel cell powered air independent propulsion boat. 2008 Fuel Cell Seminar and Exposition, Phoenix, AZ, USA, 27–30 October.

# 5
# Chemical Storage

## 5.1
## Introduction

In previous chapters, hydrogen storage as compressed gas, liquid, and in metal hydrides have been described. Basically, the techniques based on liquid hydrogen have a higher gravimetric and volumetric storage density when compared to other methods; however, a big amount of energy is needed to liquefy gaseous hydrogen and in addition, there are evaporation losses over a long time and a boil-off effect. Compressed hydrogen is easier to handle but it also has a very low volumetric storage density. Besides metal hydrides, several storage materials such as hydrocarbons, ammonia borane, and sodium borohydride are under intensive research and development. A potential hydrogen storage material must fulfill some technical requirements regarding high gravimetric and volumetric storage capacities, reversibility of the storage, cost-effective, and safety conscious performance.

Preliminary results show that ammonia borane can be thermally decomposed with very high hydrogen yield, but the reactions are not reversible and regeneration step is required. In addition, they are toxic and could possibly contaminate fuel cell catalysts.

Organic compounds such as hydrocarbons have been studied as a substances able to store hydrogen. Dehydrogenation of hydrocarbons seems to be an attractive method for hydrogen generation because only hydrogen and dehydrogenated hydrocarbons are produced in this reaction. The absence of carbon oxides eliminates the need for purification step. Another important aspect is that dehydrogenation is a reversible process which means that dehydrogenated products of cycloalkanes can be hydrogenated back to the starting cycloalkanes. In order to obtain hydrogen from hydrocarbons, their chemical binding energy must be broken. Certain interest in hydrocarbons has been risen by automotive industry like Daimler-Benz when thinking about electric cars running on fuel cells. The question is which hydrocarbon shall be considered as a source of hydrogen to feed the fuel cell. It is stated that Mercedes will use methanol, but Ford will remain with gasoline [1]. Gasoline is available everywhere and the distribution is very well established, but the advantage of methanol in comparison with gasoline is that the reforming takes place at lower temperature.

*Hydrogen Storage Technologies: New Materials, Transport, and Infrastructure*, First Edition.
Agata Godula-Jopek, Walter Jehle, and Jörg Wellnitz.
© 2012 Wiley-VCH Verlag GmbH & Co. KGaA.
Published 2012 by Wiley-VCH Verlag GmbH & Co. KGaA

## 5.2
## Materials and Properties

Hydrocarbons containing hydrogen include aromatic hydrocarbons like cyclohexane, methylcyclohexane, decaline, and heteroaromatics, such as carbazole or benzoquinone. Hydrogen storage capacities of such hydrocarbons are in the range of 6–8 wt%. General properties of various hydrocarbons are shown in Table 5.1.

Hydrocarbons are able to store hydrogen due to hydrogenation reaction of each respective bond, C=C, C=O, C=N with 1 mol of hydrogen. A useful process of hydrogen supply is the catalytic dehydrogenation of cycloalkanes (hydrogen supply) and hydrogenation of their corresponding organic liquid carriers (hydrogen storage). These reversible dehydrogenation–hydrogenation system of cycloalkanes is one of the potential to store and transport hydrogen. Exothermic hydrogenation of aromatic compounds takes place at temperatures of 100–250 °C and pressures of 1–10 bar, whereas endothermic dehydrogenation of cycloalkanes occurs by heating them above 250 °C [2]. The reactions are promoted by the catalysts based on platinum and nickel. Cycloalkanes can store 6.5 wt% of hydrogen and 60–62 kg $H_2$ $m^{-3}$ [3] and 1 mol of cycloalkane has a potential to transport from 3 to 6 mol of hydrogen.

For example, hydrogenation of benzene ($C_6H_6$) gives cyclohexane ($C_6H_{12}$) according to the following equation:

$$C_6H_6 + 3H_2 \rightarrow C_6H_{12} + 206 \text{ kJ mol}^{-1} (68.6 \text{ kJ mol}^{-1} H_2) \quad (5.1)$$

As a result of hydrogenation of naphthalene, a decalin is formed; the opposite reaction is of course the dehydrogenation of a decalin (Eq. (5.2)).

$$C_{10}H_8 + 5H_2 \rightarrow C_{10}H_{18} + 320.1 \text{ kJ mol}^{-1} (64.02 \text{ kJ mol}^{-1} H_2) \quad (5.2)$$

Hydrogen storage capacities of several cycloalkanes (in wt%) are as following: cyclohexane 7.2 (27.77 mol $L^{-1}$), methylcyclohexane 6.2 (23.29 mol $L^{-1}$), tetralin 3.0 (14.72 mol $L^{-1}$), cis-decalin 7.3 (32.4 mol $L^{-1}$), trans-decalin 7.3 (31.46 mol $L^{-1}$), cyclohexylbenzene 3.8 (17.63 mol $L^{-1}$), bicycloxehyl 7.3 (32.0 mol $L^{-1}$), cis-syn-1-methyldecalin 6.6 (29.31 mol $L^{-1}$), and trans-anti-1-methyldecalin 6.6 (28.52 mol $L^{-1}$) [3].

**Table 5.1** General properties of various hydrocarbons.

| Hydrocarbon | Chemical formula | Molar mass (g mol$^{-1}$) | Mass fraction hydrogen (%) | Melting point (°C) | Boiling point (°C) |
|---|---|---|---|---|---|
| Benzene | $C_6H_6$ | 78.11 | 7.6 | 5.5 | 80.1 |
| Toluene | $C_7H_8$ | 92.14 | 8.6 | −93 | 110.6 |
| Naphthalene | $C_{10}H_8$ | 128.17 | 6.2 | 80.26 | 218 |
| Tetralin | $C_{10}H_{12}$ | 132.20 | 9.07 | −35.8 | 206 |
| 1-Methyldecalin | $C_{11}H_{10}$ | 142.20 | 7.03 | −22 | 240–243 |
| Biphenyl | $C_{12}H_{10}$ | 154.21 | 6.48 | 69.2 | 255 |

Table 5.2 summarizes the data on hydrogen content (wt% and mol $L^{-1}$) and volume capacity showing the dehydrogenation reaction of organic hydrides.

It is said that among organic liquid carriers, methylcyclohexane, bicycloxehyl, and alkyldecalin are recommended for the efficient hydrogen storage and supply and its delivery [2]. In case of cycloalkanes as the hydrogen storage media, important issues are the temperature of the dehydrogenation process, dehydrogenation rates, reactor types for heat transfer, and the catalysts supporting efficient dehydrogenation process. The advantage of cycloalkanes dehydrogenation is that the reactions of dehydrogenation and hydrogenation occur in a reversible way producing hydrogen and aromatics without by-products like coke, polymers formation, and alkanes. The reactants and the products are recyclable in hydrogen supply and storage. The process of dehydrogenation of cycloalkanes is endothermic and the chemical equilibrium is favorable toward dehydrogenation at higher temperatures under steady-state operation in the gas phase, which may be a disadvantage as the $H_2$ storage media. With the reaction pair of decalin dehydrogenation/naphthalene hydrogenation over platinum catalysts supported on activated carbon, even 7.3 wt% of hydrogen storage can be reached (which is 64.8 kg $H_2$ $m^{-3}$) [4]. This is being close to the target storage value set by the US DOE for 2010 of 6.5 wt% (45.0 kg $H_2$ $m^{-3}$) [5]. The dehydrogenation of hydrocarbons for producing hydrogen in different reactors, ways, and using a number of metal catalysts is described below.

## 5.3
## Hydrogen Storage in Hydrocarbons

Efficient hydrogen generation of 3.0 wt% (28.2 kg $H_2$ $m^{-3}$) from tetralin dehydrogenation/naphthalene hydrogenation with "superheated liquid-film-type catalysis" has been proposed for operating proton exchange membrane (PEM) fuel cell. It was noted that even if the hydrogen storage capacity of tetralin in comparison to decalin is nearly twice lower (7.3 to 3.0 wt%), the dehydrogenation rate of tetralin and hydrogenation of naphthalene to tetralin is four to six times faster

Table 5.2 Several cyclic hydrocarbons as organic liquid carriers, reaction, values on hydrogen content, and volume capacity.

| Organic hydride | Aromatic | Reaction | $H_2$ content (wt%) | Mol $L^{-1}$ | Volume capacity ($LH_2$ $L^{-1}$) |
|---|---|---|---|---|---|
| Cyclohexane | Benzene | $C_6H_{12} \rightarrow C_6H_6 + 3H_2$ | 7.2 | 27.77 | 620 |
| Methylcyclohexane | Toluene | $C_7H_{14} \rightarrow C_7H_8 + 3H_2$ | 6.2 | 23.29 | 528 |
| Tetralin | Biphenyl | $C_{10}H_{12} \rightarrow C_{10}H_8 + 2H_2$ | 3.0 | 14.67 | 328 |
| Decalin | Naphtalene | $C_{10}H_{18} \rightarrow C_{10}H_8 + 5H_2$ | 7.3 | 32.4 | 710 |
| Bicyclohexyl | Biphenyl | $C_{12}H_{22} \rightarrow C_{12}H_{10} + 6H_2$ | 7.23 | 32.0 | 700 |

than the dehydrogenation rates for decalin at 240 °C [6]. Such high rates (conversion rate of tetraline was 95%) were obtained with carbon-supported nanosize platinum catalyst, Pt/C 5 wt. metal %, in the superheated liquid-film state in the batchwise reactor under boiling (210 °C) and refluxing (5 °C) conditions. Interesting observation was made after testing several catalysts, namely by changing catalyst from expensive platinum into carbon-supported nickel–ruthenium composite catalyst (Ni–Ru/C), and having the same process conditions, hydrogen rates obtained at 240 °C on nickel–ruthenim catalyst were similar to those obtained at 210 °C on platinum catalyst, making the process cost-effective [6]. Nonsteady spray pulse operation in thin liquid-film reactor to dehydrogenate 2-propanol on a plate $Pt/Al_2O_3/Al$ catalyst at 368 K (95 °C) was proposed by Kobayashi et al. [7]. In this process, it was possible to keep fresh catalyst surface by quick repeat between reactant feed and drying the surface to remove the reactant and products from the system, allowing to increase the reaction rate by removing adsorbed products from the catalyst before reaching the adsorption equilibrium (regeneration step). Dehydrogenation of 2-propanol has 10 times higher reaction rate than for conventional gas phase and liquid phase steady operation, proving that the nonsteady pulse operation is effective.

Through optimization of the conditions of cycloalkanes dehydrogenation process like temperature, reactant/catalyst ratio, and the support, high rate of hydrogen supply can be achieved. Japanese team developed a new type of spray-pulse mode reactor with different platinum-based catalysts (Pt and Pt–Re, Pt–Rh, Pt–Pd) for hydrogen supply from cyclohexane, methylcyclohexane, tetralin, and decalin [8]. The experimental apparatus for the nonsteady spray-pulse operation is shown in Figure 5.1.

In this reactor, the reactant in a liquid state is supplied over the catalyst from the nozzle. Catalyst is located on the heater in glass vessel with a cooling condenser to separate hydrogen, vaporized reactants, and dehydrogenated products. By feeding liquid reactant in pulses, under alternating wet and dry conditions, the production rates of hydrogen over Pt catalysts were 15–25 higher than the rates obtained in conventional gas phase and 100–1600 higher than in liquid phase reactions [2,8]. The rates of hydrogen production were dependent on the rates of the reactant feed, the highest was obtained for cyclohexane. The temperature of catalyst surface decreased with the increase of the reactant feed for all investigated hydrocarbons in the range of 287–375 °C [8]. Spray-pulsed reactor was used for the dehydrogenation of cyclohexane over Ni monometallic, Pt monometallic, and Ni–Pt bimetallic catalysts supported on activated carbon cloth [9]. It was observed that Ni monometallic catalysts supported on activated carbon with high selectivity for dehydrogenation of cyclohexane to benzene can be promising for hydrogen storage. When Ni catalyst was promoted by addition of 0.5 wt% of Pt, the activity was highest for hydrogen production rate and conversion of cyclohexane. A synergistic effect was observed for bimetallic Ni–Pt catalyst (about 60 times higher hydrogen production rates) [9]. Biniwale et al. [10,11] used spray-pulsed reactor for reforming of iso-octane ($C_8H_{18}$) at temperatures 600–700 °C over several bimetallic catalysts, Ni–Mn, Rh–Ce, and Ni–W. At 700 °C, the maximum hydrogen

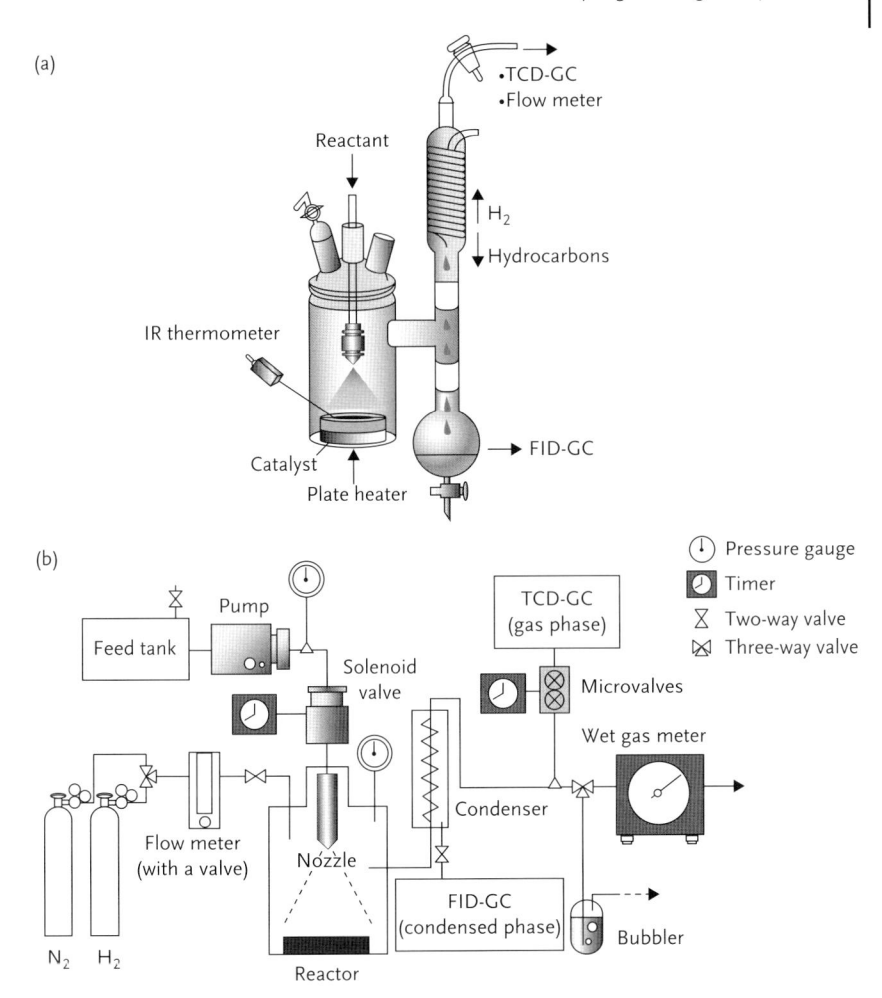

**Figure 5.1** Experimental setup of the spray-pulsed reactor for dehydrogenation reaction (a) main parts, (b) flow reactor system for dehydrogenation under nonsteady spray-pulse operation [8], reprinted by permission.

production rate over Ni–Mn catalyst was 68.27 mmol g$^{-1}$ min$^{-1}$ (partial pressure of iso-octane was 3.13 kPa with corresponding steam partial pressure 18.38 kPa). At the same temperature, hydrogen production rate over Rh–Ce catalyst was much higher, 176.50 mmol g$^{-1}$ min$^{-1}$ (98% of iso-octane conversion was reached) [10]. The alternate wet and dry conditions improved the catalyst–reactant contact and enhanced the hydrogen production rate.

High rate of hydrogen production in the dehydrogenation of cyclic hydrocarbons was obtained under "wet–dry multiphase conditions" over several platinum-containing catalysts supported on active carbon [12]. A series of catalysts have been tested in batch-type reactions, Pt–Me/active carbon, where Me was Mo, W, Re, Ru, and Pd. The catalyst surface was heated to the temperatures higher (195–400 °C) than the boiling point of respective cyclic hydrocarbons (80–197 °C); therefore, nearly all of the reactant was vaporized at the beginning of the reaction. The schematic of "wet–dry multiphase condition" is presented in Figure 5.2. The reaction takes place before all of the liquid reactant is vaporized. Unconverted reactant and the products, which are dehydroaromatized hydrocarbon and hydrogen, are vaporized due to high temperature, and the catalyst is dried and cleaned.

Due to a very dynamic change of phase, high reaction rates in the dehydrogenation process of cycloalkanes were obtained by optimizing the conditions of the reaction, such as reaction temperature, type of catalyst support, initial feeding amount of reactant, and use of bimetallic catalysts. Temperatures above 195 °C are favorable for the dehydrogenation because the temperature of the catalyst is between the boiling point of the reactants and the external heating temperature; therefore, the reaction rate is higher than the reaction rate in the liquid phase. Bimetallic catalysts (Pt–Mo/petroleum cokes carbon, Pt–W/petroleum cokes carbon, and Pt–Re/petroleum cokes carbon) showed higher reactivity than monometallic Pt/petroleum cokes carbon for the dehydrogenation of cyclohexane at 195 °C [12].

Catalysts systems with Pt/Sn on a neutral support have been found to show high dehydrogenation selectivity and catalyst stability for dehydrogenation of light paraffins at elevated temperatures. The effect was stronger after the addition of potassium to a bimetallic Sn/Pt/silica catalyst. The effect of tin and potassium over platinum catalyst for hydrogenation of ethylene and dehydrogenation of cyclohexane has been studied with a conclusion that the Pt/Sn/K catalyst might be better for dehydrogenation due to less deactivation of the surface (as it decreases the tendency of the surface to deactivate by coking) [13].

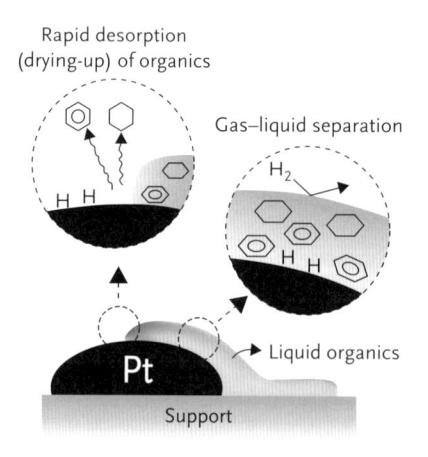

**Figure 5.2** Schematic of the "wet–dry multiphase condition" [12], reprinted by permission.

## 5.4
## Hydrocarbons as Hydrogen Carrier

Hydrogen production from hydrocarbons by reforming processes has been described in Section 2.3. Several processes are possible to convert hydrocarbon fuels to hydrogen with rather high degree of purity, acceptable even for operation in fuel cells. Disadvantage of conventional reforming processes are the relatively complex processing requirements and the operational parameters like high temperature or high pressure, adding to the overall costs of the process and certainly reducing the efficiency. In general, the main chemical processes used for hydrogen production from hydrocarbon fuels are steam reforming, partial oxidation, and autothermal reforming. Following thermodynamic and stoichiometric considerations, it was presented that steam reforming can produce a reformate with 70–80% of hydrogen on a dry basis and partial oxidation only 35–45% of hydrogen [14]. Apart from methanol reforming, other fuels require water–gas shift reaction to increase hydrogen content and reduce carbon monoxide. Partial oxidation is preferable at high temperatures of 877–1627 °C (1150–1900 K), and steam reforming being endothermic reaction takes place at 727–877 °C (1000–1150 K) and is not suitable under transient conditions. Hydrogen production from hydrocarbons can be maximized by autothermal reforming, where the energy for endothermic steam reforming is provided by exothermic partial oxidation [15]. From the results presented in the literature, it seems that in principle the catalytic reforming of methane can be a source of synthesis gas or hydrogen even on a large scale; however, mainly air or diluted feed gas was used as an oxidant causing dilution of the produced gas with nitrogen or other inert gas and showing lower energy content [16]. An autothermal low-temperature (<800 °C) catalytic partial oxidation of methane at 0.21 MPa pressure was investigated in a 1 kW reactor over ruthenium-coated monolith catalyst with pure oxygen as the oxidant medium. The produced gas showed high hydrogen yield (60.8 wt% at 1.09 kW$_{th}$), contained no nitrogen but some amount of CO (14.6% at 1.09 kW$_{th}$) [16].

Hydrogen can be produced by alcohol reforming as well. Potential good source of hydrogen is methanol because it is liquid at ambient conditions (can be used for mobile fuel cells), can be converted to hydrogen at relatively low temperatures with steam or oxygen compared to other fuels, and has high hydrogen-to-steam ratio. Reforming process of methanol results in synthesis gas with hydrogen content higher than needed for the methanol reaction. Excess hydrogen can be used for other application (as fuel in plant or can be separated in a cryogenic way for other use). $CO_2$ is added to the synthesis gas to convert some of the hydrogen to CO that increases methanol yield. Added $CO_2$ can be recovered from the reforming flue (reversible process). An apparatus and method for producing methanol by use of fuel cell to regulate the gas composition entering the methanol synthesizer have been patented [17]. In this invention, hydrogen-rich synthesis gas is passed through fuel cell that is adapted to add controlled amount of $CO_2$ to the gas stream and/or consume given amount of hydrogen in the fuel cell. Gas stream with reduced hydrogen content is then fed to methanol synthesizer to form methanol from the gas stream products.

Hydrogen production from ethanol has gained much attention worldwide [18]. Steam reforming of biomass-derived compounds, mainly ethanol, showed promising results in terms of produced hydrogen in gas or liquid phase according to the following reaction:

$$C_2H_5OH + 3H_2O \rightarrow 2CO_2 + 6H_2 \quad \Delta H = 173.1 \text{ kJ mol}^{-1} \quad (5.3)$$

It was noted that even 1% of ethanol added to water gave a substantial hydrogen evolution [19].

For vehicular applications on board reforming of gasoline or diesel is being considered since a long time. Iso-octane (2,2,4-trimethylpentane), being important component of gasoline, has been investigated as a model hydrocarbon for hydrogen production in different reforming processes like partial oxidation with various transition metals [20], steam reforming of iso-octane and methylcyclohexane over catalysts based on Ni and Fe modified with alkali earth metals [21], and autothermal reforming of iso-octane and toluene over metal-modified Ni catalysts [22]. Qi et al. [23] have built up and tested 1 kW integrated fuel processor with autothermal reformer, high- and low-water gas shift reactors for reforming of gasoline and n-octane. When using n-octane as a fuel, 1.5 mol $H_2$ per mol $C^{-1}$ was obtained and in case of commercial gasoline it was less, only 1.1 mol $H_2$ per mol $C^{-1}$.

Very interesting experiments have been performed recently by Wang et al. [24] on dehydrogenation of Jet A fuel to generate CO-free hydrogen which could be used directly in PEM fuel cell, without purification step. In addition, the same experiments were applied to cyclohexane and decalin. The dehydrogenation process was carried over $Pt/\gamma\text{-}Al_2O_3$ and $Pt/\gamma\text{-}Al_2O_3\text{-}ZrO_2/SO_4^{2-}$ catalysts at atmospheric pressure. In case of dehydrogenation of surrogate Jet A (30 wt% decane, 35 wt% dodecane, 14 wt% methylcyclohexane, 6 wt% decalin, 10 wt% t-butyl benzene, and 5 wt% 1-methyl naphthalene) in the presence of $Pt/\gamma\text{-}Al_2O_3$, the maximum hydrogen yield (hydrogen produced per hydrogen content in the Jet A) was 11.6% at 480 °C, lower than in case of pure cyclohexane (47.5%) and decalin (52.0%) [24]. It was noted that the use of bifunctional catalysts with metal and acid sites is important to increase the hydrogen yield.

## 5.5
### Application: Automotive

Organic liquids can be used indirectly to store hydrogen in a liquid form; however, several steps are needed when thinking, for example, about onboard vehicle application: first, the dehydrogenation step of the organic hydride to produce hydrogen gas followed by the transportation of the dehydrogenated product from the tank of the vehicle to a central plant (processing) and refilling the tank with fresh organic hydride and finally regeneration of the hydrogen-depleted liquid to the starting compound and back to the filling station. It seems that the infrastructure shall be very well organized. Another important point is that organic hydrides are toxic; therefore, a special care must be taken during their handling.

At Catalyst Research Center, Hokkaido University (Japan) a rechargeable direct PEM fuel cell (D-PEMFC) has been developed, using organic hydrides such as cyclohexane, methylcyclohexane, 2-propanol, or decalin [2,25]. The advantage of such a fuel cell is that it could work directly with organic fuels without the need of complex reforming, meaning significant weight and volume advantage. Instead of hydrogen, such fuel cell generates electricity and heat as a result of dehydrogenation (electro-oxidation) of organic hydrides and secondary alcohols, and regenerates hydrides (electroreduction) by supplying electricity to the fuel cell. Schematic of direct-cyclohexane fuel cell is presented in Figure 5.3 [26].

Experiments have been performed with 4 $cm^2$ single cell with Pt(Pt–Ru)/carbon as anode and Ir (Ir–Pt)/carbon as the cathode. Vaporized cyclic hydrocarbons were supplied to D-PEMFC with nitrogen gas carrier saturated with water vapor to the anode compartment. A saturated mixture of oxygen and nitrogen was supplied to the cathode side. In order to rehydrogenate aromatics by electrolysis,

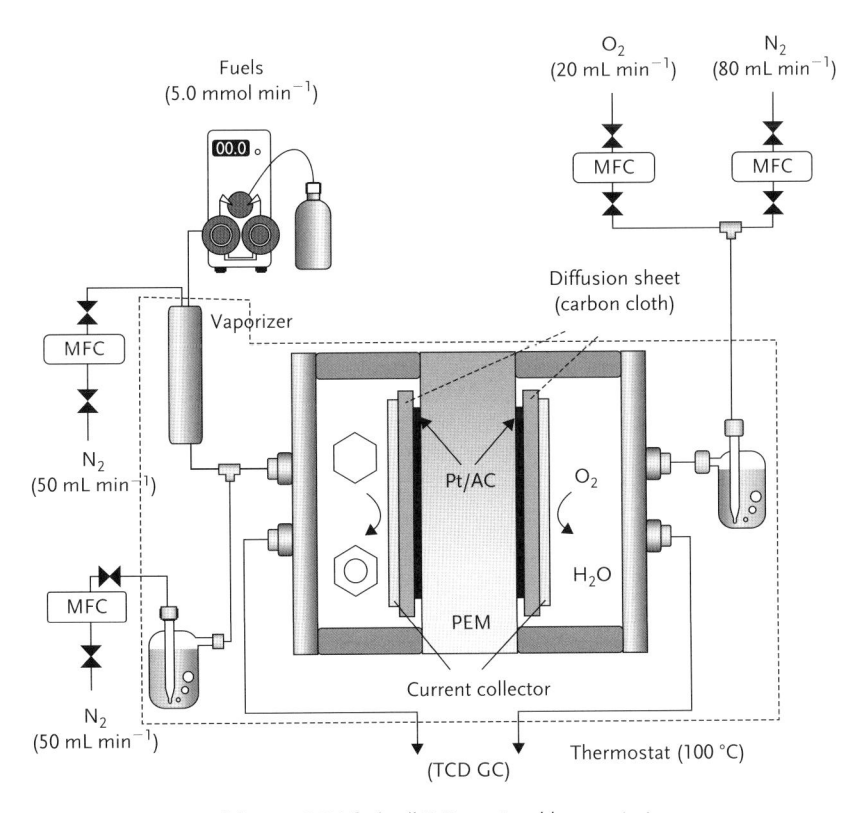

**Figure 5.3** Direct-cyclohexane PEM fuel cell [26], reprinted by permission.

water vapor was fed to the anode side together with $N_2$ gas and an acetone (or benzene) mixed with water vapor to the cathode side. Polarization curves ($I-V$) of cyclohexane, methylcyclohexane, and tetralin of D-PEMFC are presented in Figure 5.4 [26]. The best performance was achieved for cyclohexane with OCV of 920 mV and maximum power density 14–15 mW cm$^{-2}$ at 100 °C. Using 2-propanol, it was possible to reach OCV 790 mW and maximum power density of 78 mW cm$^{-2}$ [25]; however, in another report, Kariya gives value of maximum power density for 2-propanol of 100–150 mW cm$^{-2}$ measured by half-cell at 80 °C [27].

A concept for using decalin dehydrogenation/naphthalene hydrogenation as a source for fuel cell vehicle has been proposed by Hodoshima et al. [28]. It was estimated that for 50 kW fuel cell system with 45% efficiency, hydrogen supply would be of 37 8 Nm$^3$ (0.469 mol s$^{-1}$). Decaline was chosen due to high hydrogen storage density of 7.3 wt%, higher than the storage densities targeted by US DOE. Estimated reactor size at the 50 kW power was 3.0 m$^2$ for Pt/C catalyst and 0.6 m$^2$ for Pt–W/C catalyst (reaction rate with Pt–W/C catalyst was 4.88 times higher than in case of Pt/C). Hodoshima et al. [6] presented an idea of energy cogeneration for domestic use (Figure 5.5). In this model, heat and electricity are supplied also from fuel cells, not only from conventional heating. Hydrogen produced by electrolysis or as by-product from soda or oil industry can be converted and stored as tetralin by the catalytic hydrogenation of naphthalene. Hydrogen supply from tetralin to the fuel cell can be performed with domestic thermal exhaust under superheated liquid-film-type catalysis, and organic residues like naphthalene oil are recirculated to the hydrogenation sites by fuel-selling facilities. No $CO_2$ emission occurs by this kind of hydrogen circulation [6].

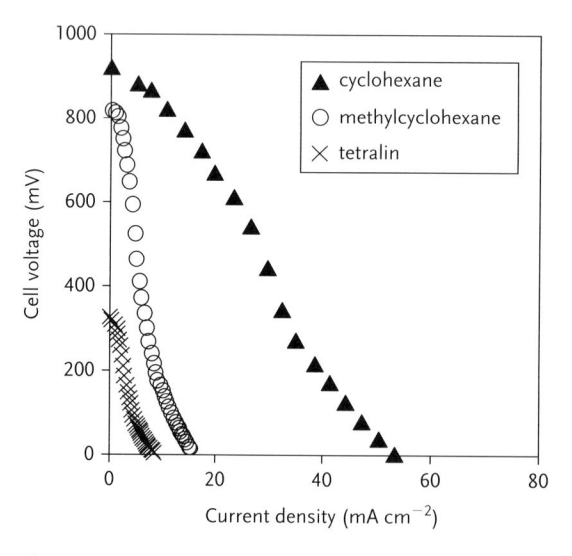

**Figure 5.4** $I-V$ curves of cyclohexane, methylcyclohexane, and tetralin in Pt–Pt D-PEMFC at 100 °C (feeding rate of fuels is 5.0 mmol min) [26], reprinted by permission.

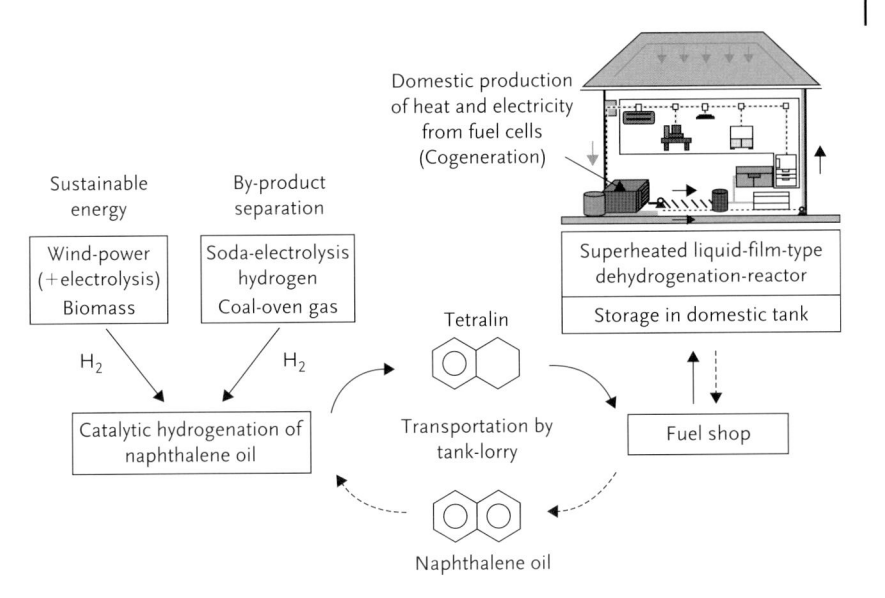

**Figure 5.5** A model for energy network for domestic production of heat and electricity from fuel cells with use of the tetralin–naphthalene pair [6], reprinted by permission.

## 5.6 Ammonia

Chemical compounds from hydrogen and nitrogen could be used as hydrogen carriers as well as the compound from hydrogen and carbon. Ammonia ($NH_3$) is the corresponding compound to methane ($CH_4$). Replacing the carbon by nitrogen has some advantages. Ammonia may hold the key to solving the storage and density issues associated with hydrogen. Ammonia actually is denser than liquid hydrogen. Ammonia can be conveniently stored in liquid form, has a high-energy density (10.8 MJ $kg^{-1}$), and the safety issues concerning storage and handling are well established. Whenever hydrogen is released from a carbon-based material, usually carbon dioxide is ejected. This carbon dioxide release is in discussion because of its rumoured role in the climate change. Whenever nitrogen compounds are used, the typically by-product is nitrogen, which is absolutely harmless for the environment. In the last years, the interest in utilizing ammonia as part of the hydrogen solution has increased.

### 5.6.1 Properties

The properties of ammonia in comparison with the properties of methane are shown in Table 5.3.

**Table 5.3** Properties of ammonia and methane.

| Properties | Ammonia | Methane |
|---|---|---|
| Molecular formula | $NH_3$ | $CH_4$ |
| Molar mass | 17.03 g mol$^{-1}$ | 16.04 g mol$^{-1}$ |
| Mass fraction hydrogen | 17.6% | 24.9% |
| Aggregation state | Gaseous | Gaseous |
| Melting point | −77.7 °C | −182 °C |
| Boiling point | −33.4 °C | −162 °C |
| Vapor pressure | $8.58 \times 10^5$ Pa (20 °C) | Not applicable |
| Octane number | 110–130 [29] | 120 |
| Heat of evaporation | 1.371 MJ kg$^{-1}$ | |
| Solubility | 541 g L$^{-1}$ water (20 °C) | |
| Density | 0.7714 g m$^{-3}$ gas (1.013 × 10$^5$ Pa at 0 °C) 681.9 kg m$^{-3}$ at −33.3 °C (liquid) 820 kg m$^{-3}$ at −80 °C (crystal solid) 817 kg m$^{-3}$ at −80 °C (transparent solid) | 0.722 kg m$^{-3}$ (1.013×10$^5$ Pa at boiling point) 415 kg m$^{-3}$ at −162 °C (liquid) |
| Ignition temperature | 630 °C | 595 °C |
| Storage pressure | $8.58 \times 10^5$ Pa (20 °C) | $200 \times 10^5$ Pa (20 °C) |
| Caloric value | 22.7 MJ kg$^{-1}$ | 55.7 MJ kg$^{-1}$ |
| heating value | 18.6 MJ kg$^{-1}$ | 50.4 MJ kg$^{-1}$ |
| Volumetric energy density (GJ m$^{-3}$) | 15.4 at −33 °C | 10.5 at 250 bar@20 °C |

### 5.6.2
### Application Areas of Ammonia

- Fertilizer
  More than 80% of ammonia is used as fertilizers either as its salts or as solutions. The production of ammonia is a significant component of the world energy budget (consuming more than 1% of all human-made power).
- Precursor to nitrogenous compounds
  Ammonia is the precursor to most nitrogen-containing compounds directly or indirectly. Virtually, all synthetic nitrogen compounds are derived from ammonia.
- Cleaner
  Household ammonia is a solution of $NH_3$ in water used as a general purpose cleaner for many surfaces.
- Minor uses
  Refrigerant – R717
  Because of its favorable vaporization properties, ammonia is an attractive refrigerant.

- Reducing agent

    Ammonia is used as a reducing agent for reducing the emission of nitrogen oxides ($NO_x$) by power plants (powered by fossil fuel) or diesel engines. This technology is called selective catalytic reduction (SCR).
- As a fuel

An overview of how ammonia can be used as fuel is shown in Figure 5.6.

Ammonia was used to power combustion engines in buses in Belgium. Liquid ammonia was used as the fuel of the rocket plane X15 in the 60s of last century. Ammonia has also been proposed as a fuel for solid oxide fuel cell (SOFC) [30] or after decomposition for alkaline fuel cell (AFC) or proton exchange membrane (PEM) fuel cell.

In case of SOFC, it was shown that ammonia can replace hydrogen as fuel without a loss of efficiency, even an increase in efficiency was detected.

Two types of electrolytes are under investigation, the oxygen-ion-conducting electrolyte, such as yttria-stabilized zirconia, and proton-conducting electrolyte [31]. The proton-conducting electrolyte shows advantage in respect to behavior with gas impurities as carbon dioxide.

For the use of ammonia as fuel in low-temperature fuel cells the decomposition of ammonia has to take place. In case of the PEM fuel cell, the potentially remaining ammonia gets separated from the produced hydrogen-rich gas, because ammonia can poison the proton exchange membrane.

In case of using ammonia as fuel for vehicular applications, cooling with ammonia is a valuable side benefit. This will allow to downsize the engine cooling system and will allow to obtain some air conditioning. Taking this in account, the overall efficiency will be improved.

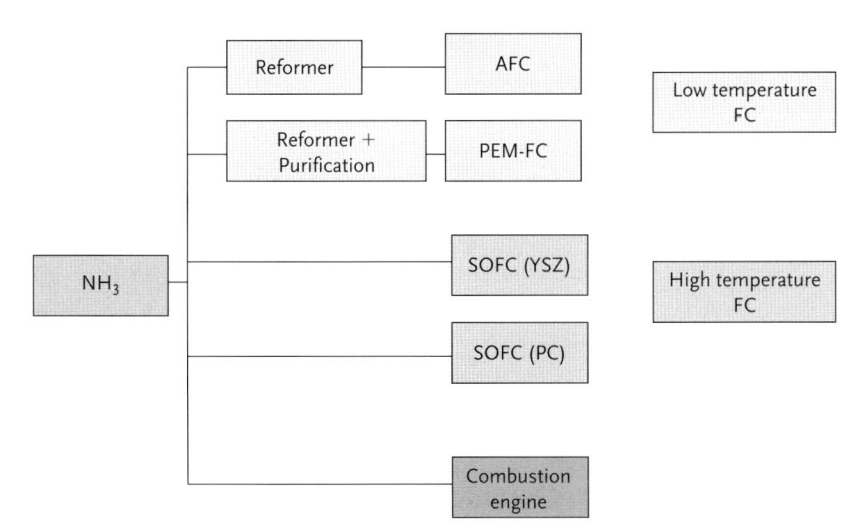

**Figure 5.6** The possible pathways of ammonia as fuel.

### 5.6.3
### Production

#### 5.6.3.1 Production from Nitrogen and Hydrogen

Ammonia is the most widely produced chemical product. Mainly, it is produced directly from nitrogen and hydrogen. The nitrogen is part of the air and could be separated from the air by liquefaction. There is no natural reservoir for hydrogen; hydrogen must be produced from suitable sources. The main process for hydrogen generation is the steam reforming using natural gas as source. (For details see Chapter 2.3.1.1)

$$CH_4(g) + H_2O(g) = CO + 3H_2(g) \quad \Delta H = +206 \text{ kJ mol}^{-1} \quad (5.4)$$

The ammonia could be produced by the Haber–Bosch process.

The final stage is the synthesis of ammonia using a form of iron oxide, as the catalyst.

$$N_2(g) + 3H_2(g) = 2NH_3(g) \quad \Delta H = -92.4 \text{ kJ mol}^{-1} \quad (5.5)$$

Typical process conditions are a pressure at 15–25 MPa and temperatures between 300 and 550 °C, passing the gases over four beds of catalyst, with cooling for the separation of liquid ammonia between each pass to maintain a reasonable recovery. On each pass, only about 15% conversions occur, but unreacted gases are recycled, so that eventually an overall conversion of 98% can be achieved.

In 2007, worldwide 131 million tons of ammonia has been produced. The steam reforming of hydrocarbons is the main source for hydrogen for the ammonia production. Steam reforming for ammonia production started in 1930. Since then, the technology has been gradually improved and energy consumption decreased from an early level of 80 MJ kg$^{-1}$ to a level of 28 MJ kg$^{-1}$ today [32].

#### 5.6.3.2 Production from Silicon Nitride

Silicon can also be used as an energy carrier. The reactions of silicon with oxygen or nitrogen are exothermic, which means they release thermal energy like other combustion processes. In case of the combustion with nitrogen, the product will be silicon nitride. Silicon nitride can be usually obtained by direct reaction between silicon and nitrogen at temperatures between 1300 and 1400 °C by using a copper catalyst; metallic silicon powder reacts with nitrogen under relatively mild conditions (about 600 °C) [33]. The thermal energy released is comparable to that of carbon or carbon-based fuels, which is 32.4 MJ kg$^{-1}$ Si.

$$3Si + 2N_2 \rightarrow Si_3N_4 \quad \Delta H = +750 \text{ kJ mol}^{-1} \quad (5.6)$$

Adding steam or a base to the silicon nitride results in the release of ammonia and other products of the reaction as silicon oxide or metal silicates could be used, for example, for the production of glass. The reaction of silicon nitride with steam is shown in the following equation.

$$Si_3N_4 + 6H_2O \rightarrow 3SiO_2 + 4NH_3 \quad (5.7)$$

The silicon nitride reacts with a base already at 350 °C and with steam at higher temperatures.

### 5.6.4
### Methods for Storing Ammonia

#### 5.6.4.1 Liquid Dry Ammonia
There are three different methods known for the storage of liquid ammonia:

- at −33 °C and atmospheric pressure,
- at −5 °C to +5 °C and at medium pressure,
- at ambient pressure and at equilibrium pressure.

Ammonia with different purities is available in steel bottles with 1, 5, or 26.5 kg liquid. Ammonia transport with pipelines is also established since many years. Steel and stainless steel are resistant against ammonia and are used for bottles, valves, tubing and pipelines.

#### 5.6.4.2 Solid-State Ammonia Storage
The main concerns against ammonia are safety related. The main safety problem in the transport sector is that there should be no risk of poisoning due to an ammonia spill. This risk is not only limited to a possible accident but also for improper handling at the filling station or in case of any leakage. To overcome the main concern against ammonia as a fuel, a possible solution is the solid storage in salts like MgCl, CaCl and so on. This technology is known from the developments of periodical refrigeration processes with ammonia and these salts have been investigated at the beginning of last century. For a good overall efficiency, the waste heat from the fuel cell has to be taken to release the ammonia from the salt. High temperatures for the desorption are required to achieve the high storage densities; this could be a limitation in some cases. Specific values are given in Table 5.4.

**Table 5.4** Specific values for different chlorides for $NH_3$- storage.

| Salt | Magnesium chloride | Calcium chloride | Strontium chloride |
|---|---|---|---|
| Molar mass (g mol$^{-1}$) | 95.21 | 110.98 | 158.52 |
| Possible utilizable molar concentration | 1; 2, 6 | 1; 2; 4;8 | 1;8 |
| Maximum possible molar concentration | 6 | 8 | 8 |
| Mass ratio of ammonia/salt (kg kg$^{-1}$) | 1.07 | 1.23 | 0.86 |
| [34], Gravimetric hydrogen storage density maximum (%) |  | 9.7 | 8.1 |

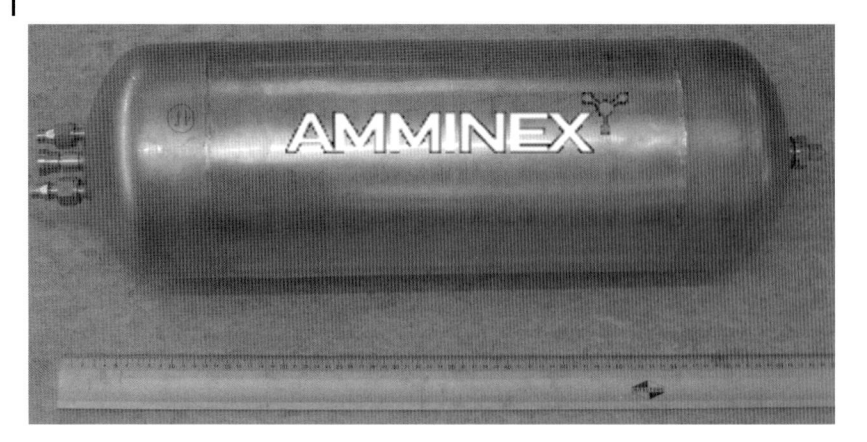

**Figure 5.7** Canister from AMMINEX with solid ammonia.

For the selection of the best salt, not only the maximum molar ratio is of interest but also the required energy to release the ammonia as well as the requested temperature.

There is a company in Denmark (AMMINEX A/S; Soborg, Denmark) working on commercialization of this technology under the trade name Hydrammine$^{TM}$. The used material allows a gravimetric hydrogen storage density of 9% and a volumetric density of 110 kg hydrogen m$^{-3}$. The ammonia released from the solid can also be used as agent for the reduction of nitrogen oxide emissions from diesel engines by selective catalytic reduction (SCR) (Figure 5.7).

## 5.6.5
### Use of Ammonia as Fuel in High-Temperature Fuel Cells

A tubular SOFC fed by ammonia was compared with the fuel cell fed by pure hydrogen [35]. For the electrolyte, the composition was 8 mol % $Y_2O_3$, balance $ZrO_2$. The best results could be obtained by using a silver anode combined with an iron catalyst. With this combination, ammonia gives an efficient performance similar to that obtained by equivalent supply of pure hydrogen. The experiments were performed at 973, 1073, and 1173 K. It could be shown that ammonia can be used directly as fuel in an SOFC system.

Fuerte *et al.* [36] investigated a tubular SOFC with different ammonia concentrations and pure hydrogen for 400 h. The tests were performed with an Ni/YSZ anode, without additional catalyst. The cell was tested at the same temperatures. From these tests, it is assumed that oxidation of ammonia in an SOFC is a two-stage process, decomposition of ammonia to nitrogen and hydrogen initially, followed by hydrogen oxidation to water. It was demonstrated that SOFC can actually utilize ammonia as fuel to produce electric power.

Haldor Topsoe AS reported that with direct ammonia, the efficiency is the same or even better compared with a hydrogen/nitrogen mixture (3/1) [37].

Thin proton-conducting electrolyte with composition $BaCe_{0.8}Gd_{0.2}O_{3-\delta}$ was prepared over substrates composed of $Ce_{0.8}Gd_{0.2}O_{1.9}Ni$ by dry-pressing method [31]. The cathode had the composition $Ba_{0.5}Sr_{0.5}Co_{0.8}Fe_{0.2}O_{3-\delta}$. The performance of a single cell was tested at 873 and 923 K with ammonia as fuel. The achieved open-circuit voltage was 1.12 and 1.1 V at 873 and 923 K, respectively. A maximum power density of 200 W cm$^{-2}$ was obtained at 923 K. While YSZ-based SOFCs are already in a mature state, $NO_x$ may be produced in the direct ammonia conventional SOFC. The high-temperature proton-conducting electrolyte is a promising alternative electrolyte for direct use of ammonia. Nitrogen and water are the only chemical products. The typically operational temperature of a proton-conducting electrolyte fuel cell fits perfectly with the required temperature for the ammonia decomposition. This will finally improve the overall system efficiency.

## 5.6.6
### Hydrogen from Ammonia

To release the hydrogen from the ammonia, first the ammonia has to be evaporated. The heat of evaporation is relatively high with 1.262 kJ kg$^{-1}$; this high value makes the ammonia attractive as a cooling agent. There are several methods described to release hydrogen from ammonia.

#### 5.6.6.1 Ammonia Electrolysis
The feasibility of the onboard hydrogen storage and production via electrolysis was evaluated [38]. Theoretically, ammonia electrolysis requires only 2.79 kJ mol$^{-1}$ $H_2$. The expected anode and cathode reactions are given in Eqs. (5.8) and (5.9)

$$2NH_3(aq) + 6OH^- = N_2(g) + 6H_2O + 6e^- \quad -0.77 \text{ V(SHE)} \quad (5.8)$$

$$6H_2O + 6e^- = 3H_2(g) + 6OH^- \quad -0.83\text{V (SHE)} \quad (5.9)$$

It means theoretical and overall voltage of 0.06 V are required. This value is very low compared with the voltage of 1.23 V required for the electrolysis of water. Starting from the 0.06 V, the required energy for 1 mol $H_2$ can be calculated to 10.8 kJ mol$^{-1}$ $H_2$. If the hydrogen is used in a fuel cell, the theoretical energy could be calculated to 241.8 kJ mol$^{-1}$ $H_2$. Assuming an efficiency of 65% (cell potential $V_{cell}$=800 mV) for the fuel cell, the produced energy can be calculated to 160 kJ mol$^{-1}$ $H_2$. It is obvious that the energy required for the electrolysis of ammonia is playing a minor role assuming high efficiencies for the electrolysis of ammonia.

#### 5.6.6.2 Catalytic Decomposition
Another way to release hydrogen from ammonia is thermal decomposition. For the decomposition of ammonia, the same catalysts as for the ammonia production could be used. $NH_3$ decomposition is endothermic and the equilibrium conversion at 100 kPa is 99.1%. Figure 5.8 shows the equilibrium conversion of $NH_3$ at different temperatures and 100 kPa. A detailed analysis of the ammonia decomposition

kinetics over Ni–Pt/Al$_2$O$_3$ was performed [39]. It was shown that a first-order rate expression provides an adequate fit for the experimental data. The activation energy was found to be in the range of 210 kJ mol$^{-1}$. A result of a review on ammonia decomposition catalysts found out that ruthenium is the most active catalyst [40], carbon nanotubes are the most effective support, and KOH is the most effective promoter for onsite generation of hydrogen from ammonia decomposition.

Several small-scale ammonia reformers (crackers) have been built in the last years. The reaction was investigated between 750 and 950 K and between 57 and 1000 kPa. In Graz, Austria, an ammonia reformer in the 10 kW class was built and tested [41,42]. For the heating of the reactor, a hydrogen burner was used with hydrogen as fuel. Another reformer was built with electrical heating. With ammonia flow rate of 661 L h$^{-1}$ and temperatures between 1073 and 1123 K, ammonia concentrations in the product stream of 150 and 300 ppm were achieved. The mass of the reactor without heat exchanger and adsorber unit was 5.6 kg. When a PEM fuel cell is fed by such a reformate, the ammonia traces have to be removed.

A small reformer for a hydrogen production of about 1 L h$^{-1}$ was built by Amminex and is shown in Figure 5.9. This reformer is heated by combustion of ammonia. For the removal of ammonia, several processes are possible: absorption, adsorption, selective membranes, etc. Typically, an adsorption process is used by applying molecular sieves. The molecular sieve can be regenerated by heat supply

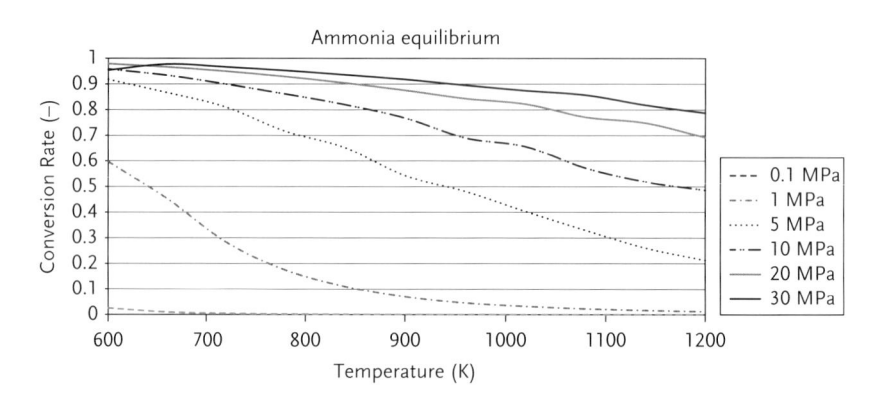

**Figure 5.8** Equilibrium conversion of NH$_3$ at different temperatures and pressures.

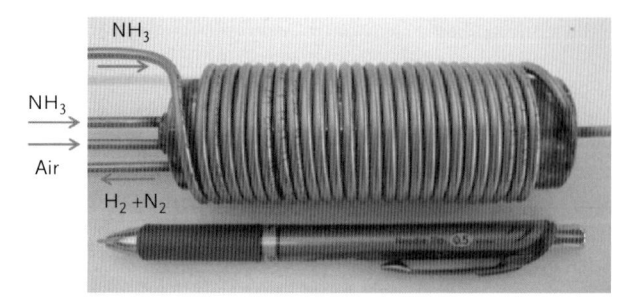

**Figure 5.9** Ammonia reformer built by Amminex.

and reduction of pressure. Heat supply by a hydrogen burner instead of electrical heating results in a better overall efficiency.

In 1977, together with Dr. Karl Kordesch, Apollo Energy Systems, Inc., located in Pompano Beach, Florida, developed alkaline fuel cells and ammonia reformers. The improved alkaline fuel cells, as well as ammonia reformers, for different power classes, are produced by Apollo Energy Systems, Inc.

When an ammonia reformer is combined directly with an alkaline fuel cell, the traces of ammonia in the hydrogen have no negative influence. In an experimental setup, the influence of the amount of ammonia in the reformate on the performance of an alkaline fuel cell was tested [44]. The tests were performed with a single-cell alkaline fuel cell with 12.5 $cm^2$ active area. A platinum catalyst was used on anode and cathode side. The cell was fed with a mixture of hydrogen/nitrogen/ammonia (0.8/0.24/0.09 mL $s^{-1}$), which means an ammonia concentration of 8%. The test duration was limited to 100 h. The measured cell voltage data showed that alkaline fuel cells are tolerant to ammonia, compared to PEM fuel cells.

### 5.6.7
### Hydrogen from Ammonia and Metal Hydride

A method has been suggested in order to overcome the high required temperatures for the decomposition of ammonia when pure hydrogen is requested. The new approach for the hydrogen release from ammonia at ambient or even lower temperatures is the ammonolysis, reacting ammonia with metal hydrides.

Experiments to generate hydrogen from ammonia and magnesium hydride ($MgH_2$) were conducted at 348–423 K [45]. The following reaction was investigated:

$$2NH_3 + 3MgH_2 \rightarrow 3Mg_3N_2 + 6H_2 \quad \Delta H = -140.34 \text{ kJ mol}^{-1} \quad (5.10)$$

Ammonia chloride is able to enhance the hydrogen release. With $PdCl_2$ or $PtCl_4$, doped $MgH_2$ in combination with the $NH_4Cl$ further increased the hydrogen formation rate. The feasibility of hydrogen production from ammonia-based reactions was demonstrated.

With this process, high-energy densities could be realized, but for the next hydrogen release, the $Mg_3N_2$ must be regenerated.

A hydrogen generator on the basis of $LiAlH_4$ shows hydrogen production at subzero temperatures.

The system with LiH was reacted at room temperature and 0.5 MPa [46]. It was found that the hydrogen can be reversibly stored according to the following equation:

$$NH_3 + LiH \rightarrow LiNH_2 + H_2 \quad \Delta H = -43.1 \text{ kJ mol}^{-1} \quad (5.11)$$

This results in a gravimetric storage density of 8.1%.

Small hydrogen generators for the use in military environments based on lithium aluminum hydride have been built [47]. The hydrogen production between 303 and 323 K takes place according to the following equation:

$$4NH_3 + LiALH_4 \rightarrow 3AlN + Li_3N + 12H_2 \quad (5.12)$$

A system for 1.44 MJ was installed. An energy density of 1.728 MJ kg$^{-1}$ was reached. The hydrogen generator was combined with a PEM fuel cell from Ball Aerospace with a power of 50 W; therefore, the ammonia had to be put out of the produced hydrogen to prevent the fuel cell from poisoning. Other possible combinations with other hydrides (like $AlH_3$) are possible and under investigation. These kinds of systems have the advantage that there is no loss of hydrogen during the storage time, but the hydride cannot easily be regenerated for each system. For smaller applications, for example, portables, this could be an effective solution.

### 5.6.8
**Energetic Consideration**

For the integral examination of the use of ammonia as hydrogen storage medium, an energetic consideration is self-evident. Here only the storage of the hydrogen is considered. The production of the hydrogen is a big issue. Actually the main hydrogen source is hydrogen from natural gas produced by steam reforming. This process needs a lot of energy and produces carbon dioxide. For the Haber–Bosch process also, hydrogen from other heritage can be used. There are different sources of renewable hydrogen possible. For the consideration here, the starting point is the availability of hydrogen at low pressure. For the production of ammonia, the Haber–Bosch process together with the nitrogen separation has to be taken into account. The product of the Haber–Bosch process is pure liquid ammonia that could be stored easily at low pressure. During the storage, no loss of ammonia is expected. The energy consumption to release the ammonia, and if required, the energy consumption for the ammonia decomposition, has to be considered.

The energy demand for the nitrogen separation from air is about 36 kJ mol$^{-1}$ nitrogen.

For 1 mol hydrogen for the Haber–Bosch process, 1/3 mol nitrogen is required. Therefore, the energy demand for providing the nitrogen is 12 kJ mol$^{-1}$ hydrogen.

The energy demand for the production of ammonia is approximately 29 kJ mol$^{-1}$ ammonia. The main effort is the compression of the gases to a high pressure for the catalytic conversion.

For the conversion of 1 mol hydrogen with the Haber–Bosch process, the energy demand is therefore 20 kJ mol$^{-1}$ hydrogen.

For the ammonia production, energy demand is 32 kJ mol$^{-1}$ hydrogen. This results in just 13% of the energy content of hydrogen. In summary, this means that only 13% of the energy content of the hydrogen must be spend for the conversion to hydrogen. This number has to be compared with the value for other hydrogen storage technologies.

For the release of the hydrogen, the energy demand is different for different processes.

First, there is the potential of the direct use of the ammonia, for example, in SOFC. In this case, no additional energy is needed. In case of ammonia electrolysis, the theoretical value is 2.79 kJ mol$^{-1}$ hydrogen.

## 5.7
## Borohydrides

In this chapter, following two borohydrides are considered:

- sodium borohydride
- ammonia boranes.

The sodium borohydride was deeply investigated by a Chrysler in the United States together with Millenium Cell. They have built one demo car together. The system called Natrium is based on the Chrysler Town and Country. Sodium borohydrides are so attractive because these can be stored as solid or as a stabilized aqueous solution without releasing hydrogen at room temperature.

The interest on the ammonia boranes has been increased in the past years.

### 5.7.1
### Sodium Borohydride

Sodium borohydride can be used in several ways:

- direct use as fuel in a PEM-based fuel cell;
- hydrogen generation by hydrolytic release; and
- thermal dehydrogenation.

Only the first two pathways are more investigated and are of interest. The general properties of sodium borohydride are given in Table 5.5.

#### 5.7.1.1 Direct Use of Sodium Borohydride as Fuel in a PEM-Based Fuel Cell

The direct use of sodium borohydride as fuel has some advantages regarding complexity of the system as well as for the energy density. With direct use of the liquid solution of sodium borohydride, simple and highly efficient fuel cells can be realized. The use of liquid fuel has also an influence on the thermal system. The fuel could directly be used to remove the heat. The liquid fuel could be advantageously combined with a liquid oxidant. Solutions of hydrogen peroxide

**Table 5.5** General properties of sodium borohydride.

| | |
|---|---|
| Chemical formula | $NaBH_4$ |
| Molar weight | 37.831 |
| Density (kg $L^{-1}$) | 1.074 |
| Melting point | 400 °C |
| Ignition temperature | >220 °C |
| Freezing point of 30% aqueous solution | −25 °C |

have been investigated. The anode reaction and the cathode reaction for an alkaline fuel cell can be described as follows:

$$NaBH_4 + 8OH^- \rightarrow NaBO_2 + 6H_2O + 8e^- \quad (5.13)$$

$$4HO_2^- + 4H_2O + 8e^- \rightarrow 12OH^- \quad (5.14)$$

The cathode reaction for an acid fuel cell can be described as follows. For the anode reaction, only the alkaline reaction has been considered since otherwise the hydrolysis reaction is favored.

$$4H_2O_2 + 8H^+ + 8e^- \rightarrow 8H_2O \quad (5.15)$$

These kinds of fuel cells working with a liquid fuel as well as with a liquid oxidant are of great interest for application in submarines and UAVs. These fuel cells have higher energy densities compared with batteries and are totally air-independent systems.

### 5.7.1.2 Hydrogen Generation by Hydrolytic Release

Hydrolytic decomposition of stabilized sodium borohydride is a highly investigated process. Sodium borohydrides release hydrogen by pumping the solution (water plus sodium borohydride) through a catalyst bed. Ruthenium was found to be one of the best catalysts. The reaction is highly exothermic. The dehydrogenated product is usually a solution of sodium borate and is not easily converted back to chemical hydrides. The produced hydrogen is highly pure and can therefore be fed directly to a PEM-based fuel cell.

The hydrogen generation from sodium borohydride follows Eq. (5.16)

$$NaBH_4 + 2H_2O \Rightarrow 4H_2 + NaBO_2 + 220 \text{ kJ} \quad (5.16)$$

The hydrolysis reaction needs water. The additional water increases the outcome of hydrogen. One way, which is the common way, is starting from an aqueous solution of sodium borohydride. The concentration of the solution has to be optimized. The more is the water in the solution the lower are the energy densities, and if less is the water in the solution then bigger is the problem of sulfuring at lower temperatures. The sodium borohide solution could be stabilized by the addition of, for example, sodium hydroxide. With different concentrations of the sodium borohydride solution, different energy densities can be achieved. The referring numbers are given in Table 5.6. From 1 kg 20% solution, 473 L hydrogen can be generated, and from 1 kg 30% solution, 710 L hydrogen can be generated.

The other way is to start with the solid sodium borohydride and adding steam for the hydrolysis.

Unlike the reaction in liquid water, up to 95% yield of hydrogen is obtained with pure steam without a catalyst [48]. It was found that adding 1 mol% of acetic acid promotes the hydrolysis reaction.

A process to release hydrogen by hydrolysis from complex hydrides was proposed at high temperature and high pressure [49]. The idea of this process is to use the heat of reaction to vaporize the water after the pressure is decreased to finally generate dry borates.

**Table 5.6** Gravimetric and volumetric storage densities for different sodium borohydride solutions.

| Solution | Gravimetric storage density (%) | Volumetric storage density (kg/100 L) |
| --- | --- | --- |
| $NaBH_4$-20 wt% solution (20 wt% $NaBH_4$, 3 wt% NaOH, 77 wt% $H_2O$) | 4.3 | 4.4 |
| $NaBH_4$-25 wt% solution (25 wt% $NaBH_4$, 3 wt% NaOH, 72 wt% $H_2O$) | 5.3 | 5.5 |
| $NaBH_4$-30 wt% solution (30 wt% $NaBH_4$, 3 wt% NaOH, 67 wt% $H_2O$) | 6.4 | 6.6 |
| $NaBH_4$-35 wt% solution (35 wt% $NaBH_4$, 3 wt% NaOH, 62 wt% $H_2O$) | 7.5 | 7.7 |

The high-energy density of sodium borohydride can be shown with the following values: 1 kg of solid $NaBH_4$ reacts with 950 g of water; the resulting hydrogen is 213.5 g, this results in an energy of 25.560 MJ.

### 5.7.2 Ammonia Borane

Ammonia borane ($H_3BNH_3$), a product of the reaction between borane and ammonia, is a solid soluble in water and strong polar solvent and could be used as hydrogen carrier. The molar mass is 30.87 g mol$^{-1}$. The chemical state is solid and the melting point is 370.76 K. Ammonia borane has high hydrogen content. Ammonia borane could be produced direct from ammonia and diborane. On a large scale, this production is carried out at 195 K in liquid ammonia. There are several options to release the hydrogen:

- by hydrolysis in an acid solution (similar to the hydrolysis of sodium borohydride);
- by adding a strong oxidizing agent;
- by hydrolysis with a homogeneous or heterogeneous catalyst;
- by thermolysis/thermal decomposition; and
- by direct use as fuel in a fuel cell electrocatalytic oxidation.

It strongly depends on the selected process and the proportion of hydrogen which is released and this determines the resulting storage densities.

For the thermally induced hydrogen release, the temperature and the energy demands are given in Table 5.7.

The electrochemical oxidation of ammonia borane was investigated on gold electrode [50]. Following cell reaction was expected:

$$NH_3BH_3 + \tfrac{3}{2}O_2 = BO_2^- + NH_4^+ + H_2O \tag{5.17}$$

**Table 5.7** Thermally induced hydrogen release from ammonia borane.

| Temperature | $H_2$ released (mol mol$^{-1}$) | Energy demand (kJ mol$^{-1}$ $H_2$) | Storage density (kg $H_2$ kg$^{-1}$ AB) |
|---|---|---|---|
| 370 | 1 | −26.6 | 0.0648 |
| 400–440 | 1 | +133.3 | 0.1297 |
|  | 1 | +562.3 | 0.1944 |

Before using ammonia borane as a storage media, the regeneration of the material has to be proofed. The regeneration is one of the biggest problems because dehydrogenation results in various different products and the energy requirement of the regeneration process is another issue.

Ammonia itself or chemical compounds from ammonia are a very interesting group of chemical compounds for hydrogen storage. Especially the compounds for long-term storage as well as the direct use of ammonia in high-temperature fuel cells make ammonia so attractive.

# References

1 O'Bockris, J. (1999) Hydrocarbons as hydrogen storage materials. Short communication. *Int. J. Hydrogen Energy*, **24**, 779–780.
2 Ichikawa, M. (2008) Organic liquid carriers for hydrogen storage in *Solid-State Hydrogen Storage, Materials and Chemistry* (ed. G. Walker), Woodhead Publishing Limited and CRC Press LLC, Cambridge, UK, 500–531.
3 Biniwale, R.B., Raylu, S., Devotta, S., and Ichigawa, M. (2008) Chemical hydrides: a solution to high capacity hydrogen storage and supply. *Int. J. Hydrogen Energy*, **33**, 360–365.
4 Sebastian, D., Bordeje, E.G., Cavillo, L., Lazaro, M.J., and Moliner, R. (2008) Hydrogen storage by decalin dehydrogenation/naphtene hydrogenation pair over platinum catalysts supported on activated carbon. *Int. J. Hydrogen Energy*, **33**, 1329–1334.
5 Targets for Onboard Hydrogen Storage Systems for Light-Duty Vehicles, US DOE, http://www1.eere.energy.gov/hydrogenandfuelcells/mypp/pdfs/storage.pdf, (September 2009).
6 Hodoshima, S., Nagata, H., and Saito, Y. (2005) Efficient hydrogen supply from tetralin with superheated liquid-film-type catalysis for operating fuel cells. *Appl. Catal. A*, **292**, 90–96.
7 Kobayashi, I., Yamamoto, K., and Kameyama, H. (1999) A proposal of a spray pulse operation for liquid film dehydrogenation – 2-propanol dehydrogenation on a plate catalyst. *Chem. Eng. Sci.*, **54**, 1319–1323.
8 Kariya, N., Fukuoka, A., Utagawa, T., Sakuramoto, M., Goto, Y., and Ichikawa, M. (2003) Efficient hydrogen production using cyclohexane and decalin by pulse-spray mode reactor with Pt catalysts. *Appl. Catal. A*, **247**, 247–259.
9 Biniwale, R.B., Kariya, N., and Ichikawa, M. (2005) Dehydrogenation of cyclohexane over Ni based catalysts supported on activated carbon using spray-pulsed reactor and enhancement in activity by addition of a small amount of Pt.*Catal. Lett.*, **105** (1–2), 83–87.
10 Biniwale, R., Kariya, N., and Ichikawa, M. (2005) Production of hydrogen-rich gas via reforming of iso-octane over Ni–Mn

and Rh–Ce bimetallic catalysts using spray pulsed reactor. *Catal. Lett.*, **100** (1–2), 17–25.

11 Biniwale, R., Mizuno, A., and Ichikawa, M. (2004) Hydrogen production by reforming of iso-octane using spray-pulsed injection and effect of non-thermal plasma. *Appl. Catal. A*, **276** (1–2), 169–177.

12 Kariya, N., Fukuoka, A., and Ichikawa, M. (2002) Efficient evolution of hydrogen from liquid cycloalkanes over Pt-containing catalysts supported on active carbons under "wet-dry multiphase conditions". *Appl. Catal. A*, **233**, 91–102.

13 Park, Y.-K., Ribeiro, F.H., and Samorjai, G.A. (1998) The effect of potassium and tin on the hydrogenation of ethylene and dehydrogenation of cyclohexane over Pt(111). *J. Catal.*, **178**, 66–75.

14 Brown, L.F. (2001) A comparative study of fuels for on-board hydrogen production for fuel-cell-powered automobiles. *Int. J. Hydrogen Energy*, **26**, 381–397.

15 Onsan, Z.I. (2007) Catalytic processes for clean hydrogen production from hydrocarbons. *Turk. J. Chem.*, **31**, 531–550.

16 Rabe, S., Truong, T.-B., Vogel, F. (2006) Autothermal reforming of methane: design and performance of a kW scale reformer using pure oxygen as oxidant. *WHEC*, Lyon, France, 16/13–16 June.

17 Farooque, M. (1988) Apparatus and method for methanol production using a fuel cell to regulate the gas composition entering the methanol synthesizer. Patent 4 772 634.

18 Chen, Y., Shao, Z., and Xu, N. (2008) Ethanol steam reforming over Pt catalysts supported on $Ce_xZr_{1-x}O_2$ prepared via a glycine nitrate process. *Energy Fuels*, **22** (3), 1873–1879.

19 Strataki, N., Bekiari, V., Kondarides, D.I., and Lianos, P. (2007) Hydrogen production by photocatalytic alcohol reforming employing highly efficient nanocrystalline titania films. *Appl. Catal. B*, **77**, 184–189.

20 Moon, D.J., Ryu, J.W., Lee, S.D., Lee, B.G., and Ahn, B.S. (2004) Ni-based catalyst for partial oxidation reforming of iso-octane. *Appl. Catal. A*, **272** (1–2), 53–60.

21 Murata, K., Wang, L., Saito, M., Inaba, M., Takahara, I., and Mimura, N.i. (2004) Hydrogen production from steam reforming of hydrocarbons over alkaline-earth metal-modified Fe- or Ni-based catalysts. *Energy Fuels*, **18** (1), 122–126.

22 Kim, D.H., Ryu, J.W., Choi, E.H., Gong, G.T., Lee, H., Lee, B.G. et al. (2008) Production of synthesis gas by autothermal reforming of iso-octane and toluene over metal modified Ni-based catalyst. *Catal. Today*, **136** (3–4), 266–272.

23 Qi, A., Wang, S., Fu, G., and Wu, D. (2006) Integrated fuel processor built on autothermal reforming of gasoline: a proof-of-principle study. *J. Power Sources*, **162** (2), 1254–1264.

24 Wang, B., Froment, G.F., and Goodman, D.W. (2008) CO-free hydrogen production via dehydrogenation of a jet a hydrocarbon mixture. *J. Catal.*, **253**, 239–243.

25 Kariya, N., Fukuoka, A., and Ichikawa, M. (2006) Direct PEM fuel cell using "organic metal hydrides" with zero-$CO_2$ emission and low-crossover. *Phys. Chem. Chem. Phys.*, **8**, 1724–1730.

26 Kariya, N., Fukuoka, A., and Ichikawa, M. (2003) Zero-$CO_2$ emission and low-crossover "rechargeable" PEM fuel cells using cyclohexane as an organic hydrogen reservoir. *Chem. Commun.*, 690–691.

27 Kariya, N. and Ichikawa, M. (2006) Novel infra-structured technology for hydrogen storage, transportation and supply using "organic hydrides". *Catalysts Catal.*, **4** (5), 339–345.

28 Hodoshima, S., Arai, H., Takaiwa, S., and Saito, Y. (2003) Catalytic decali dehydrogenation/naphtalene hydrogenation pair as a hydrogen source for fuel-cell vehicle. *Int. J. Hydrogen Energy*, **28**, 1255–1262.

29 Thomas, G. and Parks, G. (2010) *Potential Roles of Ammonia in a hydrogen economy*, http://www.hydrogen.energy.gov/pdf/nh3_paper.pdf 17 February.

30 Hansen, J.B. (2005) Oxygenates and ammonia as SOFC fuels. Proceedings of the First European Fuel Cell technology and Applications Conference; Rome, Italy, December.

31 Zhang, L. and Yang, W. (2008) Direct ammonia solid oxide fuel cell based on thin proton-conducting electrolyte. *J. Power Sources*, **179**, 92–95.

32 Rafiqul, I., Weber, C., Lehmann, B., and Voss, A. (2005) Energy efficiency improvements in ammonia production-perspectives and uncertainties. *Energy*, **30**, 2487–2504.

33 Auner, N. and Holl, S. (2006) Silicon as energy carrier – facts and perspectives. *Energy*, **31**, 1395–1402.

34 Chakraborty, D., Petersen, H.N., and Johannessen, T. (2009) Solid ammonia as energy carrier: current status and opportunities. Proceedings of European Fuel Cell Forum, Lucerne/Switzerland, 29 June–2 July.

35 Wojcik, A., Middleton, H., Domopoulos, I., and Van Herle, J. (2003) Ammonia as a fuel in solid oxide fuel cells. *J. Power Sources*, **118**, 342–348.

36 Fuerte, A., Valenzuela, R.X., Escudero, M.J., and Daza, L. (2009) Ammonia as efficient fuel for SOFC. *J. Power Sources*, **192**, 170–174.

37 Hansen J.B. (2009) Oxygenates and ammonia as SOFC fuels. Proceedings of the First European Fuel Cell Technology and Application Conference, Rome, Italy, December.

38 Bloggs, B.K. and Botte, G.G. (2009) On board hydrogen storage and production: an application of ammonia electrolysis. *J. Power Sources*, **192**, 573–581.

39 Chellappa, A.S., Fischer, C.M., and Thomson, W.J. (2002) Ammonia decomposition kinetics over Ni–Pt/$Al_2O_3$ for PEM fuel cell applications. *Appl. Catal. A*, **227**, 231–240.

40 Yin, S.F., Xu, B.Q., Zhou, X.P., and Au, C.T. (2004) A mini-review on ammonia decomposition catalysts for on-site generation of hydrogen for fuel cell applications. *Appl. Catal. A*, **277**, 1–9.

41 Bachhiesl, U. and Enzinger, P. Drei Platten-Ammoniak-Cracker (2002) zur Wasserstoffherstellung, Energieinnovation.

42 Stigler, H., Gutschi, C., and Bachhiesel, U. (2004) Ammoniak als Wasserstofflieferant für Brennstoffzellen und mögliche industrielle Anwendungen. Innovations Symposium ENO04, Graz, 4–5 February.

43 www.Apolloenergysystems.com, (20 Feb 2010).

44 Hejze, T., Besenhard, J.O., Kordesch, K., Chifrain, M., and Aronsson, R.R. (2008) Current status of combined system using alkaline fuel cells and ammonia as a hydrogen carrier. *J. Power Sources*, **176**, 490–493.

45 Li, L. and Hurley, J.A. (2007) Ammonia-based hydrogen source for fuel cell applications. *Int. J. Hydrogen Energy*, **32**, 6–10.

46 Yoshitsugu, Kojima, Satoshi, Hino, Chie, Omatsu, Tange, Kyoichi, and Ichikawa, Takayuki (2008) Novel hydrogen storage system with metal hydride and ammonia.International Symposium on Metal-Hydrogen Systems, Reykjavik, Iceland, 24–28 June.

47 Sifer, N. and Kristopher, G. (2005) An analysis of hydrogen production from ammonia hydride hydrogen generators for military fuel cell environments. *J. Power Sources*, **132**, 135–138.

48 Marrero-Alfonso, E.Y., Gray, J.R., Davis, T.A., and Matthews, M.A. (2007) Hydrolysis of sodium borohydride with steam. *Int. J. Hydrogen Energy*, **32**, 4717–4722.

49 Jehle, Walter (2007) DE 10 2005 039 061 A1 Verfahren zur Herstellung von Wasserstoff aus einem komplexen Metallhydrid.

50 Zhang, Xin-Bo, Sang, Han, Jun-Min, Yan, Hiroshi, Shioyama, Nobuhiro, Kuriyama, Tetsuhiko, Kobayashi, and Xu, Qian (2009) Electrochemical oxidation of ammonia borane on gold electrode. *Int. J. Hydrogen Energy*, **34**, 174–179.

# 6
# Hydrogen Storage Options: Comparison

## 6.1
## Economic Considerations/Costs

As seen in the previous chapters, hydrogen storage is the main key to the use of hydrogen-powered vehicles. Despite different cost and environmental options for the power plant itself, storage and the management of the storage system play the key role of future hydrogen transportation options. In this field, economic considerations as well as availability of materials and the resultant costs will be the exclusive key for future operations.

There have been several projects undertaken worldwide by NHA, European Union (EU), and South-East Asia with respect to the evaluation of economic implications of the introductions of hydrogen transportation. The EU has outlined on the "European hydrogen roadmap" within the HyWays projects and different plans for coverage of fuelling stations in Europe and has also given a roadmap and action plan for future car fleets [13].

The cost of the infrastructure build up, as well as the infrastructural costs from stationary filling stations to mobile storage systems, was estimated within a time frame between 2005 and 2035 (Figure 2.12).

For the mass usage of hydrogen vehicles (ICE, F/C, or combined), it seems that the infrastructural issues as well as the demands and availability can be fulfilled within a desired time frame of about 5–10 years; this includes hydrogen refueling stations, utilization and installation of electrolyzes, and local hydrogen productions.

On the other hand, costs and availability issues are the main driver of modern storage systems for hydrogen as far as individual transport is concerned. It can be stated, from the brief outline of different storage system within this book, and pulled from other sources, that compressed $H_2$ storage would be the most feasible option for the midterm future for individual transportation, especially compared to the new-hype on e-mobility.

This example shows extremely well that space and package requirements, as well as usage of carbon fiber for high-pressure tanks, is the output of these findings. The high demand on carbon fiber plays a major role for the cost drivers of composite tanks, and availability and dependency on crude oil will have a high economic penalty and limitation. In addition to that, an aluminum liner is mandatory as a permeation barrier. The $CO_2$ and environmental outline of worldwide aluminum production also points to cost and pollution problems. The third problem is the combination of the tank system with pipes and hoses (dressings) leading to a weight penalty and an additional safety issue. This issue is related to the "hydrogen-safe" use of stainless steel and alloys for supply pipes; such systems have to provide an antibrittleness proof due to the nature of hydrogen molecular alloy contact on conventional sheet–metal–steel.

At the end of this evaluation, the usage of high-strength fiber materials such as T700 fibers is an absolute must. Cost implications from that decision would lead to a specific problem that can be observed on all composite material applications for mass production:

1. Filament winding technique is considered to be the selected manufacturing source; this would give high mass production options as seen on propeller drive shafts and various tank systems for civil aviation.
2. For a production output of more than 500 cars per day for a single OEM, the material cost issues of carbon fiber is the main driver of the feasibility of the use of $CH_2$ in individual cars.
3. Tools and investment costs for a carbon fiber tank production play a minor role compared to material costs which seems to be the limiting factor for the introduction of mass tank systems (Figure 6.1).

    It can be stated that with carbon fiber alone, as the natural source for high tensile fiber, the demand of future car development with respect to the implementation of tanks cannot be fulfilled.

    There is an economical barrier as well as a strong availability problem, so that a search for other HT fibers has to be initiated in order to meet any target (as presented by DoE, United States in several open publications).

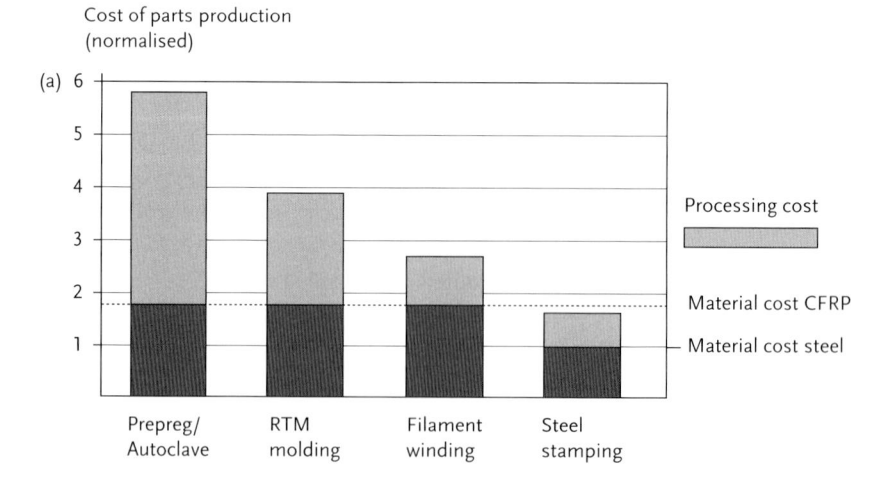

**Figure 6.1a** Material cost and production costs target outlines for carbon composite.

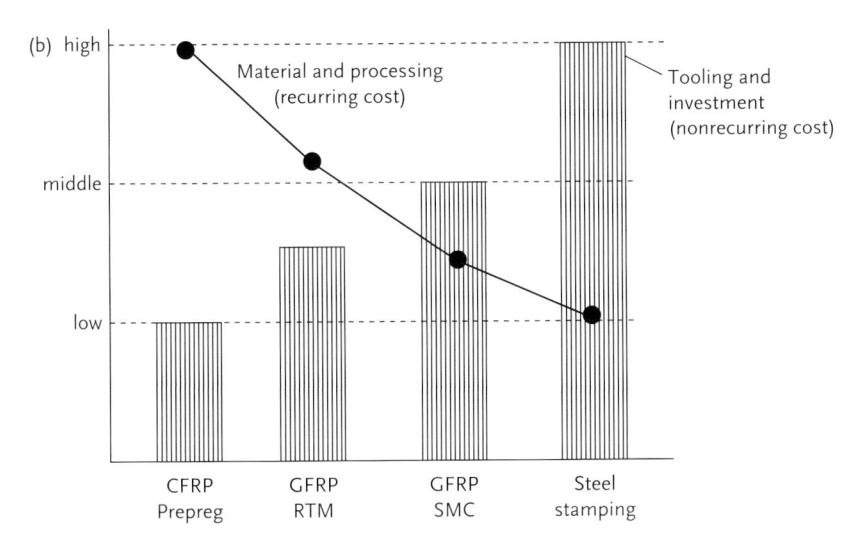

**Figure 6.1b** Nonrecurring costs and recurring costs for different materials.

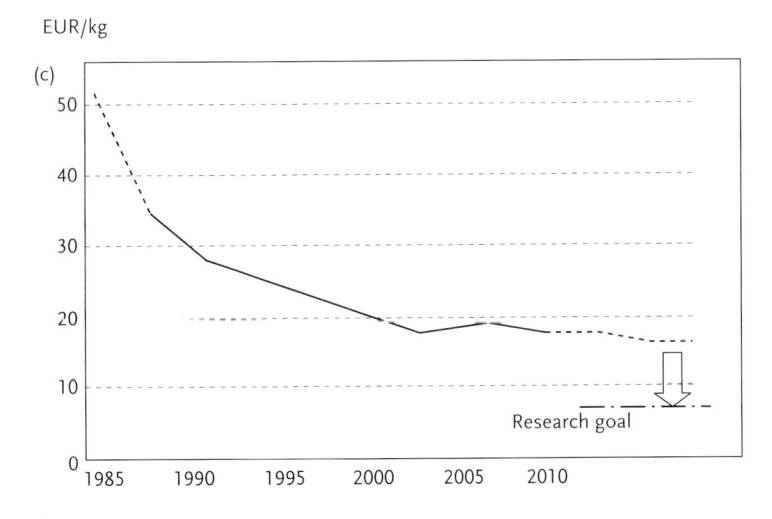

**Figure 6.1c** Cost development over the years.

In addition to this, an extraordinary demand on rolled aluminum liners would add another cost implication. Aluminum is still one of the most highly demanded and also high-cost light metal alloys produced with a high energy demand.

The storage compartment of hydrogen within vehicles including dressing lines, pressure reducers, and emergency shut-off valves is still to be designed with concepts from stationary power plant systems. Filling hoses, connectors, and the entire dressing system should be designed with a modern

hybrid composite state-of-the-art method rather than stainless steel piping system which shows compliance with very conservative design roles.
4. Pressure reduction assembly.

This outline would clearly give the indications that the main cost targets as demanded from the European Union (sixth framework program, project StorHy, [13]) cannot be met. In other words, the tank supply system will exceed the limit of 10 000 Euros per unit even under the assumption that mass production would lead to lower prices and competitive suppliers.

It is necessary to strengthen the effort in this field to relax safety requirements to a realistic level on one hand and to introduce new material and hybrid design options to lower the "per-unit" costs and availability of materials on the other (Figure 6.2).

## 6.2
## Safety Aspects

### 6.2.1
### Safety Rules and Regulations

There is a huge variety of safety rules and regulations for the use of compressed hydrogen and liquefied cryogenic hydrogen and the storage of hydrogen in a solid

**Figure 6.2** Low-pressure reducer valve with connector system and integrated shut-off mechanism.

state. These regulations are mainly driven by the demands of stationary systems (power plants and energy-packs) and are also outlined by the availability of isotropic materials and alloys such as aluminum, magnesium, and steel. In Annex 6.2.1, a complete set of European regulations and norms, especially for compressed hydrogen storage systems, is given as an example of homologation and certification issues.

The European Union has used its EN and EIHP system methodology to make a firm outline of vehicle homologation with respect to hydrogen tanks. These regulations are permanently under review and currently renewed according to general consideration from transportation issues.

As far as tanks of compressed hydrogen of any type or pressure is concerned, the design and layout of such systems is mainly referred to safety factors for:

- Daily loads, fatigue ageing, detorioration, and so on.
- Proof loads, for example, limit load, occurring several times during a lifetime, stress, and strength values have to be below yield, no plastic and permanent deformation is allowed, safe operation after occurrence of limit load case (LLC) has to be maintained.
- Ultimate loads, once in a lifetime critical load case, stresses, and strains have to be below breakage or rupture load but may be higher than yield, safe operation after ultimate load case for the rest of the mission has to be ensured, replacement repair or major inspection of desired structural parts after the occurrence of ultimate load case is mandatory.

By applying the design rules and regulations to pressurized hydrogen tank systems, for instance, these three design load cases have to be taken into account for any material selection under consideration. Daily load cases would require "design pressure" which can vary between 200 and 800 bar of atmospheric pressure for a car application. For these load cases, a brief fatigue structure analysis and component testing proof have to be performed by using S/N curves and applying the "safe-life" layout philosophy (Figure 6.3).

From the state-of-the-art design of tank systems, any crack propagation or damage-tolerant design layout strategy is inhibited by the current homologation requirements.

This is a useful, conservative, and "safe" design rule; however, it would also lead to weight implications because the fiber-reinforced design allows much higher margins with respect to damage-tolerant failure behavior (Figure 6.4).

This weight advantage, by changng the design philosophy from safe life to damage tolerant, is estimated to be greater than 20% of gross weight with the direct leverage to cost production. This can be proven by several examples of civil aviation such as helicopter main rotor blades and rear pressure bulkhead of plain fuselages.

It is clear that these components, the same as for hydrogen tank systems, are critical parts for the operation of the desired vehicles.

It is necessary to think about feasible options for the layout of tank systems with respect to aging and fatigue using new design rules.

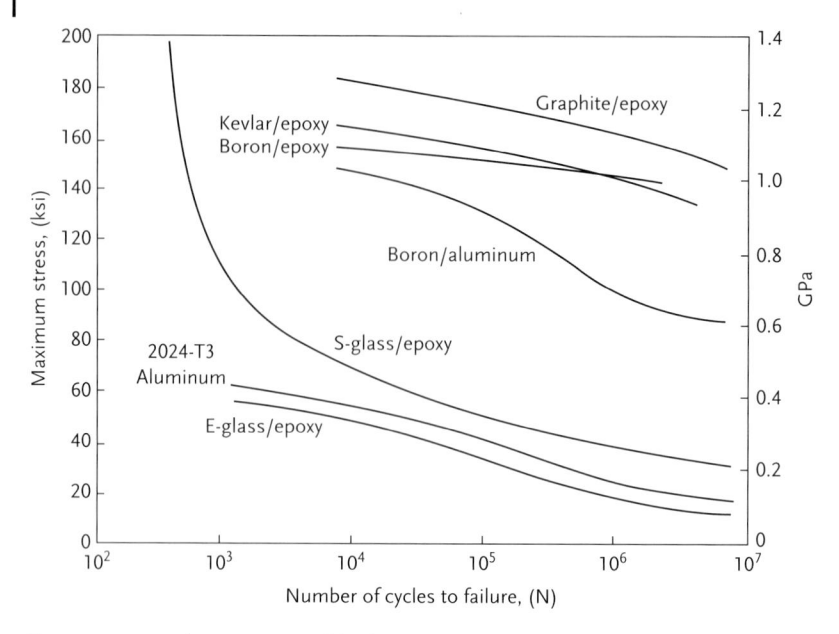

**Figure 6.3** Typical S/N curve (Wöhler diagram) for a prepreg-woven fabric carbon composite lining system.

The evaluation of tanks with respect to low- and high-cycle fatigue is a very complex issue because minor rules as well and Stromeyer/Weibull procedures for Wöhler curves are not possible for composite materials at a glance.

Following the EN design rules for pressurized tanks, the consideration of low-cycle-fatigue loading and its reserve factors will be fulfilled by the high-conservative layout of the tank against limit load. This gives automatically reliable margins of safety for any breakage of the tank, of course, but produces high-conservative design resulting in overweight, on the other hand.

Much sophisticated consideration of the structural strength of the tank has to be introduced into the EIHP in order to comply with the high capacity and design freedom of modern high-tensile composites (Scheme 6.1).

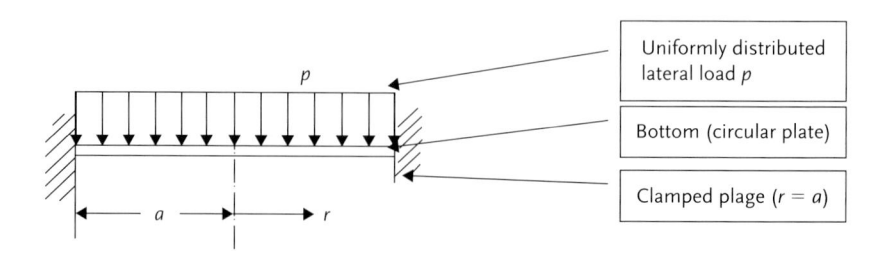

**Figure 6.4** Rear pressure bulkhead as an example of a damage-tolerant design (with kind permission from the Auto & Technik Museum Sinsheim e. V).

The deflection $w$ of the plate in polar coordinates (for symmetric boundary and loading) is given by

$$w = \frac{p}{64D}\left(r^2 - a^2\right)^2$$

$$D = \frac{Eh^3}{12(1-\nu^2)} \text{ (flexural rigidity of the plate)}$$

where
$E$ = Young's modulus
$h$ = thickness of the bottom
$w$ = deflection of the ciculate plate
$r$ = radius
$2a$ = diameter plate.

Internal forces and moments (Scheme 6.2):

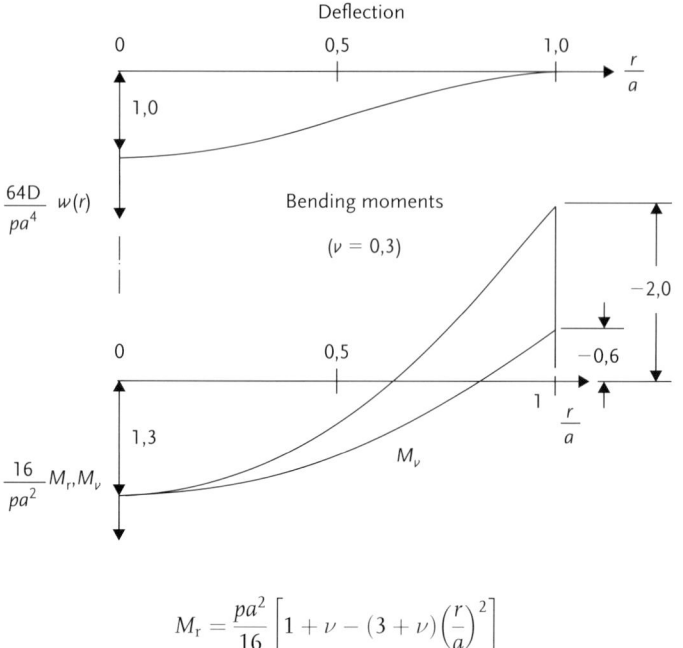

**Deflection**

$\dfrac{64D}{pa^4} w(r)$

**Bending moments** ($\nu = 0.3$)

$\dfrac{16}{pa^2} M_r, M_\nu$

$$M_r = \dfrac{pa^2}{16}\left[1 + \nu - (3 + \nu)\left(\dfrac{r}{a}\right)^2\right]$$

$$M_\vartheta = \dfrac{pa^2}{16}\left[1 + \nu - (1 + 3\nu)\left(\dfrac{r}{a}\right)^2\right]$$

$$Q_r = -\dfrac{pa}{16}\dfrac{r}{a}$$

Simple example for a circular aluminum bottom that carries a pressurized tank: $p = \text{const} = 0.001 \text{ N/mm}^2$ (assumption for a distributed load of a 50 kg tank (350 bar)).

$E = 70{,}000$ MPa;  $h = 5$ mm;  $\nu = 0.3$;  $a = 450$ mm;

$$w_{\max}(r=0) = \dfrac{pa^4}{64D} = \dfrac{0.001\,\text{N}(450\text{ mm})^4 12(1 - 0.3^2)\text{mm}^2}{64 \times 70.000 \text{ N}(5\text{ mm})^3 \text{mm}^2} = 0.8 \text{ mm}$$

$$D = \dfrac{Eh^3}{12(1 - \nu^2)} = \dfrac{70.000\,\text{N}(5\text{ mm})^3}{\text{mm}^2 12(1 - 0.3^2)}$$

$$M_{r,\max}(r=a) = \dfrac{pa^2}{16}[1 + \nu - (3 + \nu)1] = \dfrac{0.001\,\text{N}(450\text{ mm})^2}{16 \text{ mm}^2}[1 + 0.3 - (3 + 0.3)]$$

$$= -25.3 \dfrac{\text{N mm}}{\text{mm}}$$

However, changing the design rules to damage tolerant philosophy would require:

- instrumented component tests,
- extensive Wöhler curve measurements,
- measurements taken to consider rheological aging, influence of moisture, and detoriation against contact to liquids, UV rays, and gas.

For the layout of the tank system (covering fatigue!), the limit load case (LLC) is the main design criterion, giving the user a proper safety margin against yield and burst (Ultimate Load Case, ULC). For a daily load pressure of 350 bar resulting in an overall pressure reference of 370 bar to ambience, a result factor of 1.9 should be considered as a proof value.

By reaching a critical overload pressure of this kind, the tank structure may lead to yield strain of the assumed burst values.

It is demanded by the requirements that any burst case would happen after leakage of the tank and its related pipes.

This criterion leak before burst (!) is a main design concept for pressurized tanks. Leaking may occur due to microcracks due to yielding or because of the weakening of connectors, flanges, and pipes in the vicinity of the tank system. Leaking of this kind is a "structural leaking" due to yielding. "Artificial leaking" can be initiated by the introduction of safety valves which are a part of the safety equipment of the tank system (see Section 6.2.2).

Leak before burst will provide a good and safe design rule which is used for all low- and high-pressure tank systems and has given good results in recent history (Figure 6.5).

For the use of tanks, the design layout has to comply with the safety factors for proof (limit load, burst load, and ultimate loads). Reserve factors have to be 1.9 for LLC and 2.3 for ULC, resulting in a substantial margin for the "leak before burst" case. Methods of regulations, as seen above, leading to a conservative layout of the pressurized tank will not lead to a light-weight vehicle design for individual transportation.

Additional effort for new homologation rules is necessary to give the designer more degrees of freedom following the complete FMEA scenario (see Section 6.2.2).

### 6.2.2
### Safety Equipment

The design of a hydrogen storage system within vehicles, for example, ground transportation passenger cars, is linked with very strong implications with the so-called failure mode effectiveness and criticality analysis (FMECA).

FMECA or FMEA is a mandatory design step for the design of a safe tank concept within the body compartment of vehicles. The safety control of the power train and the fuelling system with all adjusted dressings should be managed by a control computing system which is also in charge of the engine or drive control.

## 6 Hydrogen Storage Options: Comparison

LH$_2$-test bench

Vacuum-fracture-trial

Crash test

Fire trial

**Figure 6.5** Typical leak before burst of hydrogen-pressurized tank in a crash environment, provided by BAM by StorHy Project.

The safety equipment of the tank system has to comply with an agreed FMEA philosophy, in which different safety events and failure modes are rated between categories such as likely remote, extremely remote (e.g., $10^{-9}$). The occurrence failure events in combination with other failure events linked to the global car system have to be rated and methods/counteractions have to be taken into account to redesign the system (Figure 6.6).

The main accessories to provide a safe operation with hydrogen can be divided into the following groups:

- safety valves and shut-off systems,
- hydrogen detection devices,
- shielding or compartment design,
- crash and impact resistant features,
- ventilation and antifire/detonation aerodynamical and structural layout methods.

1. Safety valves and shut-off systems have to be installed in the direct vicinity of the input/output pipe to the hydrogen tanks. This is relevant for low- and high-pressure lines; valves should be operated with the help of electrical features. The emergency shut-off valve systems have to be designed to provide a logical

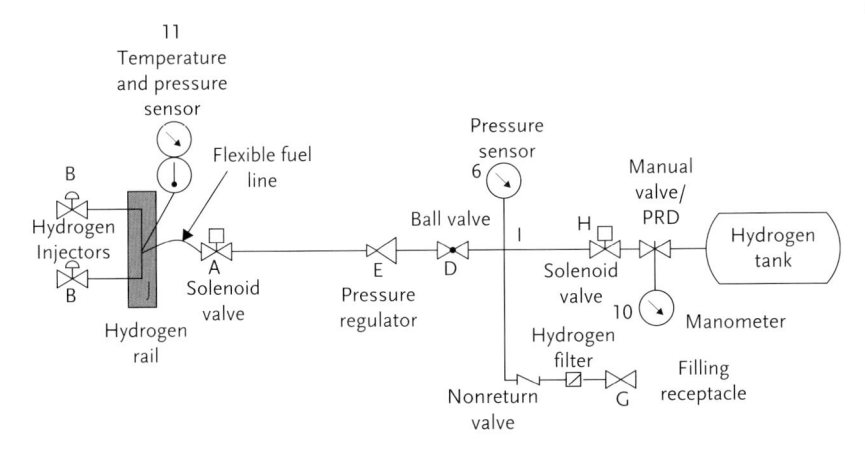

**Figure 6.6** FMEA safety design standard layout for a compressed hydrogen tank system and tank layout system.

link to pressure reducers (such as HPS). The internal system has to be designed against acceleration loads of 20 g longitudinal and 8 g transversal. This also applies to the suspension of the tank system and any framework which is related to the bearing of the tank. Valve systems have to be positioned not directly toward longitudinal crash scenario. In the case of placing valves and shut-off-systems to the outside of the body, the danger of breakage in a crash case will result in the so-called crash-recognition-sensory system which should shut-off the system within 2 s. In addition to that, the control computer should also allow a 2 s reaction time to shut-off the engine. In the low-pressure hydrogen engine supply circuit, a mechanical (hand-operateable) safety valve is necessary to avoid an engine hydrogen overflow in case of the failure of the pressure regulator (Figure 6.7)

2. Within the car body, hydrogen detection device is mandatory safety equipment which should be controlled by the safety computer system. Any leakage of hydrogen in a non- or partly-ventilated body zone should result in an emergency shut-off of the power train, with recognition of the current condition of driving (alerting the pilot). The so-called tank room should be a closed compartment with natural ventilation on two sides minimum, under recognition of the natural air flow in the car. The ventilation opening should be minimum 10% of the projected surface of the tank volume.

3. Safety compartment should be designed to comply with all conventional and required crash tests such as European New Car Assessment Program (Euro NCAP), Federal Motor Vehicle Safety Standards (FMVSS), and any other required national standards. The main threat of the tank compartment is found to be the rear crash barrier test, and any pole beside barrier crash test is proposed by FMVSS. However, the pressurized tank structure provides substantial strength against pole or back impact or contact (Figure 6.8).

**Figure 6.7** Typical pressure regulator (with kind permission from HPS).

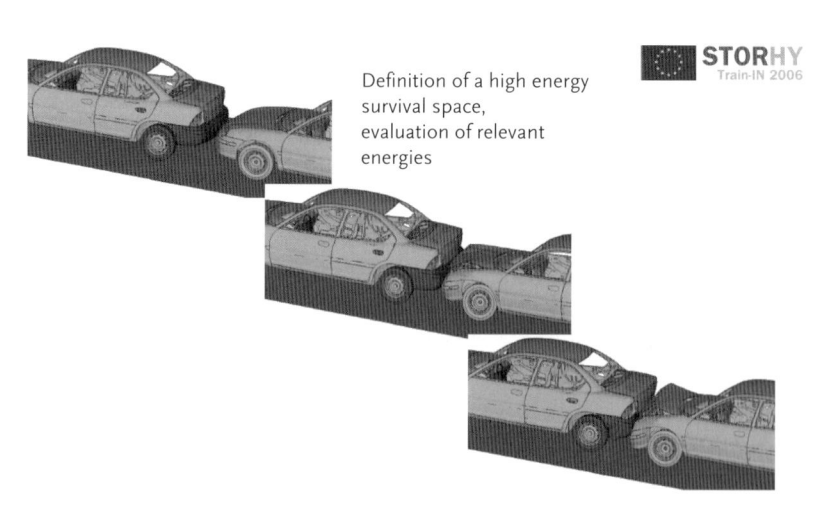

**Figure 6.8** Back crash simulation for passenger car.

4. In general, all components that may be air-washed by hydrogen should have a certification against actual homologation requirements. A proof of substantial properties for the contact with the hydrogen medium, as well as resistance against pressure and temperatures (mostly between 40 and +85 °C), must be present. In order to avoid any detoriation on fatigue loads and to supply residual strength for any crash scenario, the dressings, pipes, and hoses should be properly mounted to the fuselage of the vehicle. Standard bracket design should provide a minimum clipping distance of 15 cm for pressurized pipes (Figure 6.9).
5. The first point of ventilation is a detailed hydrogen detection sensor that alerts the driver immediately when only very few milligrams of hydrogen leak into the compartment. The entire engine unit should be investigated for any cavity, substantial blow by and fresh air exchange has to be maintained during all operational situations. In order to avoid fire or overheating or any accidental detonation of hydrogen-gas assembly, the exhaust system, in particular, has to be carefully controlled with detectors (pressure and temperature) as well as the injection/air-sucking inlet of an ICE which also has to be carefully monitored.

## 6.3
### Environmental Considerations: Waste, Hazardous Materials

For the future design of cars, individual transportation, considerations for environmental penalty as well as recycling, $CO_2$ balance, and fuel consumption will play the major role for the acceptability of vehicle families/fleets. The customer

**Figure 6.9** Bolted bracket tree on a metal high-pressure pipe.

will be aware of environmental-friendly design and power train of the selected individual transportation vehicle, so that hydrogen propulsion and hydrogen storage systems show an extraordinary compliance against these demands.

The environmental balance of a hydrogen propulsion system can be divided up into few main fields:

- hydrogen production,
- distribution and infrastructural means and methods,
- power train and engine considerations,
- $CO_2$ balance of all materials, recycling, and renewable energies,
- storage system tank, which is the main focus of this section.

Ecological environmental considerations for the storage of hydrogen are mainly related to the choice of material, productions methods, and recyclability of the design/structural materials.

Especially for aluminum, which is a good permeation barrier, the ecobalance points to high-energy sources during the raw-material production. In this field, a lot of different considerations can be given not only from industry but also from future science institutes; however, aluminum production can be related to high primary energy need, 5–10 times higher than other metal alloys. Recycling of aluminum is mainly related to secondary aluminum smelting; aluminum pellets are a main source for deoxidation of steel alloys in high oven furnaces (Figure 6.10).

The life-cycle chain of aluminum has been a well-known process for more than 150 years; however, the need for sheet metal and rolled aluminum structural pellets (shells) would give another high demand from the manufacturing part of view, as far as ecological requirements are concerned. There is still a strong link for the use of crude oil products to fire powerplants, providing the high demand of

**Figure 6.10** Aluminum, recycling of aluminum waste to provide pellets for deoxidation.

## 6.3 Environmental Considerations: Waste, Hazardous Materials

electrical power for aluminum production sites. This will give an additional $CO_2$ penalty to the aluminum life-cycle assessment.

Tank systems, especially compressed hydrogen tanks, would demand fiber-reinforced material to provide strength and structural integrity as well as stiffness. These demands lead to the obvious use of carbon fibers; high amounts of high-tensile fibers would be needed to fulfill a production line of more than 1 000 000 units/year.

Carbon fiber production is a very complex issue, mainly linked to the use of crude oil products; also PAN fibers would see some methods of alloy with crude oil. There are several serious considerations and plans for the set up of high-volume carbon fiber manufacturing such as SGL CARBON, manufacturing planed project together with BMW (mainly used for E-Car Mobility) (Figure 6.11).

Dependency on crude oil and its environmental impacts is one problem in the preface of the life cycle, recycling, and reuse of carbon fiber; (shredded) material is a problem on the other side. The carbon fiber waste, same for all other fiber wastes, is not necessarily a hazardous material but can lead to cancer and other serious diseases if the fibers are very short/chopped. In the project StorHy, the recyclability of fiber material played a major role in the feasibility of tank systems in modern cars. Environmental considerations from this study were mainly driven by the University of Nottingham. Compliance to all current legal regulations with respect to the recycling of composite material was shown in this study (Figure 6.12).

The resin system as part of the composite material is, in most cases, 2 K bonded, for example, epoxy, other thermoset resin systems such as PUR, as well as thermoplastic resins which will lead to different restriction for the reuse of the material for primary parts.

2 K matrix systems for composites will show a one-way hardening effect after curing (autoclave), so that after remelting or shredding of the composite in the

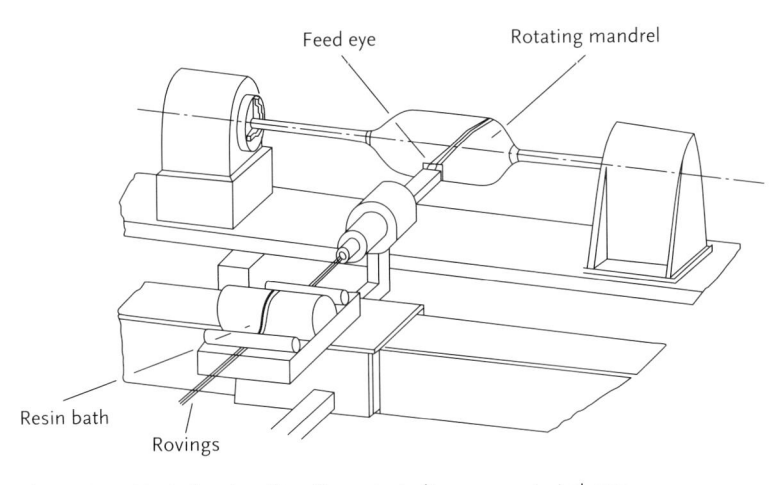

**Figure 6.11** Typical carbon fiber filament-winding process in industry.

## THE GLOBAL CARBON CYCLE

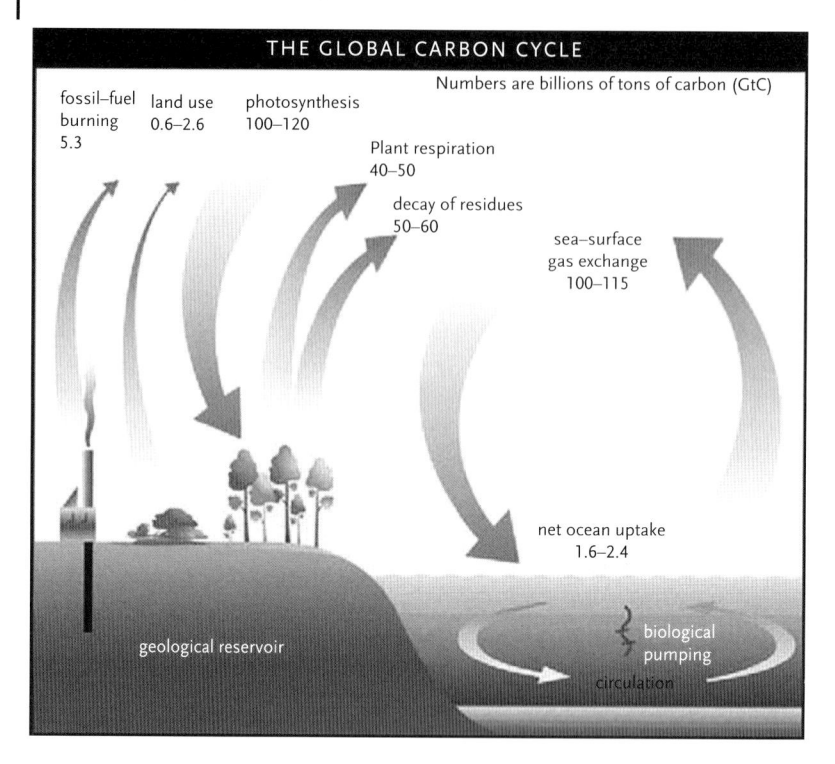

Numbers are billions of tons of carbon (GtC)

fossil–fuel burning 5.3
land use 0.6–2.6
photosynthesis 100–120
Plant respiration 40–50
decay of residues 50–60
sea–surface gas exchange 100–115
net ocean uptake 1.6–2.4
geological reservoir
biological pumping
circulation

**Figure 6.12** Recycling flow chart for carbon fiber (tank material).

process of recycling, only secondary functions of the recycled materials can be reached (filler) or the material is given in the process of "trashing/dumping or plasma-fire burning."

All metallic components of the dressings can be easily reused after melting or casting; this is mainly applicable to steel and steel alloys.

Rubber, connectors, and gasket can be environmentally recycled with the current methodology (Figure 6.13).

By setting up an ecobalance, the extremely reduced output of $CO_2$ and $NO_X$ by using ICE or F/C has to be taken into account in the calculation. Because of almost zero emissions and the substantial advantages of clean driving, any environmental considerations as far as production and recycling of the tank material is concerned play a minor role.

## 6.4
### Dimension Considerations

As seen from the previous sections, storage of hydrogen plays the key role of future mobility for ground transportation. Tank systems can be provided as state-of-the-art in a solid, liquid, and pressurized state. As seen from the argumentation,

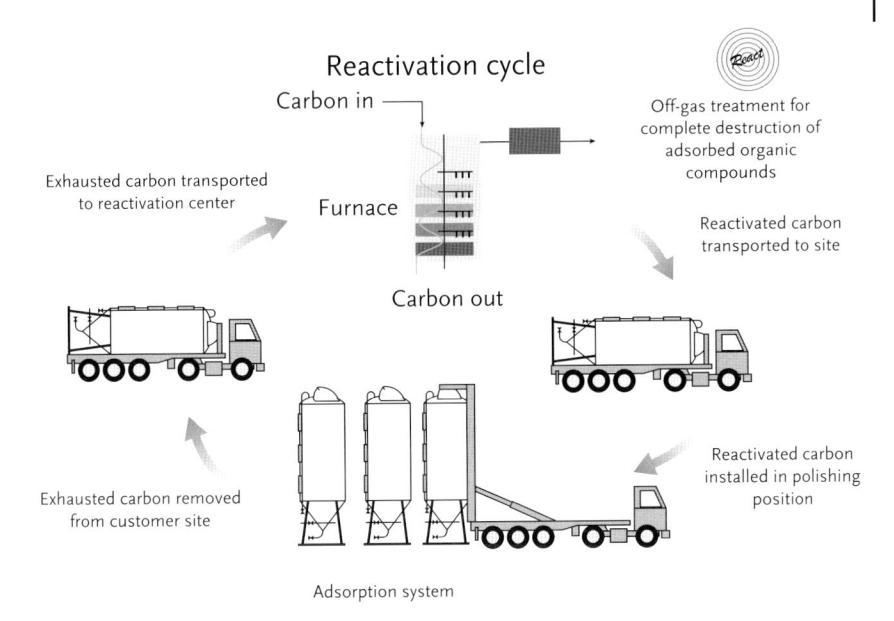

**Figure 6.13** Circle of material usage as a part of life-cycle assessment.

especially as far as cost and mass production capability is concerned, compressed hydrogen storage will play an important role for passenger cars.

Pressurized storage and cryogenic storage, however, would require a substantial space in the package compartment of the vehicle. In Figure 6.1, a typical storage system for a limousine vehicle is displayed; spatial design is basically a result of the amount of hydrogen to be carried with about a midrange of about 350 km.

The same space requirements, for instance, filling the complete tank of the car, would occur to cryogenic storage or semicryogenic-pressurized storage systems. In addition to this, a weight penalty of about 50–180 kg would spoil the practical ability of hydrogen storage for mass-produced cars.

Because of simple mathematical considerations, concerning the membrane-stress state of rotationally symmetric tanks, a spherical design is rated to be the best dimensional size of any tank design.

From this consideration, the conferential stress is only 50% of the value which will be reached, using a cylindrical or conical tank design layout (Figure 6.14).

$$\sigma_t = \frac{p_i r_i}{t} \approx \frac{p_i r}{t} \quad \sigma_a = \frac{p_i r_i}{2t} \approx \frac{p_i r}{2t}$$

$$\rightarrow \sigma_t = 2\sigma_a!!$$

where
$\sigma_t$ = tangential stress (stress in the circumferential direction),
a = axial stress (longitudinal direction),
t = wall thickness

## 6 Hydrogen Storage Options: Comparison

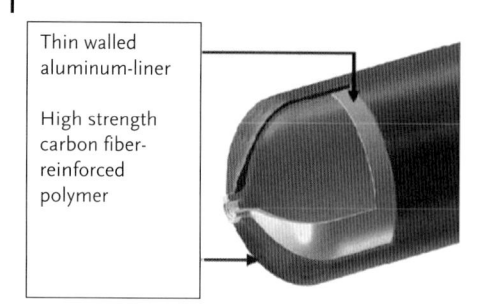

Thin walled aluminum-liner

High strength carbon fiber-reinforced polymer

**Figure 6.14**

$P_i$ = internal gauge pressure,
$r_i \approx r$ = inner (or middle) radius of the cylinder.

Numerical example (Scheme 6.3):

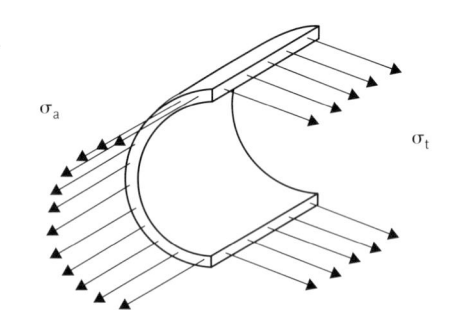

$$\sigma_t = \frac{20 \text{ N} \cdot 192 \text{ mm}}{\text{mm}^2 \cdot 8 \text{ mm}} \approx 480 \text{ MPa}$$

The surge for new concepts, avoiding spacious cylindrical tank design or spherical tanks, which can also be designed to be in groups of hexagonal packed spheres is one of the key issues of hydrogen storage for passenger cars.

Any other vehicles, marine vessels, stationary powerplants, or aircraft vehicle can allow the use of cylindrical tanks (mostly in compressed state) or racks of spherical tanks for quick exchange and refill (Figure 6.14).

For cryogenic storage tanks, the necessity of rotation-symmetrical tanks can be waived, this is mainly due to much lower pressure requirements (operating pressure between 7 and 12 bar). This will lead to free form, composite light-weight tank design, as seen in the StorHy project and explained in previous chapters. Figure 6.15 shows one of these concepts that will fit perfectly in the mid-body-passenger vehicle.

**Figure 6.15** Typical CH$_2$ tank rack from a European OEM (with kind permission from Daimler AG).

Even this progress in modern cryogenic tank design would violate the tank cost targets as well as manufacturability of the system in high production volumes.

In the end, CH$_2$ systems with high pressure (up to 750 bar) in cylindrical form are the state-of-the-art and most promising approach as a compromise of all demands and requirements.

However, the search for new materials, (fiber-reinforced design) and geometrical layouts are the most important move in engineering science in this field for the future.

For instance, tank design using combinations of hose-systems and small spherical tanks would allow a perfect placement in different areas of the car package, as seen in Figure 6.16.

The use of hydrogen tanks in vehicles would definitively lead to new package designs, such as double-shell floors (sandwich design) compromising the limousine design with van-bus design. With this strategy, storage of hydrogen will also be the driver of new passenger car designs; in combination with F/C and batteries (e-mobility), there will be a lot of interesting opportunities for the future. In Figure 6.17, two typical packages for bus and a passenger car are displayed showing high integrity of the tank system with the undercarriage compartment of the vehicle (Figure 6.18).

**Figure 6.16** Free-form tank design and placement of cryogenic storage tank in the mid-compartment of a BMW vehicle (with permission from BMW Group).

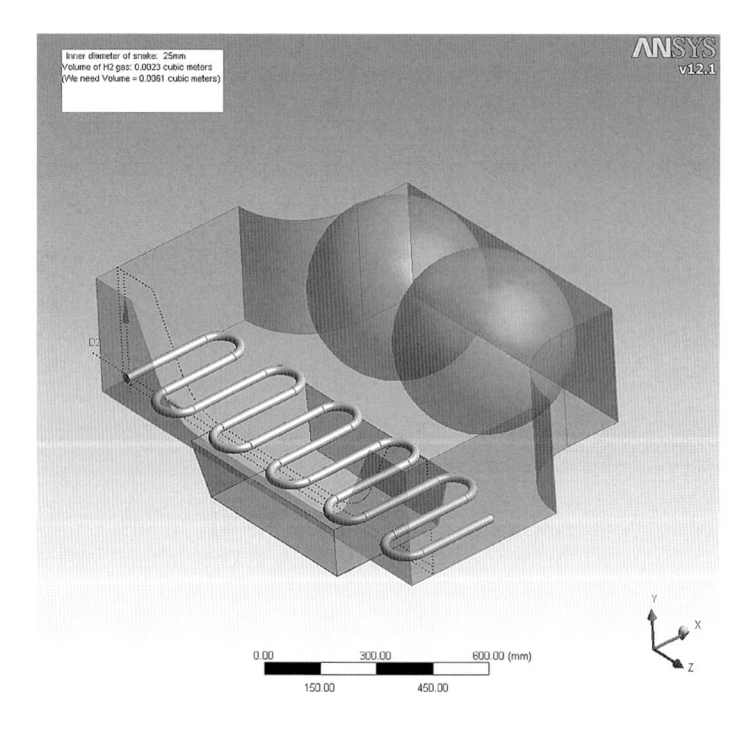

**Figure 6.17** New-age tank design, using a combination of pipes and small packs of spherical tanks.

## 6.5
### Sociological Considerations

There are many social considerations with respect to hydrogen and hydrogen storage from the early days of history to new applications in modern times. Main points of the social aspects are "bad experiences" and lots of accumulated observations on hydrogen behavior under fire and under pressure from public sciences. As learned from experiences while working with students and young people, who deal with

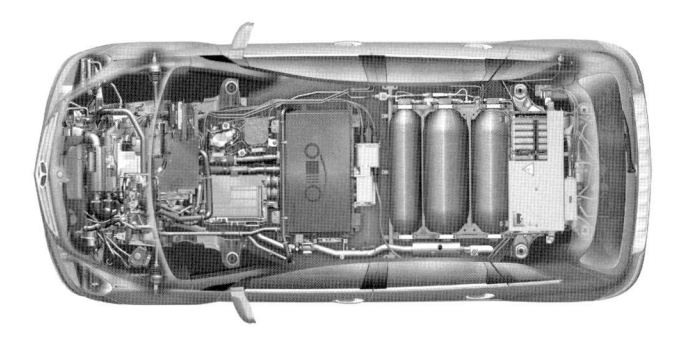

**Figure 6.18** Package proposal of an A/B-Class Daimler.

compressed gases, especially hydrogen, "natural fears" create while handling, refueling, and "touching" dangerous gases. The main considerations with respect to hydrogen in the social field of the community are pointing to dangers and cautions, more than to necessity, benefits, and advantages. From the author's point of view, the main fields of this social environment are driven by the following features:

1. Hydrogen is a dangerous gas. It may may explode or detonate.
2. High-pressurized storage of "dangerous gas" will lead to a natural fear against burst or rupture of the design structure.
3. Few bad experiences from the past especially the Hindenburg accident are still in the minds of the public.
4. Hydrogen may permeate and escape very quickly through the wall of tanks and any vehicle compartment, leading to undetected accumulation of hydrogen in nonventilated compartments, which will cause danger of detonation.
5. A lot of safety requirements and regulations will lead to scared minds in the community because "if something is tightly regulated, it must be a dangerous device."

The chemical color "red" for hydrogen as the first and lightest element is also considered to alert people, because red, as a color, is showing: DANGER, CAUTION, and often leads to emotional aggression, as seen from sociological observations. Within this insight of the sociological boundary conditions, even handling, such as fueling, refueling, touching valves and connectors will be rated as dangerous by most of the people who have never had any experiences with gasses, especially hydrogen (Figure 3.5).

The natural fear of touching and handling devices which are under high pressure will be even more increased if the natural gas is hydrogen, because of the above reasons.

In order to tackle the problem, the hydrogen community has, over the past few years, changed the strategical color from red to blue, as seen in the clean energy campaign of BMW.

To change the attitudes and alleviate anxieties, especially on hydrogen handling, the HY-RACER project of the University of Applied Science Ingolstadt was one of the main steps in the last years when dealing with students and young people.

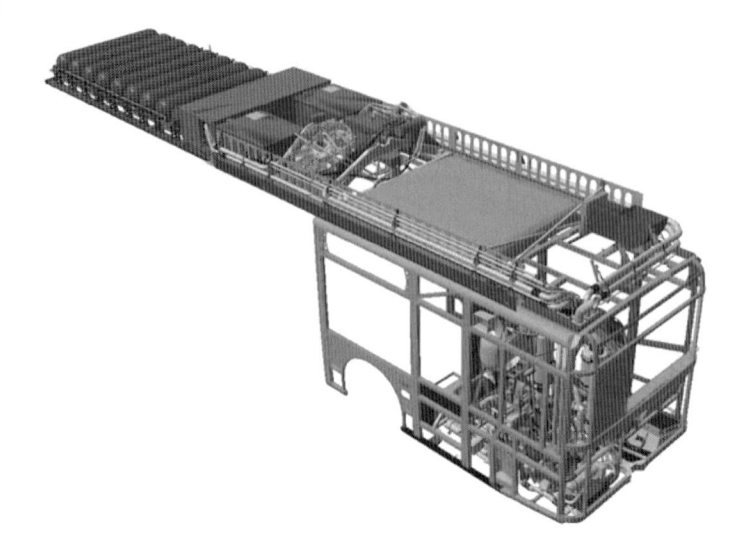

**Figure 6.19** Body-in-white-design of a bus is a good example for double-shell-flattened structure in roof design.

This project was mainly based on showcasing a small ICE car driven by compressed hydrogen on the base of a Graupner model car series (Mini-SUV).

This project mainly allowed the students/people to get a deep insight into the handling of hydrogen, and the authors were able to remove lots of fears and thoughts on the feasibility of hydrogen storage. In addition to that, the EU and project members of the STORHY project, such as Daimler, Ford, and BMW, decided to launch conferences on hydrogen together with special training on hydrogen filling, handling, and touching. In 2006, the first experimental conference of this kind was launched in Bavaria. Almost 50% of the conference time was used to put the participants in the hands-on environment and laboratories were dealing with hydrogen structure components.

The acceptability of hydrogen and its storage needs can also be encouraged by extensive use and display on vehicles for public transportation, so that people see and deal with hydrogen on a daily basis. This will lower fears and negative thoughts. With this background, any hydrogen project for ground transportation, especially for passenger cars, must have a social aspect. Handling and talking about these issues play a very important role and is very often underestimated.

## 6.6
### Comparison with Other Energy Storage System

Energy storage systems for vehicles, passenger cars, and public transportation are mainly focused on the storage of liquefied crude oil products as well as the storage of energy provided by electric current in batteries or adjusted fuel cell systems.

## 6.6 Comparison with Other Energy Storage System

Because of the importance of the storage of hydrogen for individual transportation/ground transportation, this chapter mainly focuses the emphasis of energy storage systems on the field of ground transportation. It is the strong indication from the previous deduction of hydrogen storage that solving this main problem would lead to a new generation of vehicles/passenger cars.

Thus, a direct comparison of hydrogen storage in pressurized, liquefied, or solid-state tank systems with other energy storage systems would also lead to a discussion on the future prospects on crude oil products.

With this background, energy storage for vehicles can be mainly divided into two groups.

1. Storage of electric energy with the help of batteries.
2. Storage of crude oil products in a liquefied state (conventional gasoline, diesel) or gaseous or liquefied natural gas products.

In Figure 6.20, a conventional gasoline storage system for midclass passenger car is displayed which shows the high integrity and state-of-the-art compact design of crude oil product storage.

It is obvious that a "no-pressurized" tank system can be designed in a free-form state that again allows much better integrity in degrees of freedom with respect to a package consideration.

The storage of hydrogen in the manner explained in the previous chapters would require a pressurized tank system; this would again demand a rotational symmetrical tank design. In addition to that, high-strength material is necessary to cope with the safety demands in the case of the conventional free-form gasoline tank, a low-cost, thermoplastic solution can be picked resulting in high mass-productional rates.

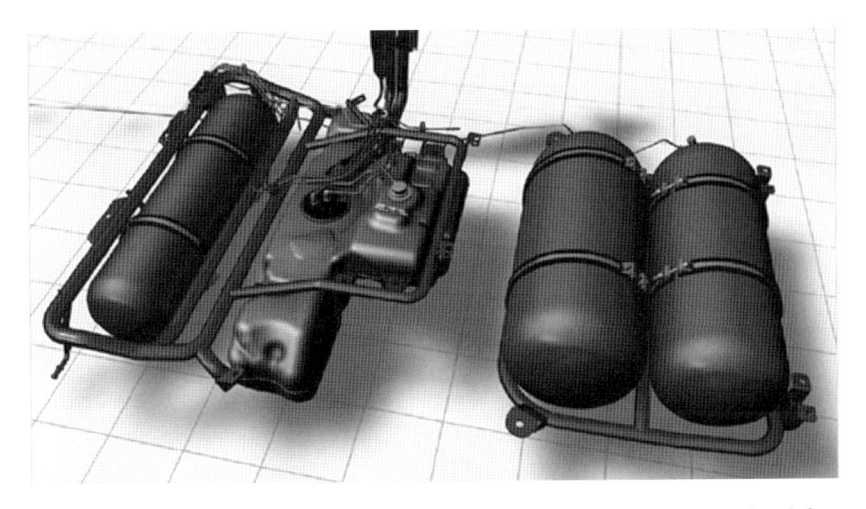

**Figure 6.20** Conventional gasoline fuel tank state-of-the art for passenger cars and tank for compressed natural gas.

## 6 Hydrogen Storage Options: Comparison

Therefore, comparison of hydrogen and hydrogen storage systems with other energy sources for vehicles is also a matter of design and structural mechanics rather than comparing energy density and efficiency of the storage system versus required space.

The source of hydrogen is a future energy carrier for passenger cars, and individual transportation would require a complete new tank design philosophy for the future with all implications shown in Chapters 3 and 5.

The storage of energy with batteries as a major key to future electric mobility plays a very important role for passenger car concepts in the moment and is to a certain extent in competition to the storage of hydrogen in combination with ICE. The use of electric battery systems in combination with combustion engines as a range extender or in combination with F/C cars is an interesting current development between different OEMs (Figure 6.21).

The storage of electricity in high-capacity batteries like lithium ion or carbon nanotubes will give a new and very promising approach to e-mobility at all which would spoil the necessity of hydrogen, hydrogen filling, and hydrogen storage in passenger vehicles.

The development of F/C vehicles with or without combustion range extenders would, of course, need hydrogen tank solution. On the other hand, this tank solution is also the key role as far as cost, safety, and availability are concerned to the feasibility of any F/C vehicle concept.

It plays a minor role if the storage density or efficiency of electric storage system is better than hydrogen storage. It can be observed that the current development in this direction is mainly driven by strategic thoughts and the current "electric mobility hip" which started in mid-2009 and has affected almost all major OEMs (Figure 6.22).

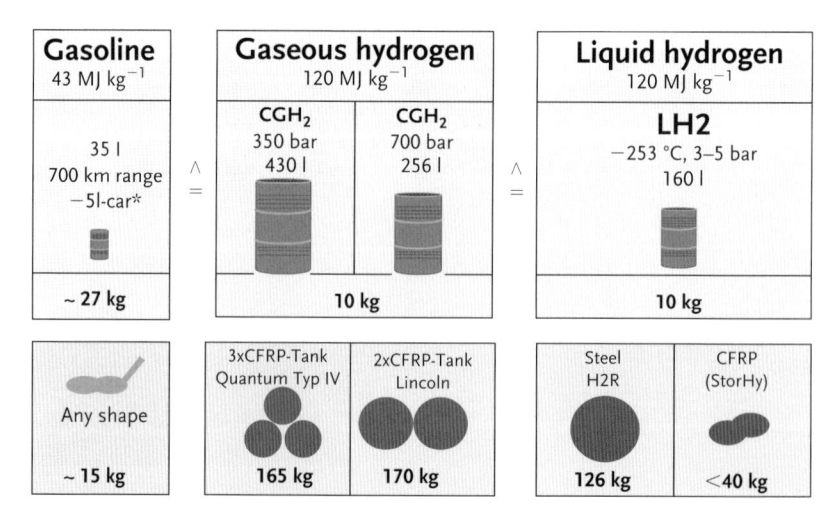

**Figure 6.21** Energy storage compared to conventional crude oil fuels.

## 6.6 Comparison with Other Energy Storage System | 221

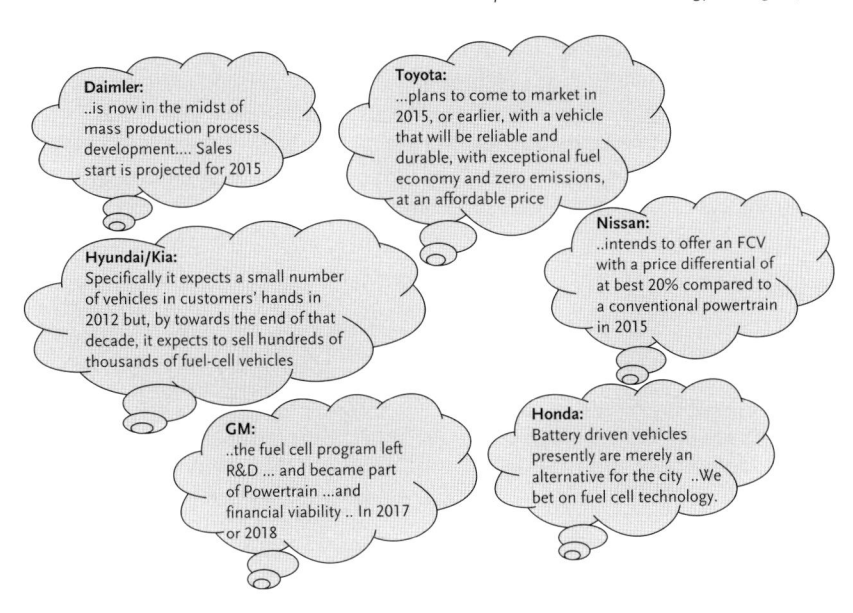

**Daimler:**
..is now in the midst of mass production process development.... Sales start is projected for 2015

**Toyota:**
...plans to come to market in 2015, or earlier, with a vehicle that will be reliable and durable, with exceptional fuel economy and zero emissions, at an affordable price

**Hyundai/Kia:**
Specifically it expects a small number of vehicles in customers' hands in 2012 but, by towards the end of that decade, it expects to sell hundreds of thousands of fuel-cell vehicles

**Nissan:**
..intends to offer an FCV with a price differential of at best 20% compared to a conventional powertrain in 2015

**GM:**
..the fuel cell program left R&D ... and became part of Powertrain ...and financial viability .. In 2017 or 2018

**Honda:**
Battery driven vehicles presently are merely an alternative for the city ..We bet on fuel cell technology.

**Figure 6.22** Statements of different OEMs regarding future vehicle concepts showing high diverging strategies (ICSAT 2010 [6]).

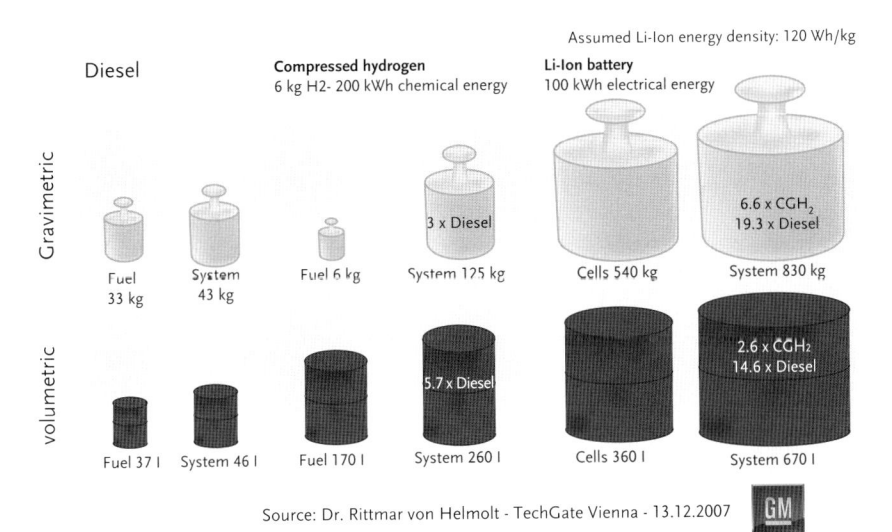

Assumed Li-Ion energy density: 120 Wh/kg

Diesel | Compressed hydrogen 6 kg H2- 200 kWh chemical energy | Li-Ion battery 100 kWh electrical energy

Gravimetric

Fuel 33 kg | System 43 kg | Fuel 6 kg | 3 x Diesel System 125 kg | Cells 540 kg | 6.6 x CGH$_2$ 19.3 x Diesel System 830 kg

volumetric

Fuel 37 l | System 46 l | Fuel 170 l | 5.7 x Diesel System 260 l | Cells 360 l | 2.6 x CGH$_2$ 14.6 x Diesel System 670 l

Source: Dr. Rittmar von Helmolt - TechGate Vienna - 13.12.2007    GM

**Figure 6.23** Weight implication of future battery systems on the cross-weight of car bodies (ICSAT 2010 [6]).

Figure 6.23 shows that there are substantial problems with the overall weight of passenger cars regarding the use of battery systems/packs as a main propulsion source. The advantages of hydrogen as a energy source are clearly in the field of low vehicle weight, high range availability, and durability of the propulsion system for more than 10 years. The use of hydrogen and hydrogen storage systems in combination with ICE is clearly in favor of all other propulsion systems as far as weight, cost, and high-volume production availability are concerned.

However, hybrid concepts such as F/C and ICE or hydrogen-powered ICE and auxiliary electric power system can play an important role in ground transportation in the field of urban mobility or short-distance drivers.

## References

1 Carlson, E.J. and Kopf, P. et al. (2005) *Cost Analysis of PEM Fuel Cell Systems for Transportation*, National Renewable Energy Laboratory, Lakewood, CO.
2 http://www.h2euro.org/2009/02/888
3 http://www.usfcc.com/
4 http://www.hydro.com/en/
5 http://www.uscar.org/guest/index.php
6 (2008) Design Guidelines for Hydrogen Piping and Pipelines. American Society of Mechanical Engineers, New York, NY, ISBN 079183137X, 9780791831373.
7 Thomas, G.J. and Jones, R.H. (2007) Materials for the Hydrogen Economy, CRC Press Inc. ISBN 10:0-8493-5024-7, EAN 9780849350245.
8 Burmeister, W. (2002) *Energiewelt Wasserstoff*, TÜV Süddeutschland Holding AG, Munich.
9 Mayer, R. (ed.) (2008) *Hydrogen Fuel: Production, Transport, and Storage*, CRC Press, Inc, Boca Raton, FL, ISBN 10: 1-4200-4575-X, EAN 9781420045758.
10 http://www.eihp.org/
11 Pehr, K. (1996) Aspects of safety and acceptance of LH$_2$ tank systems in passenger cars. *Int. J. Hydrogen Energ.*, **21** (5), 387–395.
12 Brewer, D.G. (1991) *Hydrogen Storage for Aircraft* Routledge Chapman & Hall, London
13 (2030) HyWays – H$_2$ Well-To-Wheel pathway portfolio for 10 EU member states.
14 http://www.springer.com/engineering/mechanical+eng/book/978-3-642-10796-2
15 Hordeski, M.F. (2009) *Hydrogen & Fuel Cells: Advances in Transportation and Power*, Fairmont Press, Lilburn, GA, Technology & Engineering, ISBN 0881735612.
16 http://www.hydrogenandfuelcellsafety.info/
17 von Cohen, Y.K., Gay, D., and Hoa, S. (2002) *Composite Materials: Design and Applications*, CRC Press Inc., Boca Raton, FL, ISBN 10:1-58716-084-6.
18 Chung, D.D.L. (2010) *Composite Materials. Science and Applications*, Series of Engineering Materials and Processes, Springer, Berlin,
19 Schmidtchen, U., Gradt, T., and Würsig, G. (1993) Safe handling of large quantities of liquid hydrogen. *Cryogenics*, **33** (8), 813–817.
20 http://www.storhy.net/pages.php?page=E02
21 http://www.hysafe.org/
22 http://ieahia.org/
23 http://www1.eere.energy.gov/hydrogenandfuelcells/
24 Steyr, M. (2007) Perspectives of probabilistic approach on hydrogen storage systems on vehicle level, presented by BAM. Scientific Session on Probabilistic Approach: Potentials and Procedures of the Probabilistic Approach. StorHy Annual Meeting, Berlin, Germany, February 27–28.
25 http://www.hydrogensociety.net/hydrogen_safety_issues.htm
26 http://www.dlr.de/fk/desktopdefault.aspx/tabid-2845/4374_read-6412/
27 http://www.fuelcellstandards.com/international_loc.html
28 http://www.hydrogen.energy.gov/codes_standards.html
29 http://www.hydrogenandfuelcellsafety.info/hipoc/index.asp
30 http://www.pnl.gov/fuelcells/permit_guide.stm
31 http://www.hydrogenandfuelcellsafety.info/
32 http://www.harmonhy.com/
33 http://www.hyapproval.org/

34 http://www.hyperproject.eu/
35 http://hcsp.ansi.org/
36 http://www.ch2bc.org/
37 http://www.hamburg-messe.de/H2Expo/h2_de/start_main.php
38 http://ec.europa.eu/research/fch/index_en.cfm
39 http://www.eucar.be/
40 http://www.hydrogen-engine.org/
41 http://eur-lex.europa.eu/RECH_naturel.do
42 http://h2bestpractices.pnl.gov/
43 http://www.cen.eu/
44 http://www.cenelec.eu/Cenelec/Homepage.htm
45 http://www.iccsafe.org/Pages/default.aspx
46 http://hydrogen.dot.gov/safety/
47 http://www.eihp.org/unece/docs/index.html
48 http://www.fuelcellstandards.com/portable.htm
49 http://www.nfpa.org/index.asp?cookie_test=1
50 http://www.hyways.de/
51 http://www.h2euro.org/2010/03/2337
52 von Sammes, N. (2006) *Fuel Cell Technology. Engineering Materials and Processes, Reaching Towards Commercialization*, Engineering Materials and Processes, Springer, Berlin, ISBN 10:1-85233-974-8.
53 Von Rajalakshmi, N. and Dhathathreyan, K.S. (2008) *Present Trends in Fuel Cell Technology Development*, Nova Science Pub Inc, Huntington, NY, ISBN 10: 1-60456-211-0.
54 Mobbs, P. (2007) *Energy – Beyond Oil*, Oxford University Press, Oxford, UK, ISBN 978-0-19-920996-5, ISBN 0-19-920996-0.
55 Tanakaa, T., Azumaa, T., Evansb, J.A., Croninb, P.M., Johnsonb, D.M., and Cleaverb, R.P. (2007) Experimental study on hydrogen explosions in a full-scale hydrogen filling station model. *Int. J. Hydrogen Energ.*, **32** (13), 2162–2170, Available online 7 June 2007.
56 http://www.worldenergy.org/focus/fuel_cells/377.asp
57 http://www.fuelcells.org/basics/benefits.html
58 von Kostikov, V.I. (1995) *Fibre Science and Technology*, Springer, Berlin, ISBN 10:0-412-58440-9, EAN 9780412584404.
59 http://www.dwv-info.de/
60 http://www.storhy.net/
61 Von Geitmann, S. (2006) *Wasserstoff und Brennstoffzellen*, 2. überarbeitete, aktualisierte u. erweiterte Auflage, Hydrogeit Verlag, Brandenburg, Germany, ISBN 10:3-937863-04-4, EAN 9783937863047.
62 Jessop, P.G. (2007) *Handbook of Homogeneous Hydrogenation* (eds. J.G. de Vries and C.J. Elsevier), Wiley-VCH, Weinheim, Germany, pp. 489–511.
63 Ahluwalia, R.K., Wang, X., Rousseau, A., and Kumar, R. (2004) *Fuel Economy of Hydrogen Fuel Cell Vehicles*, Argonne National Laboratory, Argonne, IL, USA.
64 http://www.greencar.com/articles/chrysler-ecovoyager-electric-car-uses-hydrogen-fuel-cell.php
65 http://www.hyracer.de/hyracer/home.html
66 von Romm, J.J. (2004) *The Hype about Hydrogen: Fact and Fiction in the Race to Save the Climate*, Island Press, Washington, DC, ISBN 10:1-55963-703-X, EAN 9781559637039.
67 O'Garra, T., Mourato, S., and Pearson, P. (2004) Analyzing Awareness and Acceptability of Hydrogen Vehicles: A London Case Study, Department of Environmental Science & Technology, Imperial College London, London, UK.
68 http://www.electricitystorage.org/ESA/home/
69 http://www.sandia.gov/ess/index.html
70 Marbán, G. and Valdés-Solísa, T. *Towards the Hydrogen Economy?*, Instituto Nacional del Carbón (CSIC), C/Francisco Pintado Fe, 26, 33011 Oviedo, *Spain*, Available online **2** February 2007.
71 Ehsani, M., Gao, Y., and Emadi, A. (2009) *Modern Electric, Hybrid Electric, and Fuel Cell Vehicles: Fundamentals, Theory, and Design*, CRC Press, Inc, Boca Raton, FL, *ISBN* **10**:1-4200-5398-1.
72 http://www.whec2010.com/
73 http://www.springer.com/engineering/mechanical+eng/book/978-3-642-10796-2

# 7
# Novel Materials

Hydrogen storage represents a high technological challenge to be solved for a commercial development. It is obvious that the search for alternative energy sources as well as their development is of high importance. Considering the access to the economic and technological infrastructure needed for energy storage, it is obvious that it is more and more important to find a way of efficient secondary energy. Among different hydrogen storage possibilities such as liquid hydrogen, compressed hydrogen gas, and absorption and adsorption in metal hydrides a group of novel materials has to be investigated. It is known that hydrogen storage technologies are rather difficult for proper evaluation; the best accuracy can be expected with well-known and developed technologies. Among such possible novel technologies, one can include hydrogen storage in carbon materials, storage in microspheres, and in solid hydrogenated polysilanes (HPS).

## 7.1
### Silicon and Hydropolysilane (HPS)

Since several years, an intensive research has been dedicated into a development of high-purity silicon production ways [1] at the NEXTSTEP GmbH and its affiliated companies at Bitterfeld-Wolfen, near Leipzig. After insolvency of the NEXTSTEP GmbH and negotiations, an Indian company Nagarjuna overtook people, patents, assets; now the names of the two new companies are: Spawnt Production and Spawnt Research. It is assumed that "In the near future perhalogenated polysilanes – accessible from silicon tetrahalide and hydrogen in plasma reactor – can serve as a link between solar energy and decentralized hydrogen generation. Silicon produced from perhalogenated polysilanes can be processed to modules of photovoltaics and solid hydrogenated polysilane (HPS) can be applied as a permanent hydrogen storage medium" [2].

A very interesting and innovative material, hydropolysilane HPS, has been synthesized from perchlorinated polysilanes in a following three-step way:

$$SiO_2 \rightarrow SiX_4 \rightarrow (X_2Si)_n \rightarrow (H_2Si)_n \qquad (7.1)$$

A simplified schematic of the process is shown in Figure 7.1.

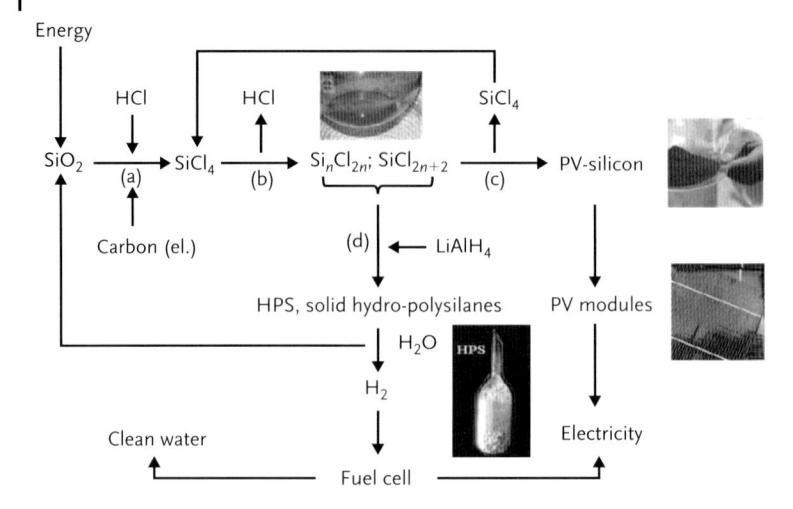

**Figure 7.1** Simplified generation of hydropolysilane from $SiO_2$ and solar energy: (a) carbochlorination, (b) plasma process, (c) pyrolysis at about 350 °C, and (d) hydrogenation.

As indicated in the above scheme, chlorine or preferably HCl gas is conducted over a solid mixture of sand/coal under access of microwaves to form silicon tetrachloride $SiCl_4$. The carbochlorination process also can be performed with conventional heating; microwave irradiation is only an efficient way to heat the C-containing material mixture to the required temperature level [2].

Under high reaction temperatures (above 1400 °C) initially formed water react with carbon, so the back reaction is suppressed and the formation of chlorosilanes is not hindered. At lower microwave energies depending on the reaction conditions, the reactor is coated with either solid or viscous perchlorinated polysilanes, soluble in $SiCl_4$ and consisting of cyclic or open chain compounds $Si_nCl_{2n}$ and $Si_nCl_{2n+2}$. It shall be noted that carbochlorination production and polysilane production are independent processes. $SiCl_4$ is first isolated from the carbochlorination product gas mixture and then polymerized.

In pyrolytic reaction, polysilanes are transformed to silicon and silicon tetrachloride, which can be recycled and passed back into the polymerization step [1]. Presently, instead of microwaves, large volume plasma excitation are applied, leading to the production of polysilanes at nearly room temperature. Starting material is silicon tetrachloride (process without step (a) in Figure 7.1) because it is inexpensive, easy accessible, and available with great purity with the purification technique developed to industrial maturity for the Siemens process [1]. Severe quantities of organic residues or hydrogen can be introduced into perhalogenated polysilanes $(X_2Si)_n$ and $Si_nX_{2n+2}$ without destroying the polymer structure (X is H or halo, preferably, F or Cl, more preferably Cl, that is, do not cause a splitting of Si−Si linkages) [3]. Hydrogenation with metal hydrides or metalloid hydrides (e.g., $LiAlH_4$, $NaAlH_4$, and $NaBH_4$) [3] leads to generation of hydropolysilane HPS

[2]. In addition, crystalline silicon produced from perhalogenated polysilanes can be molten and processed to wafers and photovoltaic modules.

According to the first spectroscopic analyses of the hydropolysilane, a linear structure of at least 20–40 silicon atoms in the polymer structure has been suggested. It is already known that the chain-like dodecasilanes $Si_{12}H_{26}$ is solid at ambient temperature [3]. HPS is a white/yellowish solid powder, odorless, and insoluble in water. Exposed to the air, it is slowly covered by a layer of $SiO_2$ and decomposes under inert gas atmosphere, starting from about 300 °C without melting. HPS can ignite spontaneously at above 80 °C only when it is piled up with a large surface area [2]. Hydropolysilane reacts with water according to the following idealized equation:

$$\frac{1}{x}(SiH_2)_x + 2H_2O \rightarrow SiO_2 + 3H_2$$

The reaction shows the capability of HPS to release 20.1 wt% of hydrogen of the material weight. HPS itself does not contain 20.1 wt% hydrogen (gravimetric hydrogen content in HPS is 6.7 wt%), but exhibits the capability to release the amount of hydrogen that equals 20.1% of the material weight [4]. Addition of another reactive hydrogen carrying material (e.g., water) is needed to utilize the complete hydrogen capacity of HPS. The production of hydrogen from HPS and water is an exothermal process, releasing about 10.8 MJ of reaction heat per kg of HPS [4]. Exothermal reaction on one hand leads to self-acceleration of the reaction as soon as it has started, on the other hand may cause temperature control challenges.

One of the safety properties of hydropolysilane is that it does not react rapidly with water under standard environmental conditions (0.1 MPa, $T=-20$ °C to 50 °C, no oxygen). It distinguishes the material from a number of similar metal hydrides, which vigorously react with liquid water or humidity. HPS can be safely transported without hydrogen losses and stored permanently. Storage of an HPS pellets on air for more than 2 years resulted in a white, apparently nonflammable product, presumably $SiO_2$. The pellet stayed hard and compact during the oxidation process and showed no significant changes in dimensions [4].

For the time being, a technically sophisticated method of producing hydrogen from HPS has not been developed. The material is not yet commercially available; however, an intensive research is ongoing at the NEXTSTEP GmbH [4]. There are many possible applications of HPS. Usually HPS is considered as a hydrogen storage material in combination with a fuel cell, because fuel cells possess a better conversion efficiency in comparison to internal combustion engines. Released hydrogen can be supplied to PEM fuel cells [5]. The overall idea is to transfer and convert the solutions selected from the group of silanes and polysilanes into hydrogen in the reaction chamber in the presence of water vapor or aqueous solution. The next step is to remove the conversion products from the chamber and to transfer generated hydrogen to the PEM fuel cell.

A potential application of hydropolysilanes could also be a new class of high-performance precursor materials for many silicon- and silicide-based thin film technologies and surface modification [3]. They could be applied onto surfaces as solutions or dispersants and to obtain silicon-based structures or layers.

They are safe in handling and can be made up into transport forms in suitable containers. Low molecular polysilanes could be used as solvents for solid polysilane mixtures as well as solvents or in solvent mixtures for preparation of solutions or dispersions of solid polysilane mixtures. Solid polysilane mixtures can be applied onto surfaces as solids and later can be melted to generate liquid polysilane structures or polysilane films to produce silicon through decomposition methods. The solutions or dispersions of solid polysilane mixtures mixed with additional components containing one or more elements from group III or V from the Mendeleev table, could be used for the production of silicon structures or layers with certain electronic properties [3].

## 7.2
### Carbon-Based Materials – General

Carbon-based materials have attracted a lot of attention and interest over the last decade. They seem to be an attractive solution for hydrogen storage when considering carbon as a cheap and light storage medium. Several years ago very high storage capacities of carbon-based materials at room temperature have been reported, even to 67 wt% for graphitic nanofibers at ambient temperature and pressure of 11.35 MPa [6], but it was never repeated in laboratories and today the opinion is that such values were impossible and came as a result of measurement errors [7]. Experiments performed at National Renewable Energy Laboratory (NREL) showed a maximum capacity for adsorption of hydrogen on single-walled carbon nanotubes is about 8 wt% [8]. It has to be mentioned again that the US DOE energy density goal for vehicular hydrogen storage for systems is 6.0 wt% and 45 kg $H_2$ $m^{-3}$. The truth to be said, the performance of carbon-based materials is rather poor and the possibility of using them for hydrogen storage depends on the dominating carbon–hydrogen interactions. In order to use carbon-based materials for hydrogen storage, the magnitude of the dominating interactions with hydrogen molecules shall be in between condition for physisorption and chemisorption [9]. Physisorption (or physical adsorption) of molecular hydrogen in nanomaterial is a nonactivated and reversible process where the hydrogen remains in its molecular state, whereas chemisorption includes dissociation of hydrogen molecules and chemical bonding of the hydrogen atoms by interaction in the lattice of a metal, alloy, or by formation of a new chemical compound [10]. Typical binding energy of physisorption is about 0.01–0.1 eV (1–10 kJ $mol^{-1}$). Chemisorption usually forms bonding with energy of 1–10 eV (100–1000 kJ $mol^{-1}$) and often involves activation energy. The physisorption of gas molecules on the surface are van der Waals interactions. In the physisorption process a gas molecule interacts with several atoms at the surface of the solid. Here the hydrogen acts as neutral but polarizable adsorbate. The hydrogen molecule is small, having two electrons (molecule radius is about 0.2 nm);

therefore, it is hard to polarize it, creating van der Waals forces is rather week. In the case of chemisorption, molecules are adsorbed on the surface by valence bonds and monolayer adsorption is created. The chemisorbed hydrogen atoms may have a high surface mobility and can interact with each other forming surface phases. The hydrogen atoms integrate with the lattice of the metal, alloy or can form a new chemical compound. According to Fichtner [10], an alternative for hydrogen storage by physical adsorption is hydrogen storage by encapsulation or trapping in microporous substances, where hydrogen trapped inside the pores by cooling can be released by raising the temperature.

Benard and Chahine [11] discussed the use and the limitations of physisorption as a storage media for hydrogen. The interest in carbon-based materials is that they can be optimized for hydrogen storage through several physical and chemical procedures. There are three following essential optimizing parameters in order to determine the best adsorbent: (i) the characteristic binding energy of the hydrogen molecule with the material, (ii) the available surface for adsorption processes, and (iii) the bulk density of the adsorbate. The binding energy is the factor that determines the operating temperature of a hydrogen storage system based on the solid. Ideally, the binding energies of carbon-based materials shall be between 0.2 and 0.8 eV (19.29–77.1 kJ mol$^{-1}$) [9], making possible that hydrogen remains at operational conditions without the need of applying high pressures and can be released after moderate heating, supplied by different sources, among them fuel cell. Ideal hydrogen storage material would have very high surface area, over 2000 m$^2$ g$^{-1}$.

## 7.2.1
### Carbon Nanotubes (CNT), Activated Carbon (AC), Graphite Nanofibers

Carbon-based nanoporous materials such as activated carbon (AC), single-walled carbon nanotubes (SWNTs), multiwalled carbon nanotubes (MWNTs), and metal-organic frameworks (MOFs) have been investigated as promising adsorbents for hydrogen storage applications by many authors. Pure carbon materials have binding energy in the range of 4–15 kJ mol$^{-1}$; lower values are typical for graphite and activated carbon whereas the higher are usual for SWNTs and MWNTs. The example of the structure of carbon nanotubes (MWNT and SWNT) is presented in Figure 7.2.

Fichtner [10] presents the overview of experimental data by several authors on hydrogen storage capacities in SWNTs over the last decade (Figure 7.3). It can be seen that depending on the chosen method and experimental conditions, the hydrogen storage capacities vary from 0.02 to nearly 10 wt%.

Study performed by Yamanaka et al. [12] showed that for carbon nanotubes prepared by the plasma-assisted chemical vapor deposition (CVD) method hydrogen content was about 8.6 wt% and in case of graphite nanofibers prepared by the thermal CVD hydrogen content was 0.77 wt%. The maximum storage capacity observed by Gundiah et al. [13] was 3.7 wt% on MWNTs prepared by the pyrolysis of ferrocene and acetylene, additionally treated with acid. Other carbon samples prepared in a different way showed much lower values, for example

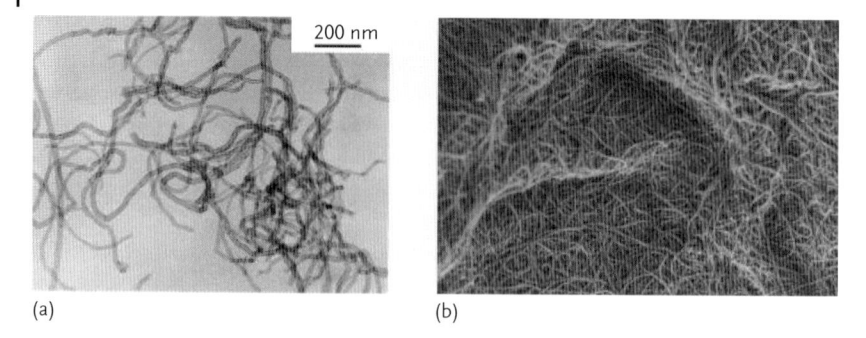

**Figure 7.2** (a) MWNT (Special surface area: 40–300 m$^2$ g$^{-1}$, thermal conductivity: ~2000 W m$^{-1}$ K$^{-1}$); (b) SWNT (Special surface area 500–700 m$^2$ g$^{-1}$, purity: >60%, outer diameter: <2 nm, thermal conductivity: ~4000 W m$^{-1}$ K$^{-1}$). Products of Schenzen Nanotech Port Co., Ltd.

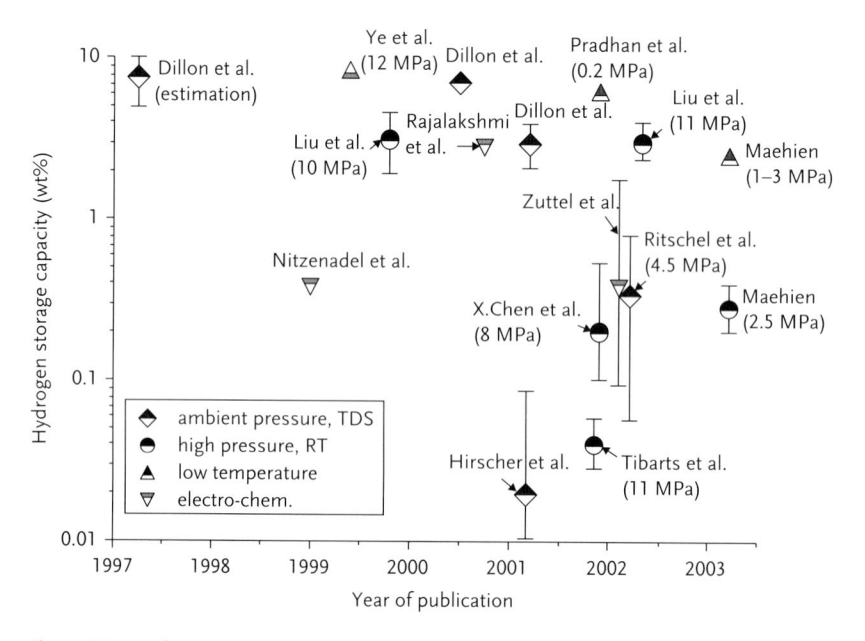

**Figure 7.3** Hydrogen storage capacities in SWNTs by different authors, methods and operational condition source: Fichtner, [10].

MWNTs synthesized by the pyrolysis of acetylene reached only 0.2 wt%, MWNTs by the pyrolysis of ferrocene showed 1.0 wt%, the value on MWNTs prepared by the arc-discharge method being well graphitized was 2.6 wt%. Most of the measurements showed that the amount of adsorbed hydrogen under ambient conditions on SWNT is relatively low, below 1 wt% [14], but under cryogenic conditions (77 K) the amount of adsorbed hydrogen increase from about 1 to 2.4 wt%. Anson et al. [15] report 0.01 wt% of hydrogen uptake at atmospheric pressure and room temperature and 1 wt% at 77 K ($-196\ °C$). At cryogenic conditions Pradhan *et al.* [16] obtained 6 wt% on SWNTs, but Gundiah *et al.* [13] report much lower values, from 0.2 to 0.4 wt%. Poirier *et al.* [17] under 77 K and 4 MPa reached maximum excess adsorption from 1.5 to 2.5 wt%, but at room temperature (pressure 4 MPa) it was less than 0.4 wt%. Reported adsorption enthalpy was about 3.6–4.2 kJ $mol^{-1}$, usual for carbon adsorbents. Dillon *et al.* in 2000 showed that at room temperature, room pressure densities of hydrogen from 3.5 to 4.5 wt% could be obtained on SWNT prepared by laser vaporization; in 2001 after optimization of the cutting procedure (in order to avoid introduction of potential impurities to SWNTs) hydrogen storage densities up to 7 wt% were achieved [18]. The authors also found that the nature of the interaction between the hydrogen molecule and SWNT is in between conventional week van der Waals adsorption and chemical bond formation.

Activated carbon is a bulky carbon having high surface area 500–2500 $m^2\ g^{-1}$, able to adsorb hydrogen in the microscopic pores. Such a great surface area of the activated carbon results from the presence of the internal structure of different pores. Most of them are strongly adsorbing micropores having diameter less than 2 nm and mesopores with diameter from 2 to 50 nm, performing mainly the role of transport channels. Macropores with diameter 100–200 nm are the widest pores in the activated carbon, the adsorption process on its surface is negligible, and similar to mesopores they act as transporting channels. The basic structure of activated carbon is shown in Figure 7.4 [19].

David [20] reports that under ambient temperature and pressure of 6 MPa, the storage capacity of hydrogen on activated carbon was only 0.5 wt%, but at cryogenic temperature and pressures 4.5–6 MPa the value increased up to 5.2 wt%.

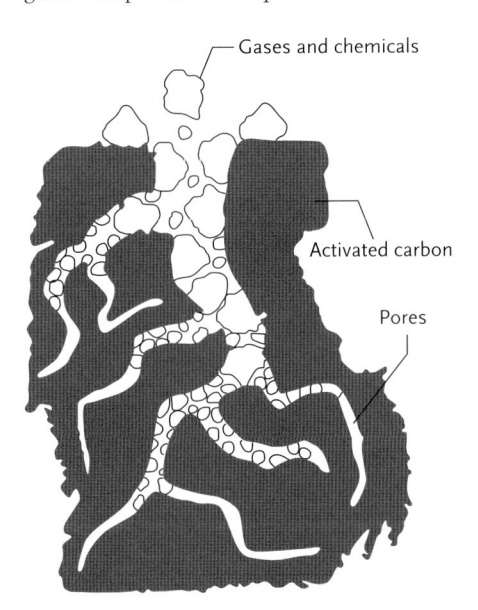

**Figure 7.4** Activated carbon structure with the pores inside (source:http://pages.total.net/ ~espitech/carbonpower.html).

Experiments performed on activated carbon obtained by carbonization of walnut shells and sawdust mixed with phosphoric acid showed that 2.2–2.8 wt% of hydrogen has been adsorbed at 77 K and pressures 1.2–1.5 MPa. Zhou *et al.* have compared hydrogen adsorption on superactivated carbon and nanotubes [21] and observed that the amount of hydrogen adsorbed on MWNT was three to five times less than on activated carbon, but the surface concentration of hydrogen on MWNT was four to six times higher than on activated carbon. Also it was noted that the influence of the temperature on the adsorption process of hydrogen is stronger in the case of the adsorption of $H_2$ on activated carbon than on MWNT. According to the authors, it seemed that the interaction between hydrogen molecules and the surface of MWNT was weaker than in the case of activated carbon, and the specific surface area of nanotubes is limited by the geometry and cannot be compared with very high surface of activated carbon. It was concluded that carbon nanotubes seemed not to be a promising hydrogen carrier for practical application [21]. A similar conclusion was also drawn by Li *et al.* [22]. Jurewicz *et al.* [23] investigated the mechanism of electrochemical storage in nanoporous-activated carbon by cathodic decomposition of water in acidic ($H_2SO_4$) and alkaline conditions (KOH) stating that in KOH electrolyte-activated carbon has significant higher hydrogen capacity due to high values of polarization.

An example of the hydrogen adsorbed on activated carbon and MWNT is presented in Figure 7.5.

Carbon nanofibers (Figure 7.6) can be synthesized through several methods such as chemical vapor deposition (CVD), hot-filament-assistant sputtering, and

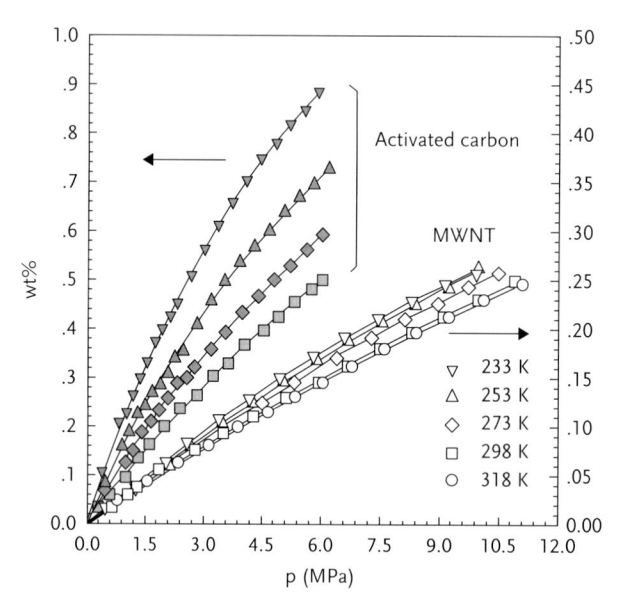

**Figure 7.5** Weight percentage of hydrogen adsorbed on activated carbon and on MWNT [21].

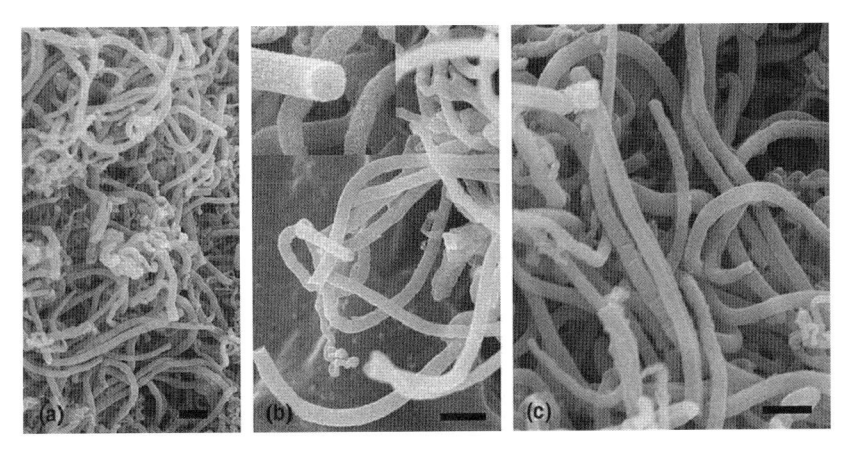

**Figure 7.6** (a–c) FESEM (field emission scanning electron microscopy) images of nanofibers, (b) high-magnification image of carbon nanofiber tip, scale 200 nm [24], (reprinted by permission.)

template-assisted methods [24]. Nanofibers are in general layered graphitic nanostructures; the reported values of hydrogen storage capacities are in the range from about 1 wt% to several tens of wt% at moderate temperatures and pressures. Such difference may result from different sample preparations at various laboratories and experimental procedures. The reproducible results at temperatures from 77 K (−196 °C) to 295 K (22 °C) and pressures varying from 0.01 to 10.5 MPa on various types of nanofibers performed by Benard and Chahine showed that the maximum obtained storage density was 0.7 wt% at pressure 10.5 MPa [11]. Li *et al.* on series of experiments reached values in the similar range, from 0.85 to 0.92 wt% [22]. Study on the interaction of hydrogen with activated charcoal, carbon nanofibers, and nanotubes (SWNT) has shown that SWNTs due to its structure have a relatively low accessible surface area (bundling of the tubes). This is why the hydrogen molecule is not entering the space between the tubes in the bundles. The hydrogen storage capacity for SWNT of 60 mL g$^{-1}$ equals 0.54 wt% of hydrogen [25]. By extrapolating to 300 K, these values have been compared with the results of Gundiah *et al.* [13] who obtained, as mentioned already, hydrogen storage capacities in the range of 0.2 wt% at 300 K and 14 MPa to 3.7 wt%. The conclusion by the authors was that at ambient conditions, the adsorption of hydrogen is negligible resulting from a very short residence time of hydrogen molecule and that: "hydrogen does not enter the interrestial channels of the SWNT because channels between nanotubes are much too small for hydrogen molecules to enter" [25].

## 7.2.2
### Other High-Surface Area Materials

Apart of carbon-based nanomaterials for hydrogen storage application a certain interest has been given to other high-surface materials such as zeolites, metal organic frameworks (MOF), covalent organic frameworks (COF), and clathrate hydrides. An overview about potential abilities to store hydrogen in these materials is presented.

### 7.2.3
### Zeolites

Zeolites have been discovered in 1756 by Swedish mineralogist Axel Fredrik Cronstedt and can be defined as a large group of aluminosilicate minerals with different chemical composition, properties, and crystalline form. Mainly they are hydrated sodium and calcium aluminosilicates. Zeolites possess a unique crystalline structure and ion-exchange capacity having a large number of pores in their structure, mainly micro- and mesopores. Zeolites have many industrial applications; they are used as ion-exchangers, for example, in water softeners or clean-up of radioactive wastes. Zeolites are applicable in drying gases and liquids, oxygen enrichment in air, filtration of drinking water, wastewater treatment, and many others.

Ideal chemical formula of natural zeolites is as follows [26]:

$$M_{x/n}\left[Al_xSi_yO_{2(x+y)}\right]pH_2O$$

where M stands for monovalent (Na, K, Li) and/or divalent (Ca, Mg, Ba, Sr) cations $n$ – cations number, $x/y = 1 \div 6$, $p/x = 1 \div 4$.

The example of zeolite structure is shown in Figure 7.7.

Two possibilities to store hydrogen in zeolites are by physisorption (already described) and by encapsulation of molecular hydrogen. Below and at room temperatures hydrogen molecules do not enter into some of zeolite pores, but at elevated conditions they are forced into the cages. When returning to ambient conditions, hydrogen remains trapped in the cages and the release may occur

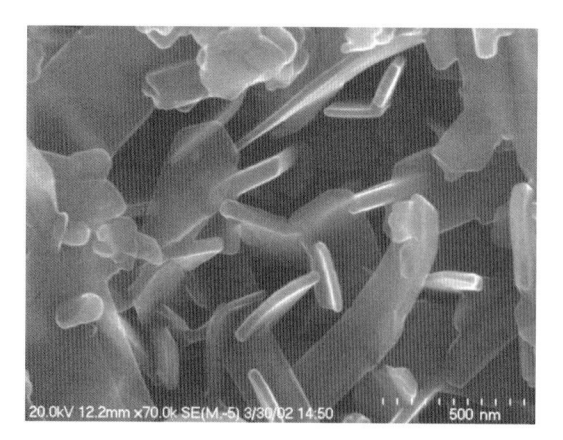

**Figure 7.7** Synthetic zeolite type hydroxycancrynite, magnification 70,000 (source: www3.uj.edu.pl).

upon reheating [27]. According to Fichtner, a factor limiting storage capacity of zeolites is the relatively big mass of the zeolitic framework containing Si, Al, O, and heavy cations [10]. It may be assumed that lighter frameworks combined with improved binding sites might improve storage capacities. Different types of zeolites show different hydrogen uptake due to the framework structure and the nature of present cations.

A review of hydrogen storage capacities by Vitillo et al. [28] presenting a wide range of experimental data did not clearly indicate the potential of various zeolites to store hydrogen. Analyzed data were collected under different working conditions, with various zeolitic frameworks and chemical composition. Depending on the source, the values oscillated between nearly 0.0 and 1.8 wt%. At temperatures of 293 K (20 °C) and pressure 1 MPa obtained storage capacities were low, not exceeding 0.12 wt% (KA zeolite) 0.11 wt% for NaA, 0.1 wt% for NaLiA and LiA. With the increase of the pressure to 6 MPa the respective values were slightly higher, up to 0.49 wt% for KA [29]. This maximum storage capacity of 1.8 wt% was obtained for NaY zeolite at 77 K (−196 °C) at pressure of 1.5 MPa by Langmi et al. [30], who also proved that hydrogen uptake was strongly dependent on the temperature, framework, and cation type. Langmi et al. [31] continued working on hydrogen-storage properties of several zeolites X, Y, A, RHO containing various exchangeable cations at cryogenic temperatures (−196 °C) and pressure range from 0 to 1.5 MPa. The highest gravimetric storage capacity of 2.19 wt% was obtained for CaX, reflecting to volumetric storage density of 31.0 kg $H_2$ $m^{-3}$ (143 hydrogen molecules per unit cell). Similar volumetric density of 30.2 kg $H_2$ $m^{-3}$ was obtained in the case of KX (144 hydrogen molecules per unit cell). Du et al. [32] investigated hydrogen uptake on several zeolites A, X, and ZSM-5 at pressures up to 7 MPa and cryogenic temperature, finding the highest gravimetric storage capacity for NaX zeolite at 4 MPa of 2.55 wt%, corresponding to more than 30 kg $H_2$ $m^{-3}$. Using atomistic simulations Hirscher and Panella [33] aimed at the theoretical determination of the maximum hydrogen capacity on several types of zeolite, finding the theoretical maximum as 2.86 wt% for FAU and RHO non-pentasile zeolites (names after the IZA – International Zeolite Association).

### 7.2.4
### Metal-Organic Frameworks (MOFs)

Metal-Organic Frameworks (MOFs) can be described as networks of transition metal atoms bridged by organic ligands, having very big specific surface of 1000–6000 $m^2$ $g^{-1}$ [11] and densities down to 0.21 g $cm^{-3}$. A variety of organic ligands and metal ions that may possibly be used to synthesize MOFs create an opportunity to design these materials with wanted pore dimensions and metal centers [33]. Unlike other porous materials, MOF have pores without walls; they are made from intersections and struts and this is the reason for high-surface areas. It is reported that MOF-5 can be activated to obtain 3800 $m^2$ $g^{-1}$ and 60% of the material is an open space, where gases and organic molecules can be introduced. By complete evacuation of the pores, it was possible to obtain surface

**Figure 7.8** Scanning electron micrograph of MOF-5 crystals obtained from the optimized large-scale preparation [35].

of 5500 m$^2$ g$^{-1}$ for MOF-177 [34]. An example of metal-organic framework is shown in Figure 7.8. Synthesized MOF-5 has a surface area of 3400 m$^2$ g$^{-1}$ and show well-shaped, high-quality cubic crystals with crystal sizes between 50 and 200 nm [35].

It is reported that at room temperature, MOF less than 1 wt% of hydrogen can be reversibly stored [36]. Similar result at room temperature and pressure of 2 MPa was obtained by Rosi *et al.* [37] on MOF-5, a composition of $Zn_4O(1,4$-benzenedicarboxylate$)_3$. At 78 K and pressure of 2 MPa, the amount of adsorbed hydrogen increased to 4.5 wt% (17.2 hydrogen molecules per $Zn_4O(1,4$-benzenedicarboxylate$)_3$ formula unit). Wong-Foy *et al.* [38] measured hydrogen uptake for seven MOF compounds at cryogenic temperature and found the maximum excess hydrogen densities in the range from 2.0 to 7.5 wt%. The maximum values have been obtained for MOF-177 with 7.5 wt% [39] and IRMOF-20 with 6.7 wt%, saturation was reached between 7 and 8 MPa. A linear correlation between surface area of MOFs under investigation and hydrogen saturation uptake was shown. MOF-17 and IRMOF-20 has obtained very high volumetric densities, 32 and 34 g L$^{-1}$, respectively, being close to US DOE target of 45 g L$^{-1}$. Gravimetric (in mg g$^{-1}$) and volumetric (in g L$^{-1}$) uptake of hydrogen for several MOFs at cryogenic temperature is shown in Figures 7.9 and 7.10, respectively.

## 7.2.5
### Covalent Organic Frameworks (COF)

Very promising and indeed interesting materials are covalent organic frameworks (COFs) being presently investigated. In here the organic building units are held together by strong covalent bonds between light elements, C, O, B, Si, N rather than metal ions. It has been shown that the synthesis of ordered COFs is possible and predesigned structures and properties can be achieved by the selection of

7.2 Carbon-Based Materials – General | 237

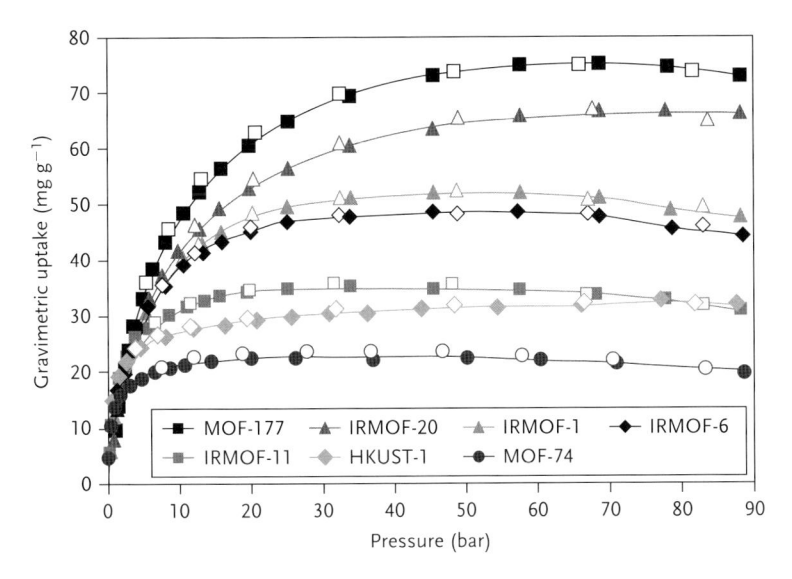

**Figure 7.9** High-pressure isotherms for activated materials at 77 K in gravimetric units representing surface excess adsorption. Filled markers represent adsorption, open markers represent desorption [38], reprinted by permission.

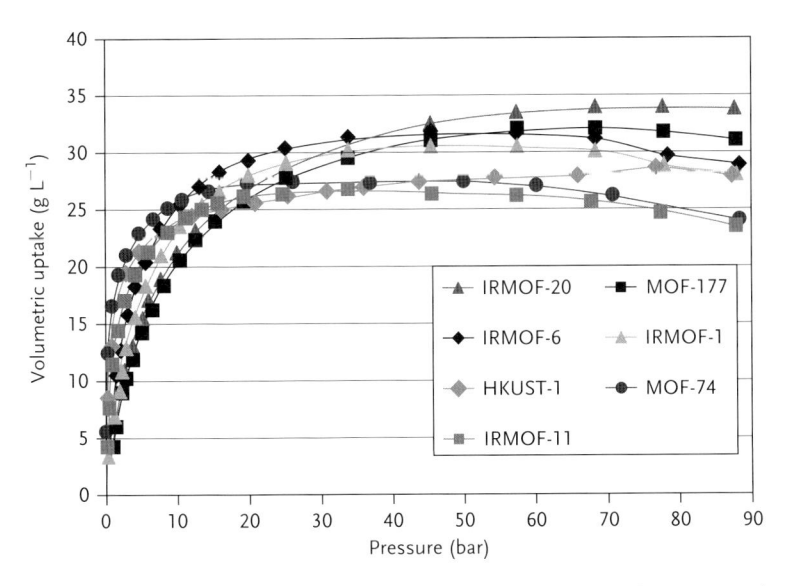

**Figure 7.10** High-pressure isotherms for activated materials measured at 77 K in volumetric units [38], reprinted by permission.

building elements and conditions during preparation [40]. COFs are characterized by high porosity and low crystal density, making them very promising candidates for hydrogen storage.

Molecular and crystal structures of COFs are presented in Figure 7.11.

Theoretical studies on prototypical COFs using Grand Canonical Monte Carlo (GCMC) simulation predicted that three-dimensional COFs are the most promising candidates for practical hydrogen storage. The highest predicted excess hydrogen uptakes at 77 K ($-196\ °C$) are 10.0 wt% for COF-105 at 8 MPa and 10.0 wt% for COF-108 at 10 MPa [41]. With these values covalent organic frameworks are much better than representative MFOs with the highest measured hydrogen uptake of 7.5 wt% for MOF-177 [38]. From the total adsorption isotherms, it is speculated that COF-108 may be able to store up to 18.9 wt% of hydrogen at 77 K and 10 MPa, followed by COF-105 with 18.3 wt%, COF-103 with 11.3 wt%,, COF-102 with 10.6 wt%, COF-5 with 5.5 wt%, and by COF-1 with 3.8 wt% [41]. The maximum volumetric hydrogen uptake is 49.9 g $L^{-1}$ for COF-102, 49.8 g $L^{-1}$ for COF-103, 39.9 g $L^{-1}$ for COF-108, 39.5 g $L^{-1}$ for COF-105, 36.1 g $L^{-1}$ for COF-1, and 33.8 g $L^{-1}$ for COF-5. To sum up, it seems that the best COF system for hydrogen storage are COF-105 and COF-108, giving the highest values reported for associative hydrogen storage of any material [41]. Isotherm measurements of hydrogen, methane, and carbon dioxide have been performed at 0.1–8.5 MPa and 77–298 K ($-196$ to $25\ °C$) at several COFs with different structural dimensions and corresponding pore sizes [42]. It was proved that three-dimensional (3D) COF-102 and COF-103 with 3D medium pores (12 Å) showed much better performance in comparison with 2D COF-1 and COF-6 with 1D small pores (9 Å) and 2D COF-5, COF-8 and COF-10 with 1D large pores (16–32 Å). Following results for the excess gas uptake have been obtained: at 3.5 MPa COF-102 showed 72 mg $g^{-1}$ for $H_2$ ($-196\ °C$), 187 mg $g^{-1}$ for methane ($25\ °C$), 1180 mg $g^{-1}$ for

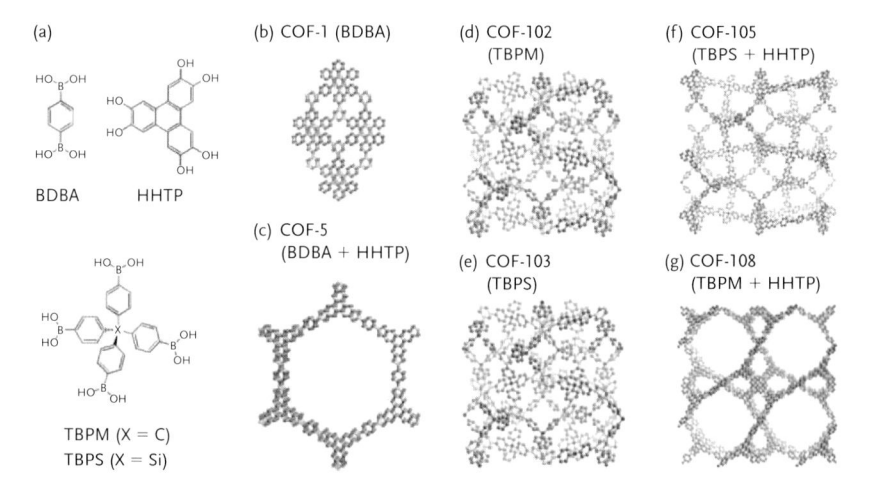

**Figure 7.11** (a) Molecular structures of building units (b–g) and crystal structures of COFs [41], reprinted by permission.

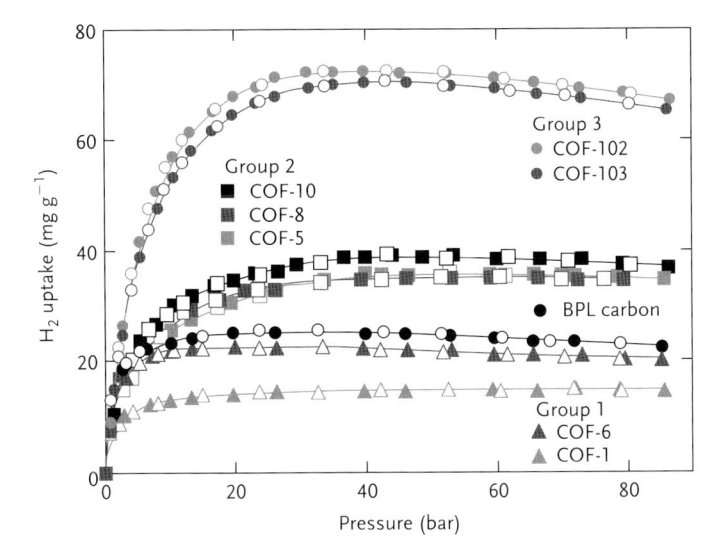

**Figure 7.12** High-pressure $H_2$ isotherms for COFs measured at 77 K data for BPL carbon (activated carbon) are shown for comparison [42], reprinted by permission.

$CO_2$. Similar performance was observed in the case of COF-103. It is stated by the authors that "These findings place COFs among the most porous and the best adsorbents for hydrogen, methane, and carbon dioxide" [42]. Experimental curves of high-pressure hydrogen isotherms for COFs at cryogenic temperature are presented in Figure 7.12. When compared this with high-pressure hydrogen isotherms for MOFs at the same temperature (Figure 7.9), it is obvious that COFs outperformed significantly MOFs.

Very detailed and comprehensive recent information about the prospects of porous inorganic solids, coordination polymers and other polymetric matrices for physical storage of hydrogen are discussed extensively in [50].

## 7.3
## Microspheres

The microspheres are also known as glass microcontainer, hollow glass-microspheres or glass microballoons. Besides different kind of glasses polymers as well as metal-coated glass microspheres are under investigation. Microspheres can also be used for storing other gases such as oxygen or helium.

Microspheres are commonly used as a cheap filler material for composite materials mainly for the automotive industry. The main advantage using the glass microspheres as filler material is the weight reduction, which is an important goal in the automotive industry. The second important application is in thermally insulating paint and tapes. For the application as low-density fillers, the glass microsphere have to be optimized regarding pressure resistance and weight.

The material as well as the size (diameter and wall thickness) have been optimized. This is already a good starting point for the use for gas storage. Different kinds of glass microspheres are commercially available. Glass microspheres are a mass product. Numerous fabrication methods have been used to produce glass microspheres: commonly the introduction of glass frit particles into an intense heat source such as a rotating arc plasma or an oxy-fuel flame. Other techniques begin with a liquid precursor that is injected as single droplet into a vertical tube furnace. A preparation method starting from sol–gel derived glass is described by Schmitt et al. [43]. This method allows to prepare a composition suitable for photo-enhanced hydrogen diffusion. The heat treated xerogel is suspended and doped and a blowing agent is added before the spray trying. The spray-dried granules are then brought in an oxy-propane flame to produce microspheres.

Microspheres are considered for gas storage since 100 years. There is a stronger interest in the last 20 years. Up to now, there is no commercial application of this technology known. There are several research institutes actually working on this topic.

The operation of microspheres for the storage of hydrogen can be described with the following steps:

- charging the microspheres at high pressure and high temperature (e.g., for glass microsphere @100 MPa and 1000 K);
- transferring the filled microsphere to a storage container;
- storing the filled microspheres in the storage container at defined conditions (temperature, the temperature determines the permeability and therefore the loss of hydrogen);
- optionally transferring the filled microsphere to the discharge reactor; and
- discharging the microspheres by heating up the microspheres.

The idea of using microspheres to store compressed hydrogen came up in the beginning of the last century. With the use of microspheres, high gravimetric and volumetric hydrogen storage densities are possible. The idea is to fill the microsphere with gaseous hydrogen at a high pressure, and then the hydrogen could be stored in spheres. The permeability of the material the spheres are made of has to be increased to achieve a fast release of the hydrogen. Therefore the temperature has to be elevated.

For this process the following main requirements for the material could be listed.

- High tensile strength for a broad temperature range.
- Low permeability for hydrogen at low temperature (storage temperature).
- High permeability for hydrogen at raised temperature (discharge temperature).
- Easy producible.

The investigated materials for this application are different kinds of glass. For three different investigated glasses the composition as well as some physical properties is given in Tables 7.1 and 7.2, respectively.

**Table 7.1** Composition of silica borosilicate and soda lime glass.

| Glass | SiO$_2$ | Al$_2$O$_3$ | K$_2$O | Na$_2$O | MgO | CaO | B$_2$O$_3$ |
|---|---|---|---|---|---|---|---|
| Silica glass | 100 | | | | | | |
| Borosilicate | 70–80 | 2–7 | 4–8 | | 5 | | 7–13 |
| Soda lime | 70–75 | | | 12–16 | | 10–15 | |

**Table 7.2** Properties of different glasses.

| | Tensile strength MPa | Density kg m$^{-3}$ | Softening point K |
|---|---|---|---|
| Silica glass | 50 | 2250 | 1870–1940 |
| Borosilicate | 282 | 2230 | 990–1150 |
| Soda lime | 19–77 | 2440 | 970–1000 |

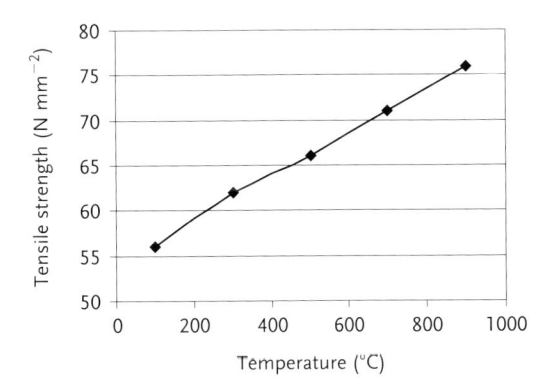

**Figure 7.13** Tensile strength as a function of the temperature for silica glass (corning 7940) [44].

For example, the tensile strength as a function of the temperature of a silica glass is shown in Figure 7.13.

The following characteristics besides material properties are useful to describe the microspheres:

- Aspect ratio: $D/H$
- $D$: diameter
- $H$: wall thickness.

The ratio of the diameter of the sphere and the wall thickness is named the aspect ratio. For some investigated microspheres the characteristic is given in Table 7.3. In Figure 7.14 for example a SEM image of microspheres is shown.

## 7 Novel Materials

**Table 7.3** Dimensions of investigated microspheres for hydrogen storage [45, 49].

|  | im 30 K (3 M) | S 38 (3 M) | S60 HS (3 M) | D32/4500 (3 M) | Kohli [49] |
|---|---|---|---|---|---|
| Diameter (µm) | 18 (average) | 40 (50%) | 29 (50%) | 40 (50%) | |
| Aspect ratio (−) | | | | | 80–180 |
| Nitrogen isostatic crush strength (MPa) | 193 | 40 | 124 | 31 | |

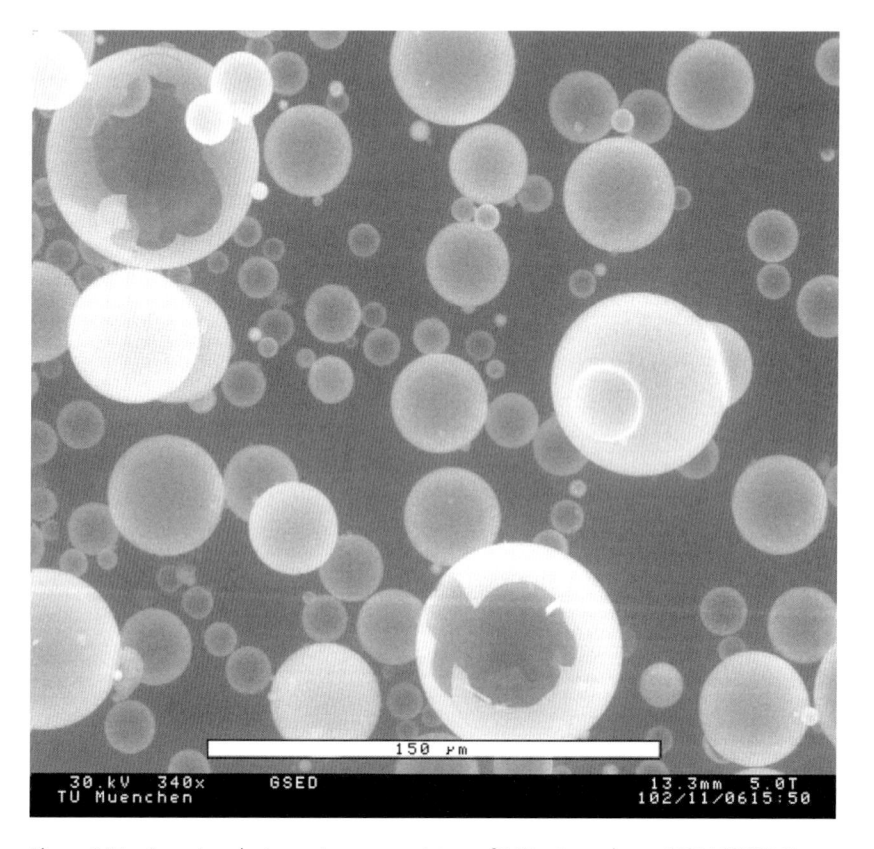

**Figure 7.14** Scanning electron microscope picture of 3 M microspheres (D32/4500)[46].

The permeability coefficient is the parameter to describe the gas transport through the walls.

The permeability coefficient has the following dimensions:

$$P = \frac{(\text{amount of permeant}) \times (\text{film thickness})}{(\text{area}) \times (\text{time}) \times (\text{pressure drop across the film})}$$

The permeability coefficient strongly depends on the temperature. The dependency is given in Eq. 7.4. Silica glass has the highest permeability. Other additives reduce the permeability. The permeability determines the charging time as well as the storage time and the extraction rate.

In Figure 7.15, the temperature dependency of the permeability coefficient for different materials is shown:

$$P = P_0 \exp\left(-\frac{E_k}{T}\right)$$

where
$P$: permeability coefficient
$P_0$: coefficient
$E_k$: gas permeation activation energy
$T$: temperature

For spheres with a high aspect ratio the following equation can be written:

$$P_{max} = 4 \times R_m \times \frac{H}{D}$$

where
$P_{max}$: maximum pressure
$D$: diameter
$H$: wall thickness
$R_m$: tensile strength.

The maximum weight content of hydrogen in spheres with a small wall thickness can be described by

$$\eta = \frac{2 \times R_m \times \delta_{H_2}(P)}{3 \times \delta(\text{mat})}$$

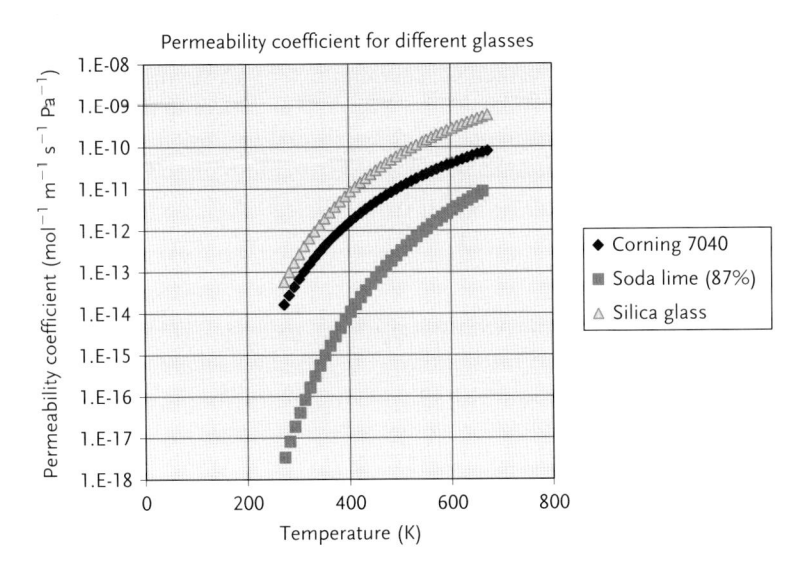

**Figure 7.15** The permeability coefficient as function of the temperature for different glasses.

where
η: weight content
$R_m$: tensile strength
P: pressure
δ: density
Index mat: microsphere material.

For the calculation of the required volume the package density has to be considered. Assuming ideal spheres with the same diameter the maximum package density that can be reached is 74%.

As for each storage method not only the storage density, but also other characteristics are important, which are as follows.

- Confinement time
- Energy consumption
- Fuel supply time
- Extraction rate
- Expenses for refueling.

## 7.3.1
### Methods for Discharging

To realize a defined extraction rate, the temperature of the walls of the microspheres have to be increased to reach the required permeability. There were several methods under investigation.

The common way to heat up the spheres is to heat the spheres by a hot gas stream. For this process, additional energy is required. This method is not very fast and is not the most effective one.

Keding et al. [47] report an innovative method for heating up the walls of microspheres. He proposed a method combining the two storage concepts of hollow glass microspheres and chemical storage. The idea is to use the generated heat by hydrolyses reaction of sodium borohydride to heat up the microspheres for releasing the hydrogen. Water will be used to transport the loaded microspheres in a reactor and in parallel sodium borohydride will be injected into the reactor. The water reacts catalytically with the sodium borohydride. To support this reaction, the spheres have to be coated with an adequate catalyst. The sodium borohydride system is described in detail in Section 5.7. The exothermal reaction of the hydrolysis of sodium borohydride is a suitable method for heating up the walls of the microspheres. The reaction in the liquid phase allows a good heat transfer to the spheres. Later on the empty microspheres have to be separated from the reaction product a solution of borates. Keding et al. report that such a system can reach hydrogen storage densities of up to 10 wt%. For the experiments he used different glass microspheres from 3 M namely S38, S60HS, and iM30 K.

Shelby [48] investigated another option for releasing the hydrogen. He compared the thermally driven with a photo-induced outgassing of the microspheres. The photo-induced hydrogen diffusion by using cobalt-doped microspheres was demonstrated. For the experiments, he used glass microspheres from Mo-Sci. Corp. and filled them up to 10.7 MPa with hydrogen. The photo-induced outgassing results in a faster response rate than that obtained by heat, even though the temperature used in the thermal outgassing is 100 K greater than which results from application of light. The first results indicate that less than 10% of the gas is lost in the first 30 days at storage temperatures between $-20\,°C$ and $50\,°C$.

Kohli *et al.* [49] reported in 2008 that they have experimentally attained a gravimetric density as 15–17% for the glass microspheres at pressure ranges between 15 and 30 MPa. For experiments, they produced the microspheres based on Corning 7040 (which is a borosilicate glass), Soda-lime, and quartz.

The hydrogen storage in microspheres has some advantages:

- cheap, plentiful raw material,
- bulk good, easy handling of the spheres,
- high storage densities are possible, and
- save handling of the microspheres.

A large number of microspheres can be bunched together in a tank that can be build very light, because the tank has not to be under a high pressure. In the case of an accident, the microsphere system would not break to release a large quantity of hydrogen. Instead, some of the microspheres would just spill onto the floor. A limited number of microspheres could possibly break, but release only small amounts of hydrogen.

To use these advantages, it is important to also know the challenges. The correct choice of the microspheres with material and dimension is very important, because hereby the storage pressure is determined. But in addition the leakage at ambient temperature is determined by the same properties. The capacity of the microspheres is not only determined by the material properties and dimensions, but also by the required hydrogen release rates. The depletion follows an exponential function. This means that the spheres never will be really empty, but depleted to a defined pressure. The leakage rate gives a limitation to the storage time. Also the method used to increase the permeability for discharging has to be carefully selected to finally have a successful system. With the investigated system, the resulting charging times are in the range of 0.5–20 h with corresponding discharging times from 0.2 to 10 h.

### 7.3.2
**Resume**

Glass microspheres could be an attractive solution for gas storage. The mass-related storage density could be up to 15% ore even more. The resulting storage density is between the energy density of compressed and of liquid hydrogen. The volume-related storage density is the same as for compressed hydrogen. Storing gas in microspheres is storing compressed hydrogen; therefore, this is obvious. Before the implementation of this technology, for example, fuelling cars, the whole chain from filling to transportation finally to the engine has to be considered. For a future application, the whole system has to be optimized, starting from the material, dimensions and finally the process for the discharging.

A breadboard validation in laboratory environment should be the next step, then a breadboard validation in the relevant environment could follow.

**References**

1 Auner, N. (2009) Method for the production of silicon from silyl halides, US 2009/0127093 A1 US.
2 Auner, N. and Lippold, G.(2007) Energy economy based on silicon: facts and perspectives. *Nachrichten aus der Chemie,* 627–633.
3 Auner, N., Holl, S., Bauch, C., Lippold, G., and Deltschew, R. (2010) Solid polysilane mixtures, US 2010/0004385 A1.
4 NEXTSTEP GmbH (2009) Wolfen, Entwicklungszentrum, Bitterfeld-Wolfen. Hydropolysilane.
5 Pretterebner, J. and Oppenweiler, D.E. (2009). Process of supplying fuel cell with hydrogen by means of silanes and polysilanes, US 2009/0136800 A1
6 Chambers, A., Park, C., Terry, R., Baker, K., and Rodriguez, N.M. (1998) Hydrogen storage in graphite nanofibers. *J. Phys. Chem. B,* **102** (22), 4253–4256.
7 Riis, T., Sandrock, G., Ulleberg, O., and Vie,P.J.S.(2005) Hydrogen Storage – Gaps and Priorities, HIA HCG Storage Paper.
8 Dillon, A.C., Gilbert, K.E.H., Parilla, P.A., Alleman, J.L., Hornyak, G.L., Jones, K.M. et al. (2002) Hydrogen storage in carbon single-wall nanotubes. Proceedings of the 2002 US DOE Hydrogen Program Review, NRE L/CP-610–32405, Vancouver, British Columbia, 1–3 June 2009.
9 Iniguez, J. (2008) Modelling of carbon-based materials for hydrogen storage in Solid-State Hydrogen Storage, Materials and Chemistry (ed. G. Walker), Woodhead Publishing Limited and CRC Press LLC, Cambridge, UK.
10 Fichtner, M. (2005) Nanotechnological aspects in materials for hydrogen storage. *Adv. Eng. Mater.,* **7** (6), 443–453.
11 Benard, P. and Chahine, R. (2007) Storage of hydrogen by physisorption on carbon and nanostructured materials. *Scr. Mater.,* **56**, 803–808.
12 Yamanakaa, S., Fujikane, M., Uno, M., Murakami, H., and Miura, O. (2004) Hydrogen content and desorption of carbon nano-structures. *J. Alloys Compd.,* **366** (1–2), 264–268.
13 Gundiah, G., Govindaraj, A., Rajalakshmi, N., and Dhathathreyan, K.S. (2003) Hydrogen storage in carbon nanotubes and related materials. *J. Mater. Chem.,* **13**, 209–213.
14 Becher, M., Haluska, M., Hirscher, M., Quintel, A., Skakalova, V., Dettlaff-Weglikovska, U. *et al.* (2003) Hydrogen storage in carbon nanotubes. *C. R. Phys.,* **4** (9), 1055–1062.
15 Anson, A., Callejas, M.A., Benito, A.M., Maser, W.K., Izquierdo, M.T., Rubio, B. et al. (2004) Hydrogen adsorption studies on single wall carbon nanotubes. *Carbon,* **42** (7), 1243–1248.
16 Pradhan, B.K., Harutyunyan, A., Stojkovic, D., Zhang, P., Cole, M.W., Crespi, V. *et al.* (2002) Large cryogenic storage of hydrogen in carbon nanotubes at low pressures. *Mat. Res. Soc. Symp. Proc.,* 706, Z10.3.1–Z10.3.6 .
17 Poirier, E., Chahine, R., Benard, P., Lafi, L., and Dorval-Douville, G. (2006) Hydrogen adsorption measurements and modeling on metal-organic frameworks and single-walled carbon nanotubes. *Langmuir,* **22** (21), 8784–8789.
18 Dillon, A.C., Gennett, T., Alleman, J.L., Jones, K.M., Parilla, P.A., and Heben, M.J. (2001) Optimization of single-wall

nanotube synthesis for hydrogen storage. IEA Task 12: Metal Hydrides and Carbon for Hydrogen Storage 2001, NREL/CH-590–31288.
19 http://pages.total.net/espitech/carbonpower.html [Online].
20 David, E. (2005) An overview of advanced materials for hydrogen storage. *J. Mater. Process. Tech.*, **162–163**, 169–177.
21 Zhou, L., Zhou, Y.-P., and Sun, Y. (2004) A comparative study of hydrogen adsorption on superactivated carbon versus carbon nanotubes. *Int. J. Hydrogen Energ.*, **29** (5), 475–479.
22 Li, Y., Zhao, D., Wang, Y., Xue, R., Shen, Z., and Li, X. (2006) The mechanism of hydrogen storage in carbon materials. *Int. J. Hydrogen Energ.*, **32** 2513–2517.
23 Jurewicz, K., Frackowiak, E., and Beguin, F. (2004) Towards the mechanism of electrochemical storage in nanostructured materials. *Appl. Phys. A*, **78**, 981–987.
24 Zou, G., Zhang, D., Dong, C., Li, H., Xiong, K., Fei, L., et al. (2006) Carbon nanofibers: synthesis, characterisation and electrochemical properties. *Carbon*, **44**, 828–832.
25 Schimmel, H.G., Kearley, G.J., Nijkamp, M.G., Visser, C.T., de Jong, K.P., and Mulder, F.M. (2003) Hydrogen adsorption in carbon nanostructures: comparison of nanotubes, fibers and coals. *Chem. Eur. J.*, **9**, 4764–4770.
26 Ciciszwili, G.W., Andronikaszwili, T.G., Kirow, G.N., and Filizowa, L.D. (1990) *Zeolity Naturalne*, Polskie Wydawnictwa Naukowo-Techniczne, Warszawa, ISBN 83-204-1283-8.
27 Anderson, P.A. (2008) Storage of hydrogen in zeolites in *Solid-State Hydrogen Storage, Materials and Chemistry* (ed. G. Walker), Woodhead Publishing Limited and CRC Press LLC, Cambridge, UK.
28 Vitillo, J.G., Ricchardini, G., Spoto, G., and Zecchina, A. (2005) Theoretical maximal storage of hydrogen in zeolitic frameworks. *Phys. Chem. Chem. Phys.*, **7** (23), 3948–3954.
29 Kayiran, S.B. and Darkrim, F.L. (2002) Synthesis and ionic exchanges of zeolites for gas adsorption. *Surf. Interface Anal.*, **34** (1) 100–104.
30 Langmi, H.W., Walton, A., Al-Mamouri, M.M., Johnson, S.R., Book, D., Speight, J.D. et al. (2003) Hydrogen adsorption in zeolites A, X, Y and RHO. *J. Alloys Compd.*, **356–357**, 710–715.
31 Langmi, H.W., Book, D., Walton, A., Johnson, S.R., Al-Mamouri, M.M., Speight, J.D. et al. (2005) Hydrogen storage in ion-exchanged zeolites. *J. Alloys Compd.*, **404–406**, 637–642.
32 Du, X. and Wu, E. (2006) Physisorption of hydrogen in A, X and ZSM-5 types of zeolites at moderately high pressures. *Chin. J. Chem. Phys.*, **19**, 457–462.
33 Hirscher, M. and Panella, B. (2007) Hydrogen storage in metal-organic frameworks. *Scr. Mater.*, **56**, 809–812.
34 Yaghi, O.M. and Li, Q. (2009) Recticular chemistry and metal-organic frameworks for clean energy. *MRS Bull.*, **34**, 683–690.
35 Stallmach, F., Groger, S., Kunzel, V., Karger, J., Yaghi, O.M., Hesse, M. et al. (2006) NMR studies on the diffusion of hydrocarbons on the metal-organic framework material MOF-5. *Angew. Chemie. Int. Ed.*, **45**, 2123–2126.
36 Poirier, E., Chahine, R., Bénard, P., Lafi, L., Dorval-Douville, G., and Chandonia, P.A. (2006) Hydrogen adsorption measurements and modelling on metal-organic frameworks and single-walled carbon nanotubes. *Langmuir*, **22**, 8784–8789.
37 Rosi, N.L., Eckert, J., Eddaoudi, M., Vodak, D.T., Kim, J., O'Keefe, M. et al. (2003) Hydrogen storage in microporous metal-organic frameworks. *Science*, **300**, 1127–1129.
38 Wong-Foy, A.G., Matzger, A.J., and Yaghi, O.M. (2006) Exceptional $H_2$ saturation uptake in microporous metal-organic frameworks. *J. Am. Chem. Soc. Commun.*, **128**, 3494–3495.
39 Mandal, T.K. and Gregory, D.H. (2009) Hydrogen: a future energy vector for sustainable development. *Proc. IMechE J. Mech. Eng. Sci.*, **224**, 539–558, doi 10.1243/09544062JMES1774.
40 Uribe-Romo, F.J., Hunt, J.R., Furukawa, H., Klock, C., O'Keeffe, M., and Yaghi, O.M.

(2009) A crystalline imine-linked 3-D porous covalent organic framework. *J. Am. Chem. Soc. Commun.*, **131** (13), 4570–4571.

41  Han, S.S., Furukawa, H., Yaghi, O.M., and Goddard, W.A. III (2008) Covalent organic frameworks as exceptional hydrogen storage materials. *J. Am. Chem. Soc. Commun.*, **130**, 11580–11581.

42  Furukawa, H. and Yanghi, O.M. (2009) Storage of hydrogen, methane and carbon dioxide in highly porous covalent organic frameworks for clean energy applications. *J. Am. Chem. Soc.*, **131**, 8875–8883.

43  Schmitt, M.L., Shelby, J.E., and Hall, M.M. (2006) Preparation of hollow glass micro spheres from sol–gel derived glass for application in hydrogen gas storage. *J. Non-Cryst. Solids*, **352**, 626–631.

44  Shakel Ford, J.F. and Alexander, W. (2001) CRC Handbook of Material Science & Engineering, CRC Press, Boca Raton, Florida, USA.

45  www.3m.com/product/information/Glas-Bubbles.html (21 May 2012).

46  Herr, M. and Lercher, J.A. (2003) Hydrogen Storage in Microspheres, ESA Contract 16292/o2/NL/PA

47  Keding, M., Schmid, G., and Tajmar, M. (2009) Innovative hydrogen storage in hollow glass-microspheres. HFC Conference+ Exhibition Conference Proceedings, Vancouver, British Columbia, 1–3 June.

48  Shelby, J.E. (2006) Glass Microspheres for Hydrogen Storage, FY Annual Progress Report.

49  Kohli, D.K., Khardekr, R.K., S., Rashmi, and Gupta, P.K. (2008) Glass microcontainer-based hydrogen storage scheme. *Int. J. Hydrogen Energ.*, **33**, 417–422.

50  Reardon, H., Hanlon, J., Hughes, R.W., Godula-Jopek, A., Mandal, T.K. and Gregory, D.H. (2012) Emerging concepts in solid-state hydrogen storage: The role of nanomaterials design. *Energy Environ. Sci.*, **5**, 5951–5979.

# Index

*Note*: Page numbers in *italics* refer to figures and tables.

$AB_2$ alloys, 136–137
$AB_5$ hydride, 136
Activated carbon (AC), 229, 231–232
Aircraft, 128–131
Alanates, 139, 141–143
Alcohols, 23
Alkaline electrolyzers, 33–34
Alkaline fuel cell (AFC), 183
Aluminum, hydrogen from, 50–51
Aluminum alloys, 108–109
Amides-imides, 143–146
Amminex
– ammonia reformer built by, *186*, 188
Ammonia, 181
– application areas of, 182–183
– energetic consideration, 190
– as fuel
–– in high-temperature fuel cells, 186–187
–– in low-temperature fuel cells, 183
– hydrogen release from, 187
–– ammonia electrolysis, 187
–– catalytic decomposition, 187–189
– hydrogen release from ammonia and metal hydride, 189–190
– methods of storage, 185
–– liquid dry ammonia, 185
–– solid-state ammonia storage, 185–186
– production, 184
–– from nitrogen and hydrogen, 184
–– from silicon nitride, 184–185
– properties, 181, *182*
Ammonia borane, 193–194
Ammonia electrolysis, 187
Ammonia reformer built by Amminex, *186*, 188
Anabaena variabilis, 48

Apollo Energy Systems, Inc., 189
Austenitic stainless steel, 108
Autoignition temperature (AIT), 56
Autothermal reforming (ATR), 20
Auxiliary power unit (APU), 90

*Bacillus*, 48
Biomass
– hydrogen from, 45
–– biological processes, 47–49
–– thermochemical processes, 47
Boil-off, 121–123
Borohydrides, 146–148, 191
– ammonia borane, 193–194
– sodium borohydride, 191
–– as fuel in a PEM-based fuel cell, 191–192
–– hydrogen generation by hydrolytic release, 191–192
Braze materials, 124
Building integrated photovoltaics (BIPV), 28–29

Calcium chloride, *185*
Carbon-based materials, 228–229
– activated carbon (AC), 229, 231–232
– carbon nanotubes (CNT), 229–233
– covalent organic frameworks (COFs), 236, 238–239
– graphite nanofibers, 229, 232–233
– high-surface area materials, 233–234
– metal-organic frameworks (MOFs), 235–236, *237*
– zeolites, 234–235
Carbon fiber reinforced plastics (CFRP), 7
Carbon fibers, 211

*Hydrogen Storage Technologies: New Materials, Transport, and Infrastructure*, First Edition.
Agata Godula-Jopek, Walter Jehle, and Jörg Wellnitz.
© 2012 Wiley-VCH Verlag GmbH & Co. KGaA.
Published 2012 by Wiley-VCH Verlag GmbH & Co. KGaA

Carbon nanofibers, 232–233
Carbonyl sulfide (COS), 25
Catalytic decomposition, 187–189
Cavendish, Henry, 11
Centre for Renewable Energy Sources (CRES), 31
Chemical storage, 171
– ammonia, 181
– – application areas of, 182–183
– – as fuel in high-temperature fuel cells, 186–187
– – catalytic decomposition, 187–189
– – electrolysis, 187
– – energetic consideration, 190
– – hydrogen from, 187
– – hydrogen from ammonia and metal hydride, 189–190
– – liquid-state storage of, 185
– – production from nitrogen and hydrogen, 184
– – production from silicon nitride, 184–185
– – properties of, 181, *182*
– – solid-state storage of, 185–186
– application, 178–180
– borohydrides, 191
– – ammonia borane, 193–194
– – sodium borohydride, 191–192
– hydrocarbons as hydrogen carrier, 177–178
– hydrogen storage in hydrocarbons, 173–176
– materials and properties, 172–173
Chemical vapor deposition (CVD), 229, 232
Chemisorption, 228, 229
Chlamydomonas reinhardtii, 49
Claude process, 117, *118*
*Clostridium*, 48
*Clostridium pasteurianum*, 49
Coal, 18
Coke formation, 20
Collins liquefier, 117–118
Combined heat and power (CHP) system, 157, *158*
Compressed hydrogen, 68, 98–100
– applications, 111–114
– materials, 106–109
– mechanical compressors, 100–101
– nonmechanical compressors, 101–106
– properties of, 98
– sensors, instrumentation, 110
– tank filling, 110–111
Covalent organic frameworks (COFs), 236, 238–239

Cycloalkanes
– dehydrogenation–hydrogenation system of, 172
– dehydrogenation of, 173, 176
– hydrogen storage capacities of, 172

Decalin dehydrogenation, 180
Dehydrogenation–hydrogenation system of cycloalkanes, 172
Dehydrogenation of hydrocarbons, 171
Desulfurization, 19
Diaphragm compressor, 101, *102*
Differential scanning calorimerty (DSC), 143, *144*
Dimension considerations, of hydrogen storage options, 212–216
Direct biophotolysis, 48
Direct PEM fuel cell (D-PEMFC), 179
Distribution of hydrogen, 87–89
DOE technical targets, *113*
Dynetek, 4

Economic Comission for Europe (ECE), 64–65
Economic considerations/costs, of hydrogen storage options, 197–200
Electrochemical hydrogen compressor, 103–104
Electrolysis, of steam, 38–40
Electrolysis technology, 27–28
*Enterobacter*, 48
European New Car Assessment Program (Euro NCAP), 207
European Wind Energy Association (EWEA), 32

Failure mode effectiveness and criticality analysis (FMECA), 205
Federal Motor Vehicle Safety Standards (FMVSS), 207
Fiber-reinforced design, 201, 215
Fiber-reinforced polymers (FRP), 127
Filament winding technique, *2*, 198
Filling stations, 86–87
Formula H racing, 9
Fossil fuels, 18
Freeze-thaw method, 120
Fuel cells (F/C), development of, 81, 84

GALILEO, 149
Gallium arsenide (GaAs), 30
Gasoline, 171

Gas storage in underground, 111–112
GenHy®, 36, *37*, 39
Glass microcontainer. *See* microspheres
Gleocapsa alpicola, 48, 49
Graphite nanofibers, 229, 232–233
Grove, William Robert, 11

Haber–Bosch process, 190
Haldor Topsře convective reformer (HTCR), 19
Half membrane theory, 5
Hazardous materials, of hydrogen storage options, 209–212
HDW (Howaldtwerke Deutsche Werft AG), 162
Heat exchanger with an auger, 120–121
Heavy metal hydrides. *See* Intermetallic hydrides
High pressure electrolyzer, 104–106
High pressure metal hydride storage tank, 109
High-surface area materials, 233–234
High-temperature electrolysis of steam (HTES), 38–40
High-Temperature Electrolysis of Steam, 38–41
High-temperature engineering test reactor (HTTR), 45
High-temperature water splitting
– and high-temperature nuclear energy and solar energy, 42–45
History/background, of hydrogen storage, 1
Hydrocarbons, 22
– containing hydrogen, 172
– dehydrogenation of, 171
– general properties of, *172*
– as hydrogen carrier, 177–178
– hydrogen storage in, 173–176
Hydrogen, 11
– ammonia production from, 184
– basic thermodynamic data of, *12*
– comparison with other fuels, 14–16
– – mass and volume energy density of, *15*
– energy content of, 12
– features of, 217
– hydrocarbons containing, 172
– parameters, *14*
– phase diagram of, 13–14
– physical properties of, *12*
– production, 16–18
– – from aluminum, 50–51
– – from biomass, 45

– – coal gasification, 25–27
– – combination with fossil fuels, 18
– – electrolyzers, types of, 33
– – hydrocarbons, partial oxidation and autothermal reforming of, 20–21
– – HyPr-RING method for, 21–23
– – natural gas, steam reforming of, 19–20
– – pathways of, 17
– – Plasma-Assisted Production from, 23–25
– – water-splitting processes, 27
– storage methods, *14*
– storage safety aspects, 53–55
– – from production to user, 65–66
– – human health impact, 62–63
– – hydrogen vehicles, 70–73
– – incidents with, 61–62
– – properties related to safety, 55–61
– – regulations, codes, and standards (RCS), 63–65
– – safe removal of, 73
– – sensors, 63
– in transportation sector, 17
Hydrogenated polysilanes (HPS), 225–228
Hydrogen attack, 107
Hydrogen carrier, hydrocarbons as, 177–178
Hydrogen chloride gas (HCl), 25
Hydrogen Components Inc., 149
Hydrogen cyanide (HCN), 25
Hydrogen embrittlement, 106–107
Hydrogen-filling station, 83, 93
Hydrogen from ammonia and metal hydride, 189–190
Hydrogen generation by hydrolytic release, 191–192
Hydrogen MASER, 150
Hydrogen permeation
– through metals, 107
– through nonmetals, 107–108
Hydrogen-powered vehicles, 96
Hydrogen Refuelling Stations (HRS), *69*
Hydrogen release from ammonia, 187
– ammonia electrolysis, 187
– catalytic decomposition, 187–189
Hydrogen slush, 120–121
Hydrogen storage assembly (HSA), 150
Hydrogen storage in hydrocarbons, 173–176
Hydrogen storage options
– comparison with other energy storage systems, 218–222
– dimension considerations, 212–216
– economic considerations/costs, 197–200
– hazardous materials, 209–212
– safety equipment, 205–209

– safety rules and regulations, 200–205
– sociological considerations, 216–218
– waste materials, 209–212
Hydrogen sulfide, 25
Hydrogen transportation, 197
Hydrogen vehicles, 70–73
HYDROSOL-2, 43
HyICE project, 85
HyPr-RING process, from hydrocarbons, *22*
HyRacer project, 85
Hystore Technologies Ltd., 149

IGCC (integrated gasification combined cycle), 25
IGFC (integrated gasification fuel cell), 25
Indirect biophotolysis, 48
Infrastructural requirements, 81, 93–96
– distribution, 87–89
– filling stations, 86–87
– military, 89–91
– portables, 92–93
– transportation, 81–85
Intermetallic hydrides, 135–137
Internal combustion engines (ICEs), 3, 5, 82, 84
International Partnership for Hydrogen and Fuel Cells in the Economy (IPHE), 55
Iodine–sulfur (I–S) cycle, 17

Joule–Brayton cycle, 118

Lavoisier, Antoine, 11
Life cycle assessment (LCA), 40
Limit load case (LLC), 201, 205
Linde–Sankey process, 117
Liners, used materials for, 109
Liquefaction, 116–120
– Claude process, 117, *118*
– Collins liquefier, 117–118
– Joule–Brayton cycle, 118
– Linde process, 116–117
– magnetic refrigeration, 118–120
– thermoacoustic, 120
Liquid dry ammonia, 185
Liquid hydrogen, 13, 68
Liquid hydrogen storage system, 124–125
– from Linde, *129*, *130*
Liquid natural gas (LNG), 126
Liquid/slush hydrogen, 114
– applications, 125–131
– – aircraft, 128–131
– – road and rail transportation, 126

– – rockets, 131
– – sea transportation, 126
– – solar power plants, 131
– – storage, 125
– – vehicles, 127–128
– boil-off, 121–123
– braze materials, 124
– hydrogen slush, 120–121
– insulation, 123–124
– liquefaction, 116–120
– ortho para conversion, 114–115
– properties, 114, *115*
– sensors/instrumentation, 124–125
– tank materials, 123
Liquid storage (LH2), of hydrogen, 1, *2*
Load vector, 5
Lower heating value (LHV), 12
Low-temperature water splitting, 27–28

Magnesium-based hydrides, 137–139, 165
Magnesium chloride, *185*
Magnetic refrigeration, 118–120
Material Safety Data Sheet (MSDS), 55
Maximum power point tracker (MPPT), 28
Mechanical compressors, 100–101
– diaphragm compressor, 101
– piston compressors, 101
Metal hydride compressor, 102–103
Metal hydride fuel-cell (MHFC) stacks, 161
Metal hydrides (MH), 68, 131–134
– application, 149–163, *164–165*
– classical
– – intermetallic hydrides, 135–137
– – magnesium-based hydrides, 137–139, *140*
– hydrogen from, 189–190
– light, 139
– – alanates, 139, 141–143
– – amides-imides, 143–146
– – borohydrides, 146–148
– outlook, 163, 165–166
Metal hydride tanks (MHT), 149
Metal-organic frameworks (MOFs), 229, 235–236, *237*
Methane, 19, 23, 171, 177
– properties of, *182*
Microspheres, 239–244
– methods for discharging, 244–245
– resume, 245–246
Microwave cavity and shield assembly (MCSA), 150
Military, 89–91
Mineral oil, 18
Mini-self-teaching robot, 92–93

Multiple interconnected solar cells, 28
Multiwalled carbon nanotubes (MWNTs), 229, 230

Nafion®, 34
Naphthalene hydrogenation, 180
Natural gas, 18
NEXTSTEP GmbH, 225
Nitrogen, ammonia production from, 184
Nonequilibrium plasma-assisted technologies
– conversion rates of, 24
– energy efficiencies of, 26
Nonmechanical compressors, 101–106
– electrochemical compressor, 103–104
– high pressure electrolyzer, 104–106
– metal hydride compressor, 102–103
Novel materials, 225
– carbon-based materials, 228–229
–– activated carbon (AC), 229–233
–– carbon nanotubes (CNT), 229–233
–– covalent organic frameworks (COFs), 236, 238–239
–– graphite nanofibers, 229–233
–– high-surface area materials, 233–234
–– metal-organic frameworks (MOFs), 235–236, 237
–– zeolites, 234–235
– microspheres, 239–244
–– methods for discharging, 244–245
–– resume, 245–246
– silicon and hydropolysilane, 225–228

Original equipment manufacturerAQ2 s (OEMs), 82

Passenger car, 208
Passive Hydrogen MASER, 149–150
Perhalogenated polysilanes, 225
Permeability coefficient, 242–243
Photoelectrochemical (PEC) device, 30
Photovoltaic-electrolysis system, 28–31
Photovoltaics (PVs), 17
Piston compressors, 101
Portables, 92–93
Pressure storage, of hydrogen, 1, 2
Pressure swing adsorption (PSA), 97
Production of hydrogen, 16–18
– from aluminum, 50–51
– from biomass, 45
– coal gasification, 25–27
– combination with fossil fuels, 18
– electrolyzers, types of, 33

– hydrocarbons, partial oxidation and autothermal reforming of, 20–21
– HyPr-RING method for, 21–23
– natural gas, steam reforming of, 19–20
– pathways of, 17
– plasma-assisted production from, 23–25
– water-splitting processes, 27
–– low-temperature water splitting, 27–28
2-Propanol, dehydrogenation of, 174
Proton exchange membrane (PEM) electrolyzers, 31, 33, 34–38
Proton exchange membrane (PEM) fuel cell, 103, 173, 183, 227
– sodium borohydride as fuel in, 191–192
Purification of hydrogen, 97

Regulations, codes, and standards (RCS), 63–65
Road and rail transportation, 112, 126
Rockets, 131

Safety equipment, of hydrogen storage options, 205–209
Safety rules and regulations, of hydrogen storage options, 200–205
Schönbein, Christian Friedrich, 11
Sealing, used materials for, 109
Sea transportation, 126
Selective catalytic reduction (SCR), 183
Sensors, 63
– compression, 110
– liquid hydrogen storage system, 124–125
Siemens, 162
Silicon nitride, ammonia production from, 184–185
Silicon production, 225–228
Single-walled carbon nanotubes (SWNTs), 229, 230
Slush guns, 121
Small solar power system (SSPS), 43
Sociological considerations, of hydrogen storage options, 216–218
Sodium borohydride, 191, 147, 244
– as fuel in a PEM-based fuel cell, 191–192
– hydrogen generation by hydrolytic release, 191–192
Solar cells, 29
Solar power plants, 131
SOLID-H™, 149
Solid oxide electrolysis cells (SOEC), 33
Solid-oxide fuel-assisted electrolysis cell (SOFEC), 40, 41, 42
Solid oxide fuel cell (SOFC), 39, 183

Solid-state ammonia storage, 185–186
Space flight programs, 1
Spawnt Production, 225
Spawnt Research, 225
Steam-reforming reaction
– of natural gas, 19
Steam-to-carbon (S/C) ratio, 19
Storage safety aspects, of hydrogen, 53–55
– human health impact, 62–63
– hydrogen vehicles, 70–73
– incidents with, 61–62
– from production to user, 65–66
– – garage for repairing cars, 70
– – production, 66–67
– – refuelling stations, 67–68
– – storage, 68–70
– – transportation, 68–70
– properties related to safety, 55–61
– regulations, codes, and standards (RCS), 63–65
– safe removal of, 73
– sensors, 63
Strontium chloride, 185

Tank filling, 110–111
– See also filling stations
Tank supply system, 200
Thermal decomposition, 187
Thermoacoustic liquefaction, 120

Thermogravimetric analyses (TGA), 142
Thin-film solar cells (TFSCs), 29
Titanium alloys, 109
Transportation, hydrogen application for, 81
TU-155 experimental aircraft, 130

U212A submarine, 162–163
United Nations (UN), 64–65
Unmanned aircraft vehicle (UAV), 90
Upper heating value (HHV), 12

Van der Waals coefficients, 14
Van der Waals equation, 13
Van't Hoff equation, 135
Vehicles, 127–128
Very high temperature gas-cooled reactor (VHTR), 39

Waste materials, of hydrogen storage options, 209–212
Water electrolysis, 17
Water gas shift reaction (WGSR), 19
Well-to-wheel assessment, 83–84, 96
Wet–dry multiphase conditions, 175–176
Wind–electrolysis system, 31–33
Wind–hydrogen generation system, *31, 32*
World Wind Energy Association (WWEA), 32

Zeolites, 234–235